Artificial Intelligence Applications and Innovation

Artificial Intelligence Applications and Innovation

Editors

João M. F. Rodrigues
Pedro J. S. Cardoso
Marta Chinnici

MDPI

Basel • Beijing • Wuhan • Barcelona • Belgrade • Novi Sad • Cluj • Manchester

Editors
João M. F. Rodrigues
NOVA LINCS and Instituto
Superior de Engenharia (ISE)
Universidade do Algarve
Faro, Portugal

Pedro J. S. Cardoso
NOVA LINCS and Instituto
Superior de Engenharia (ISE)
Universidade do Algarve
Faro, Portugal

Marta Chinnici
ICT Division-HPC Lab,
Department of Energy
Technologies and Renewable
Energy Sources (TERIN)
ENEA C.R. Casaccia
Roma, Italy

Editorial Office
MDPI
St. Alban-Anlage 66
4052 Basel, Switzerland

This is a reprint of articles from the Special Issue published online in the open access journal *Applied Sciences* (ISSN 2076-3417) (available at: https://www.mdpi.com/journal/applsci/special_issues/Artificial_Intelligence_Applications_and_Innovation).

For citation purposes, cite each article independently as indicated on the article page online and as indicated below:

Lastname, A.A.; Lastname, B.B. Article Title. *Journal Name* **Year**, *Volume Number*, Page Range.

ISBN 978-3-0365-9949-6 (Hbk)
ISBN 978-3-0365-9950-2 (PDF)
doi.org/10.3390/books978-3-0365-9950-2

Contents

About the Editors

João M. F. Rodrigues

João Rodrigues holds a Ph.D. in Electronics and Computer Engineering. He is a Coordinator Professor with Habilitation at the Institute of Engineering, University of the Algarve, Portugal, where he has lectured on Computer Science and Computer Vision since 1994. He is a member of the Research Centre NOVA LINCS. Professor Rodrigues has extensive research experience, having participated in more than 20 nationally or internationally funded scientific projects, some of which he served as the coordinator. He is a co-author of more than 200 scientific publications, belongs to the Editorial Board of several international journals, and belongs to the organization of several special issues in journals, tracks, workshops, and international conferences. He is the editor of several books on pattern recognition and developments and technologies for human–computer interaction. His main areas of interest are computer vision, human-centered AI (HCAI), human–computer interaction (HCI), affective computing, human-senses modeling, and adaptative interfaces

Pedro J. S. Cardoso

Pedro J. S. Cardoso obtained his Ph.D. in Mathematics/Operations Research from the University of Seville, Spain, an M.Sc. in Computational Mathematics from the University of Minho, Portugal, and a B.Sc. in Mathematics/Computer Science from the University of Coimbra, Portugal. Serving as a Coordinator Professor and researcher at the Universidade do Algarve for over two decades, he is also a member of the Research Centre NOVA LINCS. With extensive expertise in algorithms and data structures, Pedro specializes in machine learning and multiple objective meta-heuristics. He has been a dedicated lecturer in these fields for many years. Throughout his career, Pedro has actively participated in more than a dozen scientific and development projects, boasting a portfolio of over ninety publications. Furthermore, he has served as an editor for more than a dozen books. His primary research interests encompass machine learning, particularly in real applications, affective computing, anomaly detection, and energy efficiency.

Marta Chinnici

Dr. Chinnici graduated in Mathematics (2004) magna cum laude at the University of Naples (Italy), where she received her Ph.D. in Mathematics and Computer Science (2008) with a thesis focusing on stochastic self-similar processes and application in non-linear dynamical systems. Currently, she is a Senior Researcher at ENEA in the Department of Energy Technologies and Renewable Energy Sources, ICT Division, HPC Lab, where she conducts the study, analysis, research and development on ICT with particular reference issues relating to energy efficiency in data center (DC), high-performance computing (HPC) and data science. She is a European Commission Expert as a review/evaluator in ICT and computer science for many European programs. She is responsible and a manager for ENEA of many EU and national projects. She is also a member of the Technical Program Committee of various international conferences and workshops and an editor/referee and editor-in-chief of relevant journals. She is the author of books, relevant scientific articles, essays, and speeches for national and international conferences.

applied sciences

MDPI

Editorial

Artificial Intelligence Applications and Innovations: Day-to-Day Life Impact

João M. F. Rodrigues [1,*,†], Pedro J. S. Cardoso [1,†] and Marta Chinnici [2,†]

1 LARSYS (ISR-LX) & ISE, Universidade do Algarve, 8005-139 Faro, Portugal; pcardoso@ualg.pt
2 ICT Division-HPC Lab, Department of Energy Technologies and Renewable Energy Sources (TERIN), ENEA C.R. Casaccia, 00123 Roma, Italy; marta.chinnici@enea.it
* Correspondence: jrodrig@ualg.pt
† These authors contributed equally to this work.

Keywords: artificial intelligence; human-centered AI; data analysis; data science; big data

1. Introduction

The idea of an intelligent machine has fascinated humans for centuries. But what is intelligence? Some define it as the capacity for learning, reasoning, understanding or, from a different perspective, the aptitude to grasp truths, relationships, facts, or meanings. All these perspectives require the capacity to acquire data from the surrounding world and, possibly, act over that environment. In short, the building of more or less autonomous agents, served with sensors and actuators, capable of learning and producing educated answers has been long foreseen.

New trends in intelligente systems comprise, among other aspects, pervasive robotization, ubiquitous online data access, empowered edge computing, smart spaces, and digital ethics. These trends build the research on "Artificial Intelligence Applications and Innovation", impacting our day-to-day life, our cities, and even our free time. Nevertheless, artificial intelligence (AI) is still closely associated with some popular misconceptions that cause the public to either have unrealistic fears about it or to have unrealistic expectations about how it will change our workplace and life in general. It is important to show that such fears are unfounded and that new trends, innovations, technologies, and smart systems will be able to improve the way we live, benefiting society without replacing humans in their core activities.

Citation: Rodrigues, J.M.F.; Cardoso, P.J.S.; Chinnici, M. Artificial Intelligence Applications and Innovations: Day-to-Day Life Impact. *Appl. Sci.* **2023**, *13*, 12742. https://doi.org/10.3390/app132312742

Received: 22 November 2023
Accepted: 24 November 2023
Published: 28 November 2023

2. Artificial Intelligence Applications and Innovation

This Special Issue (SI) delves into mutually dependent subfields including, but not restricted to, machine learning, computer vision, data analysis, data science, big data, internet-of-things (IoT), affective computing, natural language processing, privacy and ethics, and robotics. The established set of papers form a comprehensive collection of contemporary "Artificial Intelligence Applications and Innovation" that serves as a convenient reference for AI experts as well as newly arrived practitioners, introducing them to different fields and trends.

In this context, the mentioned advancements and technologies have the potential to enhance our lifestyle; examples are presented by Iovane et al. [1], where a model for assessing the relevance of opinions in uncertainty and info-incompleteness conditions is proposed, and by Mndawe et al. [2], where a stock price prediction framework is introduced, supported by a sentiment classifier based on news headlines and tweets. The latter work uses four machine learning models for a fundamental analysis and six long short-term memory (LSTM) model architectures, including a developed LSTM encoder–decoder model for technical analysis. Data used in the experiments are mined and collected from news sites, tweets (from Twitter), and Yahoo Finance. A deep data assimilation (DDA) model is presented by Arcucci et al. in [3], where novel integration of data assimilation with machine learning through the use of a recurrent neural network is designed.

Other innovative advances include Domingo's deep learning and IoT application for beach monitoring [4], where a study of beach attendance prediction at Castelldefels beach is conducted. The evaluation of robotic solutions and their potential benefits, compared to conventional processes when adopted on construction, is presented by Marcher et al. [5].

Related to health and well-being, a model for differential diagnosis of Raynaud's phenomenon based on thermal hand patterns is presented by Filippini et al. [6], and the issues and needs of dyslexic students are studied by Zingoni et al. [7]. Silva et al. [8] introduce a tool for lung nodule malignancy assessment using computed tomography images.

Two studies relate to energy consumption optimization. Namely, Hong et al. [9] present a driving cycle-based state of charge prediction for electrical vehicles batteries using deep learning methods, and Hora et al. [10] introduce a metaheuristic for effective electric energy consumption prediction.

In the field of identification, again, two studies are presented, namely facial ID document validation in mobile devices addressed by Medvedev et al. [11] and a review of tools and methods for the (re-)identification in urban scenarios presented by Oliveira et al. [12].

Finally, more conceptual studies are also available. In particular, Turner et al. [13] present a modular dynamic neural network architecture for continual learning that can deal with the phenomenon of catastrophic forgetting, and Chen et al. [14] analyze the temporal structures in evolutionary networks. In the latter case, the authors propose a community detection algorithm based on graph representation learning that uses a Laplacian matrix to extract the node relationship data of the edges of the network structure that are directly connected at the preceding time slice. A deep sparse autoencoder learns to represent the network structure under the current time slice, and the *K*-means clustering algorithm is used to partition the low-dimensional feature matrix of the network structure under the current time slice into communities.

Despite the widespread use of AI in various applications, recent advancements indicate that the field is still in its early stages of development, with many opportunities for growth and innovation ahead.

3. Future of AI

Although the Special Issue is closed, much more in-depth and distinct research in AI applications is foreseeable. E.g., as AI systems become more complex, there is a growing need for transparency and interpretability, making the field of explainable AI (XAI) one of the present and also future trends. On other words, XAI focuses on making AI systems more understandable and accountable, allowing humans comprehension of the decision-making processes of AI algorithms.

AI ethics and bias mitigation, which concerns bias in AI algorithms and their ethical implications, is also a fundamental trend. This field leads to increased efforts to develop and implement ethical AI practices, addressing stricter regulations, better frameworks for ethical AI development, and increased awareness of bias issues.

Edge AI is both a current and a future area of interest. With the rise of the IoT, there is a growing trend toward deploying AI models directly on edge devices, like smartphones or IoT devices, rather than relying solely on centralized cloud servers. This can lead to faster response times, improved privacy, reduced bandwidth usage and more sustainable AI ecosystems.

Generative models, such as generative adversarial networks (GAN) and variational autoencoders (VAE), are presented in many models and becoming more sophisticated. These models are used for creating realistic synthetic data, generating content, and similar applications, with potential breakthroughs in areas like creativity and content creation.

AI in healthcare, as presented in some chapters of the SI, is currently playing and anticipated to continue playing a crucial role in personalized medicine, drug discovery, and diagnostic tools. As evidenced, integrating AI into healthcare can actually result in better treatment plans, more precise diagnoses, and better patient outcomes.

Other present and future trends include natural language processing (NLP), autonomous systems, and AI applications on fields like cybersecurity or in finance. As a matter of fact, continued advancements in NLP could lead to more natural and context-aware interactions with AI systems. This trend includes improvements in language understanding, sentiment analysis, and language generation. The development of autonomous vehicles, drones, and robotic systems is ongoing. Advancements in machine learning and sensor technologies are expected to drive progress in making these systems safer and more reliable. With the increasing sophistication of cyber threats, AI is being used to enhance cybersecurity measures. AI systems can detect anomalies, identify patterns, and respond to security incidents in real time. In the financial sector AI is likely to continue making inroads for tasks such as fraud detection, algorithmic trading, and personalized financial advice.

Nevertheless, the integration of quantum computing and AI is probably the next "jump". There is growing interest in exploring how quantum computing can be integrated with AI to solve complex problems more efficiently.

Altogether, these trends validate a future where AI becomes more integrated into various aspects of our lives, solving complex problems and improving efficiency across different industries. Keeping in mind that the AI field is dynamic, and new trends may emerge as technology continues to evolve, it is therefore fundamental to keep up with the latest developments in the field to stay ahead of the curve.

Funding: This work was supported by the Portuguese Foundation for Science and Technology (FCT), project LARSyS—FCT Project UIDB/50009/2020, and Project ECS 0000024 Rome Technopole, CUP B83C22002820006, National Recovery and Resilience Plan (NRRP), Mission 4, Component 2 Investment 1.5, funded from the European Union—NextGenerationEU.

Conflicts of Interest: The authors declare no conflict of interest.

References

1. Iovane, G.; Landi, R.E.; Rapuano, A.; Amatore, R. Assessing the Relevance of Opinions in Uncertainty and Info-Incompleteness Conditions. *Appl. Sci.* **2022**, *12*, 194. [CrossRef]
2. Mndawe, S.T.; Paul, B.S.; Doorsamy, W. Development of a Stock Price Prediction Framework for Intelligent Media and Technical Analysis. *Appl. Sci.* **2022**, *12*, 719. [CrossRef]
3. Arcucci, R.; Zhu, J.; Hu, S.; Guo, Y.K. Deep Data Assimilation: Integrating Deep Learning with Data Assimilation. *Appl. Sci.* **2021**, *11*, 1114. [CrossRef]
4. Domingo, M.C. Deep Learning and Internet of Things for Beach Monitoring: An Experimental Study of Beach Attendance Prediction at Castelldefels Beach. *Appl. Sci.* **2021**, *11*, 10735. [CrossRef]
5. Marcher, C.; Giusti, A.; Matt, D.T. On the Design of a Decision Support System for Robotic Equipment Adoption in Construction Processes. *Appl. Sci.* **2021**, *11*, 11415. [CrossRef]
6. Filippini, C.; Cardone, D.; Perpetuini, D.; Chiarelli, A.M.; Gualdi, G.; Amerio, P.; Merla, A. Convolutional Neural Networks for Differential Diagnosis of Raynaud's Phenomenon Based on Hands Thermal Patterns. *Appl. Sci.* **2021**, *11*, 3614. [CrossRef]
7. Zingoni, A.; Taborri, J.; Panetti, V.; Bonechi, S.; Aparicio-Martínez, P.; Pinzi, S.; Calabrò, G. Investigating Issues and Needs of Dyslexic Students at University: Proof of Concept of an Artificial Intelligence and Virtual Reality-Based Supporting Platform and Preliminary Results. *Appl. Sci.* **2021**, *11*, 4624. [CrossRef]
8. Silva, F.; Pereira, T.; Frade, J.; Mendes, J.; Freitas, C.; Hespanhol, V.; Costa, J.L.; Cunha, A.; Oliveira, H.P. Pre-Training Autoencoder for Lung Nodule Malignancy Assessment Using CT Images. *Appl. Sci.* **2020**, *10*, 7837. [CrossRef]
9. Hong, S.; Hwang, H.; Kim, D.; Cui, S.; Joe, I. Real Driving Cycle-Based State of Charge Prediction for EV Batteries Using Deep Learning Methods. *Appl. Sci.* **2021**, *11*, 11285. [CrossRef]
10. Hora, S.K.; Poongodan, R.; de Prado, R.P.; Wozniak, M.; Divakarachari, P.B. Long Short-Term Memory Network-Based Metaheuristic for Effective Electric Energy Consumption Prediction. *Appl. Sci.* **2021**, *11*, 11263. [CrossRef]
11. Medvedev, I.; Shadmand, F.; Cruz, L.; Gonçalves, N. Towards Facial Biometrics for ID Document Validation in Mobile Devices. *Appl. Sci.* **2021**, *11*, 6134. [CrossRef]
12. Oliveira, H.S.; Machado, J.J.M.; Tavares, J.M.R.S. Re-Identification in Urban Scenarios: A Review of Tools and Methods. *Appl. Sci.* **2021**, *11*, 10809. [CrossRef]

13. Turner, D.; Cardoso, P.J.S.; Rodrigues, J.M.F. Modular Dynamic Neural Network: A Continual Learning Architecture. *Appl. Sci.* **2021**, *11*, 12078. [CrossRef]
14. Chen, D.; Nie, M.; Wang, J.; Kong, Y.; Wang, D.; Huang, X. Community Detection Based on Graph Representation Learning in Evolutionary Networks. *Appl. Sci.* **2021**, *11*, 4497. [CrossRef]

applied sciences

Article

Assessing the Relevance of Opinions in Uncertainty and Info-Incompleteness Conditions

Gerardo Iovane [1,†], Riccardo Emanuele Landi [2,*,†], Antonio Rapuano [1,†] and Riccardo Amatore [1,†]

1 Department of Computer Science, University of Salerno, 84084 Fisciano, Italy; giovane@unisa.it (G.I.); arapuano@unisa.it (A.R.); ramatore@unisa.it (R.A.)
2 Rigenera S.r.l., Via Aventina 7, 00153 Rome, Italy
* Correspondence: riccardo.landi@rigenera2020.it
† These authors contributed equally to this work.

Abstract: Researchers are interested in defining decision support systems that can act in contexts characterized by uncertainty and info-incompleteness. The present study proposes a learning model for assessing the relevance of probability, plausibility, credibility, and possibility opinions in the conditions above. The solution consists of an Artificial Neural Network acquiring input features related to the considered set of opinions and other relevant attributes. The model provides the weights for minimizing the error between the expected outcome and the ground truth concerning a given phenomenon of interest. A custom loss function was defined to minimize the Mean Best Price Error (MBPE), while the evaluation of football players' was chosen as a case study for testing the model. A custom dataset was constructed by scraping the Transfermarkt, Football Manager, and FIFA21 information sources and by computing a sentiment score through BERT, obtaining a total of 398 occurrences, of which 85% were employed for training the proposed model. The results show that the probability opinion represents the best choice in conditions of info-completeness, predicting the best price with 0.86 MBPE (0.61% of normalized error), while an arbitrary set composed of plausibility, credibility, and possibility opinions was considered for deciding successfully in info-incompleteness, achieving a confidence score of 2.47 ± 0.188 MBPE ($1.89 \pm 0.15\%$ of normalized error). The proposed solution provided high performance in predicting the transfer cost of a football player in conditions of both info-completeness and info-incompleteness, revealing the significance of extending the feature space to opinions concerning the quantity to predict. Furthermore, the assumptions of the theoretical background were confirmed, as well as the observations found in the state of the art regarding football player evaluation.

Keywords: decision support systems; uncertainty; info-incompleteness; machine learning; artificial intelligence; football market; athlete evaluation

Citation: Iovane, G.; Landi, R.E.; Rapuano, A.; Amatore R. Assessing the Relevance of Opinions in Uncertainty and Info-Incompleteness Conditions. *Appl. Sci.* **2022**, *12*, 194. https://doi.org/10.3390/app12010194

Academic Editors: João M. F. Rodrigues, Pedro J. S. Cardoso and Marta Chinnici

Received: 25 November 2021
Accepted: 23 December 2021
Published: 25 December 2021

1. Introduction

The main concept used for estimating the possibility of the occurrence of an event is the probability, in which the certain event is the upper extreme and the impossible event is the lower one. Between these two bounds, there are more or less probable events. Probability, however, is a limited concept, as it can be affected significantly by the lack of information; for instance, in the case that all the information on the environment in which dice are thrown is known, it is possible to easily decide the exact face of the dice that will be obtained. However, there are some events in which the probability can be computed through some additional information; i.e., in a dice throw, the number of faces can be known, while, in a financial context, where it is intended to compute the probability of an asset reaching a certain price, the information is almost totally absent. This is called an *uncertain and info-incomplete* environment.

To better estimate the occurrence of an event, different definitions of what is called *plausible reasoning* have been proposed. Iovane et al. [1] have modeled decision making and reasoning in uncertainty and info-incompleteness conditions as the evaluation of *Probability*, *Plausibility*, *Credibility*, and *Possibility*, providing several models of interest (capital letters are used for specifying the concepts as defined by the authors). Probability is conceived as an estimate of evidence concerning a given phenomenon and Plausibility, Credibility, and Possibility as opinions extracted from the area of knowledge which does not regard direct evidence. In fact, in the case the probability estimation (i.e., the direct evidence) that a phenomenon occurs is weak, it is possible to reduce the uncertainty by acquiring information concerning, e.g., the opinion of experts or the sentiment of a group of people regarding the aforementioned phenomenon. When an estimate of evidence is available, i.e., a probability, we decide in conditions of *info-completeness*; when an estimate of evidence is not available, we decide in conditions of *info-incompleteness*. In a nutshell, when an event is very unlikely, or a decision based on the analysis of evidence cannot be performed, it is more promising to consider other sources of information, which may be less reliable than deciding "blindly".

In the present work, it is proposed to re-enforce the model defined by Iovane et al. [1] by using machine learning to estimate the relevance of Probability, Plausibility, Credibility, and Possibility opinions in conditions of both info-completeness and info-incompleteness. To achieve the above goal, it was decided to adopt, as we explain in the following Sections, the *best price* model, as proposed by the authors, to a real study case. In particular, we will refer to the football players' market, where each athlete is characterized by an economic evaluation. A custom dataset was built to train and test an Artificial Neural Network (ANN) in estimating the weights of Probability, Plausibility, Credibility, and Possibility in the above field of interest.

The work is organized as follows. In Section 2, a summary of the prodromic theory [1] is provided. Section 3 analyzes the state of the art and the most important studies in the field of decision making and reasoning in conditions of uncertainty and info-incompleteness, as well as in the research regarding the evaluation of athletes through Artificial Intelligence methodologies. Section 4 shows our proposed solution for weighting the opinions, while Section 5 describes the dataset and the implementation of the proposed learning model on the chosen study case. Finally, Section 6 discusses the results, and Section 7 summarizes the work and indicates the future direction.

2. Theoretical Background

In this section, a discussion of the advancements in the plausibility theory, with particular regard to the prodromic study [1], is provided.

According to Polya [2,3], the plausible reasoning is not subjective, and it is treated as a conditional probability: given two events A and B, the author conceives the plausibility as the confidence of B given that A is true. After Polya, Dempster–Shafer's theory was defined [4,5]: the plausibility is no longer intended as one-dimensional but as a series of mutually exclusive alternatives with a maximum probabilistic value. To overcome the difficulties of Dempster–Shafer's model, a new solution, called Dezert–Smarandache's theory, was proposed [6]: there, the *plausibility* becomes an upper limit of the probabilistic value concerning a given event. Iovane et al. [1] provided a further contribution by extending the concept of plausibility to *credibility* and *possibility*. The authors defined the expectation function obtained through the composition of Probability, Plausibility, Credibility, and Possibility.

The study provides seven models for computing the above function. By considering $P_1, P_2, P_3, P_4 \in \mathbb{R}^k$, with $k > 0$, as the opinions concerning Probability, Plausibility, Credibility, and Possibility, respectively, the aforementioned models can be summarized as follows.

- *Average model*: the simplest model. It computes the average of the four opinions as

$$a_1 = a_1(P_1, P_2, P_3, P_4) = \frac{1}{4}\sum_{i=1}^{4} P_i. \quad (1)$$

The model assigns the same importance to the different distributions of Probability, Plausibility, Credibility, and Possibility.

- *Product model*: the expectation function is defined by the P_i product as

$$a_2 = a_1(P_1, P_2, P_3, P_4) = \prod_{i=1}^{4} P_i. \tag{2}$$

- *Weighted average model*: a_1 and a_2 assume that all P_i have the same importance. Instead, this model extends a_1 by weighting the P_is with

$$\sum_{i=1}^{4} \alpha_i = 1, \tag{3}$$

where α_i is the weight of P_i and a_3 is defined as

$$a_3 = \sum_{i=1}^{4} \alpha_i P_i. \tag{4}$$

- *Weighted product model*: extends a_2 weighting the P_is as well. Formally,

$$a_4 = \prod_{i=1}^{4} \alpha_i P_i. \tag{5}$$

- *Overlap model with shift based on probability*: differently from other previously defined models, the overlap with shift based on probability allows a hierarchical use of the P_is. Formally,

$$a_5 = \begin{cases} P_1 & if\ 1\% < P_r \leq 100\%, \\ P_2 & if\ 0.1\% \leq P_r \leq 1\%, \\ P_3 & if\ 0.01\% \leq P_r \leq 0.1\%, \\ P_4 & if\ P_r \leq 0.01\%. \end{cases} \tag{6}$$

In this model, the expectation function is selected from Probability, Plausibility, Credibility, and Possibility. There, the selection depends on the classical probability value (P_r). Each P_i, if selected, has a coefficient

$$0\% \leq c_i \leq 100\%. \tag{7}$$

- *Overlap model with shift based on hierarchical P_i*: the probability does not have a pivotal role; this is an alternative model for a_5. There, the expectation function is defined by the authors as

$$a_6 = \begin{cases} P_1 & \text{if } \bar{P}_{P_1} - 3\sigma_{P_1} \leq P_r \leq \bar{P}_{P_1} - 3\sigma_{P_1}, \\ P_2 & \text{if } \bar{P}_{P_2} - 3\sigma_{P_2} \leq P_r \leq \bar{P}_{P_2} - 3\sigma_{P_2} \text{ and } P_r > \bar{P}_{P_2} - 3\sigma_{P_1}, \\ P_3 & \text{if } \bar{P}_{P_3} - 3\sigma_{P_3} \leq P_r \leq \bar{P}_{P_3} - 3\sigma_{P_3} \text{ and } P_r > \bar{P}_{P_2} - 3\sigma_{P_2}, \\ P_4 & \text{if } \bar{P}_{P_4} - 3\sigma_{P_4} \leq P_r \leq \bar{P}_{P_4} - 3\sigma_{P_4} \text{ and } P_r > \bar{p}_{P_3} - 3\sigma_{P_3}. \end{cases}$$

There, i.e., in a financial context like in the present work, in the case the goal is to obtain the best price of an athlete, $\bar{P}$ represents the average price in the dataset and σ the standard deviation. It is important to note that, in this context, the above P_is identify the distribution of $\bar{P}$ and σ and not the P_i value. In other words, the selection of the P_i depends on the values of σ and $\bar{P}_i$ and the P_i distribution.

- *Model based on Dempster's composition rules*: the last model the authors proposed is based on the Dempster's composition rules, which are defined only for the plausibility. Iovane et al. [1] extended those rules to Probability, Plausibility, Credibility and Possibility. Formally,
  - $m(P_1)$, relative $bel(P_1)$ and $Dpl(P_1)$;
  - $m(P_2)$, relative $bel(P_2)$ and $Dpl(P_2)$;
  - $m(P_3)$, relative $bel(P_3)$ and $Dpl(P_3)$;
  - $m(P_4)$, relative $bel(P_4)$ and $Dpl(P_4)$;

  where m is called "mass function" by Dempster (the degree of belief), *bel* is the belief function, and *Dpl* represents the plausibility.

While the described models are used to compute the most likely price, i.e., the best price, we can extend the information defining the *occurrence* as

$$O(E) = \alpha_1 P_1(E) + \alpha_2 P_2(E) + \alpha_3 P_3(E) + \alpha_4 P_4(E), \tag{8}$$

where $P_i(E) : \mathbb{R}^k \rightarrow [0,1]$, with $k > 0$, is how probable the event E is and α_i represents the weights of the $P_i(E)$, with $0\% \leq \alpha_i \leq 100\%$. Once the best price and how much this price occurs are determined, the last thing needed for describing an event is the reliability of the information. To compute the reliability of the best price, we can use the standard deviation. Formally,

$$R(E) = \sigma^2(E) = \sqrt{\frac{\sum_{i=1}^{N}(x_i - \bar{x})^2}{N-1}}, \tag{9}$$

where x_i is in the set P_1, P_2, P_3, P_4.

Therefore, the final output of the model is the triad

$$(E, O(E), R(O)). \tag{10}$$

In the prodromal study [1], the authors simulated the datasets to prove the correctness of the models. In the present work, it is intended to face a real case by using a neural network for estimating the weights of the best price associated with the expectation function a_3. In the next section, the state of the art regarding athletes' price estimation is analyzed.

3. Related Work

The economical evaluation of football players is a much-addressed issue. In particular, in the financial area, the evaluation of an asset is made by supply and demand. The financial world applied to the sport is complex; differently from traditional finance, there are only two actors in the negotiation of a player: the buyer and the seller. They can agree on any price, and this can lead to several problems from a regulatory point of view. The question in this field is: can we have a reference point for the football players' evaluation? In this context, crowdsourcing is significant through Transfermarkt, but a more reliable tool is still needed.

As the financial world behind football, as well as the sports world in general, is vast and based on complex economic models, several studies have investigated how to predict or estimate the athletes' market value. Dobson and Goddard [7] proposed an interesting and detailed study concerning the economics of professional English football at the club level. As mentioned in the previous section, Iovane et al. [1] applied the described models to two different applications to prove the validity of the theory. The two experiments consisted of two simulations regarding the probabilities fields of biometrics and sport odds; there, the authors simulated the datasets, adding uncertainty through randomness over the input space. The weights α_i were defined without a backtest; thus, a deep study on the estimation of the weights is needed. The present work is conceived to solve the above difficulty.

In [8,9], the authors tried to estimate the market value of football players. Behravan and Razavi [8] clustered the football players by roles; after that, they used a hybrid regression method involving Particle Swarm Optimization (PSO) and Support Vector Regressor (SVR) for each cluster. They obtained a final accuracy for their model of 74%.

Furthermore, the sentiment can affect the athlete evaluation: an interesting study was conducted by Singh and Lamba [10]; they described how crowdsourcing, previous year statistics, and popularity of players can affect the evaluation. Regarding crowdsourcing, in [11], the authors proved how, in the context of German soccer, a community became the main source for reporting market values to predict the actual transfer fees. The authors described the evaluation process performed by the community, together with the accuracy of the estimated market values, and which variables are important to make a price estimation. They found that the variables that are mostly correlated with the price are those of age, precision, success, assertion, and flexibility.

The importance of athletes' age is analyzed and discussed in several papers. In particular, Gonzalez et al. [12] investigated the relative age effect, which was predominant in players born in the first months of the year compared to those born in the last months. The results show that, except for the youth categories, the relative age does not affect the professional football player market evaluation but only the selection in youth categories.

Other scholars investigated the variables affecting the football players' market value. In [13], the values of Transfermarkt.de were acquired, while in [14], the authors analyzed the Football Manager game values. Felipe et al. [15] investigated the influence of team variables and the player role on the athletes' market values. Another interesting variable for the price estimation of football players is popularity; Franck and Nüesch [16], as well as Kiefer [17], investigated the influence of the players' popularity on their market value. The authors proved that the market value of the players is influenced by both talent and non-performance-related popularity. In [18], the authors proposed a decision support system for football club managers and players' agents by estimating the correct wages of football players. Player skills, performances in the previous season, age, the trajectory of the improvement, personality, and other features were considered.

Although there are several works that estimate the economical value of athletes, a decision support system merging different opinions can perform well even in conditions of uncertainty and info-incompleteness. The present work aims to investigate the roles and the weights of Probability, Plausibility, Credibility, and Possibility in the market evaluation of football players, for both improving the state of the art in the research area and providing a case study in which the assessment of the best opinion, in conditions of info-incompleteness, can be achieved.

4. Proposed Solution

Referring to the best price function defined in (3), it was intended to predict the weights α_1, α_2, α_3, and α_4, related to the opinions associated with Probability, Plausibility, Credibility, and Possibility, respectively, by extracting data from different sources. Given the occurrences of the above four opinions and the related ground truth, it is possible to approximate the best price function, i.e., the best opinion, by defining a learning model trained to predict the weights. This approach permits obtaining the relevance of opinions in conditions of both info-completeness and info-incompleteness, together with the most promising estimate regarding a given phenomenon.

To achieve the above goal, it was decided to propose the model described in Figure 1, in which an ANN receives four features associated with Probability, Plausibility, Credibility, and Possibility opinions and an arbitrary set of other relevant attributes. Each of the opinions is extracted from a dedicated source of information, as well as the additional set of attributes.

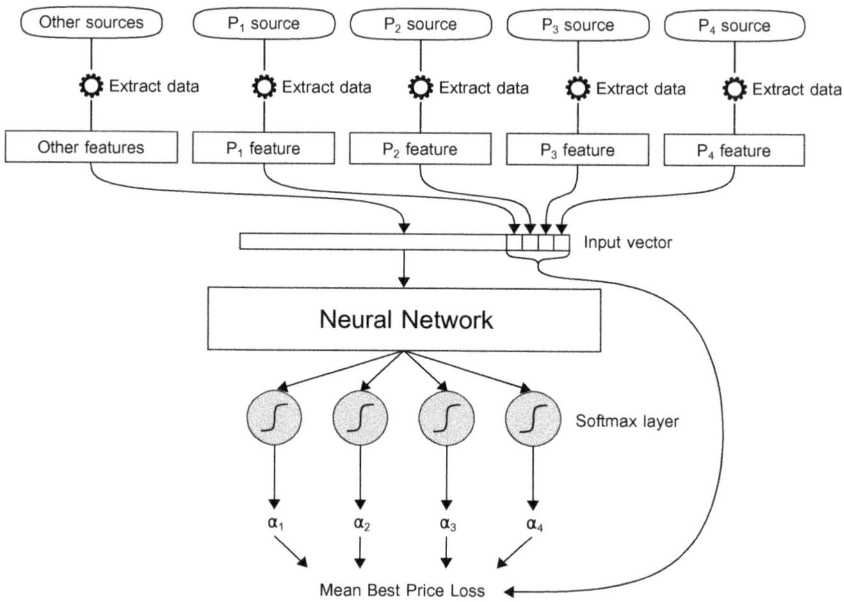

Figure 1. The proposed learning model to predict the relevance of opinions and an estimate of the best opinion in uncertainty and info-incompleteness conditions.

The model outputs four weights extracted from the components of a softmax layer to minimize a custom loss function, called *Mean Best Price Loss*, which was defined as

$$L_{MBP} = \frac{1}{N}\sum_{i=1}^{N} |P_{0_i} - (\alpha_{1_i}P_{1i} + \alpha_{2_i}P_{2i} + \alpha_{3_i}P_{3i} + \alpha_{4_i}P_{4i})|, \tag{11}$$

where N is the batch size, while P_{0_i}, P_{1i}, P_{2i}, P_{3i}, and P_{4i} are the ground truth and the Probability, Plausibility, Credibility, and Possibility opinions concerning the i-th occurrence, respectively. Similarly, the weights α_{1_i}, α_{2_i}, α_{3_i}, and α_{4_i} are the components of the softmax layer concerning the i-th occurrence. The loss function can be considered as the Mean Absolute Loss, in which the second term of the subtraction represents the prediction of the proposed model given a certain instance of features. The performance metric associated with the above loss is called *Mean Best Price Error* (MBPE).

The proposed approach above permits the ANN to learn the parameters that provide the optimal weights for minimizing the difference between the best price prediction, i.e., the estimate of the best opinion, and the true expected outcome. For instance, suppose there is some interest in evaluating the reliability in the occurrence of a given phenomenon. Suppose the phenomenon is also characterized by some specific attributes and that there exist sources of information from which one or more opinions can be extracted. In the case evidence about the phenomenon exists, a probable occurrence can be obtained; in the case some experts are involved in the study of the phenomenon, a plausible occurrence can be obtained; in the case the people discuss the phenomenon, a credible occurrence can be considered; in the case some other less relevant sources of information are available, a possible occurrence can be extracted. The acquisition of opinions related to a common domain of attributes permits an inference to be performed on the occurrence related to the given phenomenon of interest.

Under the above hypotheses and definitions, we have decided to test the proposed approach on a real study case regarding football players' market evaluation. Such a problem is ideal for studying the relevance of opinions in uncertainty and info-incompleteness conditions, as the prediction of the next transfer cost of a player is subjected to several sources of speculation. It is possible to evaluate the proposed model in conditions of both info-completeness, i.e., when the opinion concerning the Probability is available, and info-incompleteness, i.e., when the opinion related to the Probability is completely lacking. The following section evaluates the MBPE error as the subsets of opinions vary; for evaluating the performance of the model in conditions of info-incompleteness, e.g., the feature related to the Probability is set to zero, so as to test the case in which a decision should be taken when only Plausibility, Credibility, and Possibility opinions are fully or partially available.

5. Experiments on a Study Case: Football Players Evaluation

As discussed in the previous section, to provide a case study on which to verify the effectiveness of the proposed model, we performed the prediction of the transfer cost of a player starting from their attributes and the opinions associated with the Probability, Plausibility, Credibility, and Possibility, extracted from the web. Formally, in the present use case the opinions are defined as $P_1, P_2, P_3, P_4 \in \mathbb{R}$, since prices are one-dimensional quantities.

To realize the above scope, four sources of information were chosen:

1. Transfermarkt players evaluation in the year 2021/2022 for the opinion related to Probability;
2. Football Manager players evaluation for the opinion related to Plausibility;
3. Wikipedia descriptions as a sentiment, combined with Football Manager society costs of players, for the opinion related to Credibility;
4. FIFA21 players evaluation for the opinion related to Possibility.

Each source of information is normalized as a price, e.g., having a player with Transfermarkt, Football Manager, and FIFA21 evaluations of 16, 15, and 25 million of euros for the Probability, Plausibility, and Possibility opinions, respectively, while a Football Manager society has a cost of 20 million euros, combined with a sentiment score of 0.95, for the Credibility opinion. Data extraction was performed by scraping, through four distinct scripts, the web pages related to the major European football leagues, i.e., Serie A, La Liga, Premier League, Ligue 1, and Bundesliga. The ground truth was found by acquiring the cost of transfers from the same source chosen for finding the opinions related to Probability. Regarding the players' attributes, it was decided to acquire their characteristics from the same source adopted for finding the opinions related to Possibility.

The data extraction process can be summarized, for each player, as the parallel execution of the following tasks:

- Extraction of the Transfermarkt evaluation, at one year before the next transfer, to obtain the Probability feature P_{1i};
- Extraction of the Football Manager evaluation, at one year before the next transfer, to obtain the Plausibility feature P_{2i};
- To obtain the Credibility feature P_{3i}, performing the computation of a sentiment score from the related Wikipedia description, at one year before the next transfer, and weighting the result to the extracted Football Manager society cost;
- Extraction of the FIFA21 evaluation, at one year before the next transfer, to obtain the Possibility feature P_{4i};
- Extraction of the FIFA21 attributes, at one year before the next transfer, to obtain other features.

In the following sections, a custom dataset composed of data extracted from the different sources of information and the adopted feature selection methodology, together with the discussion of the experimental results, is presented. It was decided to investigate the proposed model in conditions of info-completeness, i.e., when the opinion related to Probability is considered, and under different configurations of uncertainty and info-incompleteness, i.e., when the opinions related to Plausibility, Credibility, and Possibility are completely or partially available.

5.1. Sentiment Analysis

The analysis of sentiment was conducted on the texts acquired from the Wikipedia pages concerning the players. The text sentiment classification model employed is based on BERT (Bidirectional Encoder Representations from Transformers) [19] and fine-tuned on the IBM Claim Stance Dataset [20,21]. The solution receives a text string as input, while it outputs a score in the range $[0, 1]$, in which 0 and 1 represent the most negative and positive emotions, respectively. The BERT framework permits fine-tuning a pre-trained language model for tackling several Natural Language Processing tasks, of which in the present study the sentiment analysis was valued. The model is characterized by two operational phases: (i) *pre-training*, for training, through an unsupervised approach, the model over several tasks; (ii) *fine-tuning*, for optimizing, through a supervised approach, the parameters found in the previous task to the sentiment analysis. For both processes, the same Multi-layer Bidirectional Transformer Encoder architecture was employed: it generates word embeddings, i.e., univocal probabilistic representations of words, through a bi-directional training approach; for the first task, i.e., pre-training, the network is trained for solving the two problems of Masked Language Modeling (MLM) and Next Sentence Prediction (NSP); for the second task, i.e., fine-tuning, a custom fully connected layer is added to the pre-trained network for solving the desired supervised problem. The MLM problem consists in predicting the masked words of a sentence to learn their bi-directional context, while the NSP problem regards the determination of the order through which two sentences are employed in a given context.

The classifier reaches, in terms of accuracy, 94% and provides a good estimate of the sentiment in a text. In the present study, the output of the aforementioned model is used as an opinion modulation factor characterizing the sentiment associated with a given phenomenon.

5.2. Dataset

The data were extracted from the Transfermarkt, FIFA21, and Football Manager sources available on the web. In particular, Transfermarkt was the first source explored, as the extraction of the data was made dependent on the latest transfers of the calendar year 2021. The employed scraping process extracts the names and the related market values and transfer costs from Transfermarkt; then, for each extracted name, the sources related to FIFA21 and Football Manager are considered for obtaining attributes and values concerning Plausibility, Credibility, and Possibility opinions. Meanwhile, Wikipedia pages regarding the extracted names are processed through the sentiment analysis algorithm described in Section 5.1. The players' transfer costs provided by Transfermarkt (min = 0.1 million of euros, max = 125 million of euros) serve as ground truth for the training and test processes. A total of 398 data occurrences of transferred players have been considered.

Figure 2 shows the pair distributions of Probability, Plausibility, Credibility, and Possibility features obtained by building the dataset.

The pair distributions enhance the characteristics in the population of opinions by considering two features at a time as coordinates. The Figure, on the first diagonal, also shows the density distributions related to the opinions. It can be noticed that the observations, expressed in millions of euros, related to Credibility ($m = 13.33$, $SD = 29.47$) and Possibility ($m = 26.19$, $SD = 101.72$) present the highest variance. The distribution concerning Plausibility, instead, is characterized by the lowest variance ($m = 6.62$, $SD = 10$). The Plausibility opinions, which in the present study concern the decisions of experts, tend to occupy a definite area of hypotheses characterized by less uncertainty. As the considered field of opinions expands itself towards the areas of sentiment and other less relevant sources of information, the uncertainty on the players' evaluation increases, as the opinions are more heterogeneous.

Regarding the other features considered, the set concerning skills, age ($m = 24$, $SD = 3.66$, y.o.), position (e.g., offensive guard, wide receiver, etc.), wage ($m = 14.53$, $SD = 36.18$, millions of euros), height ($m = 182.48$, $SD = 7.13$, cm), weight ($m = 76.81$,

$SD = 7.74$, kg), preferred foot (right foot, left foot), and preferred positions was adopted. Non-numeric features, such as position or preferred foot, have been enumerated.

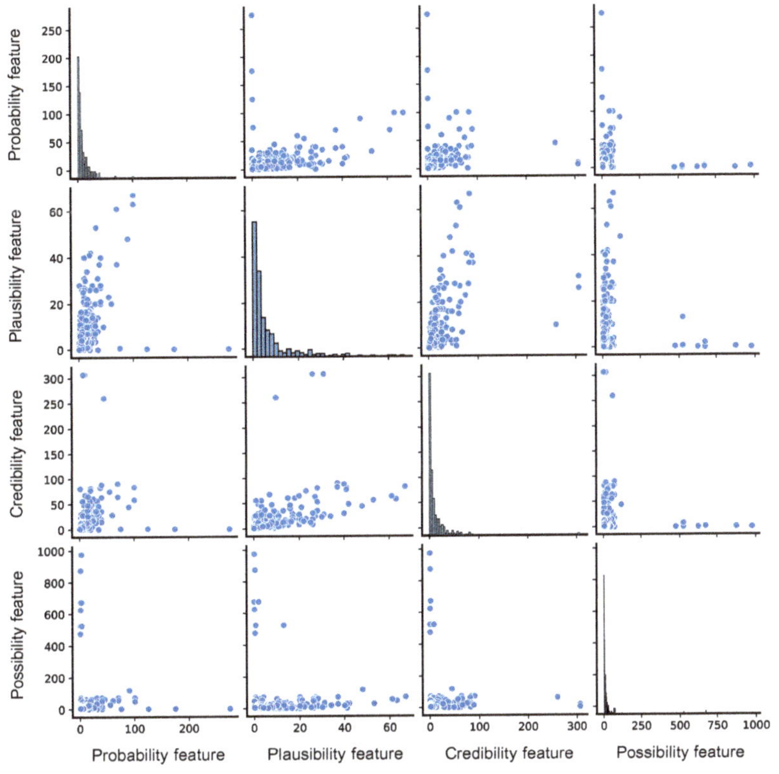

Figure 2. Pair distributions of features related to Probability, Plausibility, Credibility, and Possibility opinions.

In Figure 3, the box and whiskers plots related to the distributions of skills are shown. As organized in FIFA21, the scores for skills are specified as integer numbers in the range $[0, 100]$.

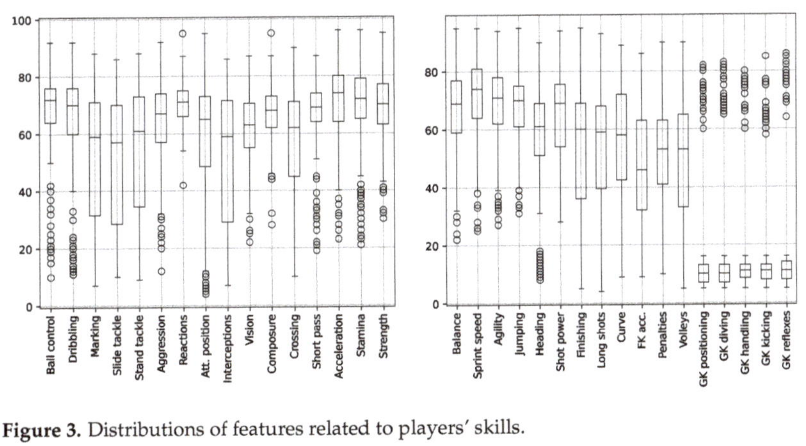

Figure 3. Distributions of features related to players' skills.

The attributes which present the highest dispersion are *making*, *slide tackle*, and *interceptions*, while those characterized by the highest amount of outliers are the dimensions reserved for goalkeepers, i.e., *GK positioning*, *GK diving*, *GK handling*, *GK kicking*, and *GK reflexes*. These last attributes present such characteristics due to the limited occurrence of goalkeepers in the dataset.

5.3. Implementation Details

The input features were pre-processed through a re-scaling in the $[-1, 1]$ range. For each input data point reserved for the backpropagation process, the related non-re-scaled values of Probability, Plausibility, Credibility, and Possibility are used by the L_{MBP} loss function to compute the prediction error. Thus, the network acquires two inputs: the first is the vector of re-scaled attributes and opinions, employed for performing the prediction; the second is the vector of non-re-scaled opinions, employed for computing the loss function. In our experiments, the network consists of three dense layers composed of 512, 256, and 16 neurons, respectively, all characterized by ReLu activation; the end-point, as already discussed in Section 4, consists of a four-neuron softmax layer.

The learning parameters of the network were initialized by sampling from a random uniform distribution.

5.4. Feature Selection

Feature selection was performed by computing feature importance through a brute force approach, as the number of features is limited. The employed process is described through the pseudo-code described in Algorithm 1, in which the $positiveImportanceScores$ represents the list of importance scores concerning the considered set of features.

Algorithm 1 The employed algorithm for computing feature importance

$importanceScores \leftarrow array[N_{features}]$
$baseScore \leftarrow evaluate(compiledModel, X_{train}, X_{test}, y_{train}, y_{test})$
$index \leftarrow 0$
while $index < N_{features}$ **do**
  $X^{new}_{train} \leftarrow deleteFeatureAtIndex(X_{train}, index)$
  $X^{new}_{test} \leftarrow deleteFeatureAtIndex(X_{test}, index)$
  $predScore \leftarrow evaluate(compiledModel, X^{new}_{train}, X^{new}_{test}, y_{train}, y_{test})$
  $importanceScores[index] \leftarrow predScore - baseScore$
end while
$lowerBound \leftarrow -min(importanceScores)$
$positiveImportanceScores \leftarrow array[N_{features}]$
while $index < N_{features}$ **do**
  $positiveImportanceScores[index] \leftarrow lowerBound + importanceScores[index]$
end while

The process starts by instantiating an array of null values characterized by $N_{features}$ dimensions, i.e., the number of considered dimensions, and by training and evaluating the compiled model on the original set of features. For each input dimension in the original set, a new set of inputs is generated by deleting the given feature associated with the considered dimension; then, the same model is trained and evaluated on the new input. At each iteration, the array initialized at the beginning is populated with the differences between the prediction scores obtained by evaluating the model with the original set of features and the temporary subsets of input dimensions. Finally, the result is obtained by adding the negation of the minimum occurrence to the elements of the array.

The procedure allows one to evaluate the importance of a given feature by computing the effect on the performances in terms of Mean Best Price Error. In the case in which the elimination of a given feature results in better performances, the related dimension is considered relevant for the decision. Its relevance is directly proportional to the difference between the errors obtained by evaluating the model with and without the given dimension.

5.5. Results

The best model obtained is characterized by online learning for backpropagation and Adam optimization at the learning rate of 10^{-2}. The proposed model was tested by considering different subsets of opinions to study the performances in different conditions of uncertainty. The experimentation was performed by considering the most important features obtained through the process described in Section 5.4.

In Table 1, the importance of the considered input dimensions is shown in the form of a ranking. In particular, the features providing an importance score lower than or equal to 0.056 contributed to a higher MBPE in the prediction.

Table 1. Results of the feature importance analysis obtained through the proposed method.

Feature	Importance	Feature	Importance
Composure	0.137	Long shots	0.073
GK reflexes	0.136	Short pass	0.072
Interceptions	0.13	Balance	0.071
Curve	0.129	Position	0.064
Acceleration	0.124	FM sentiment-society evaluation (P_3)	0.06
FIFA evaluation (P_4)	0.124	Vision	0.058
Stamina	0.102	GK handling	0.056
GK positioning	0.102	Finishing	0.05
Weight	0.1	Aggression	0.048
Volleys	0.099	FK accuracy	0.047
Marking	0.091	Crossing	0.037
Ball control	0.09	Penalties	0.037
Preferred positions	0.08	Slide tackle	0.034
Stand tackle	0.086	Heading	0.031
TM evaluation (P_1)	0.083	Dribbling	0.027
GK diving	0.078	Age	0.025
Strength	0.077	Jumping	0.023
FM value (P_2)	0.077	Agility	0.016
Reactions	0.076	GK kicking	0.014
Wage	0.075	Sprint speed	0.006
Height	0.074	Preferred foot	0.0

Table 2 shows the Mean Best Price Error, expressed in millions of euros, regarding different sets of opinions after feature selection. Except for the evaluation of the model by considering Plausibility, Credibility, and Possibility opinions individually, a significant increase in the performances was found after feature selection. However, the set of dimensions related to Credibility and Possibility increases the uncertainty in the prediction. The analysis of the differences between the groups of re-scaled input features associated with a prevalence, during the prediction phase, in one of the weights α_1, α_2, α_3, and α_4, provided significance for the dimensions of *height* ($p = 0.006$) and *marking* ($p = 0.039$). For all the experiments, the training and testing were performed on 338 and 60 samples, respectively.

The best performances were obtained by considering the set composed of Probability and Plausibility, as well as the Probability considered individually. The results identify substantial differences between the decision tests conducted using the different sets of opinions. For the problem addressed, concerning the prediction of the transfer cost of the players, we obtained that the smallest error, in the prediction phase, is reached by using the decision associated with the Probability (0.86 MBPE, 0.61% of normalized error) only. The worst performances, however, were found individually considering the opinions of Plausibility, Credibility, and Possibility. This result corresponds to the prodromal theory of decision and reasoning in uncertainty and info-incompleteness conditions [1], since the obtaining, and the use, of particularly relevant opinions concerning the sphere of Probability represents a condition of info-completeness. Conversely, by eliminating the direct evidence, i.e., by neglecting the Probability opinion, there is a larger error in the prediction phase. The decision in conditions of info-incompleteness can introduce greater uncertainty in the decision phase, as the lack of direct evidence forces the decision-maker to evaluate the opinions deriving from experts, sentiments, and subjects of weaker relevance. The prediction problem addressed in the present study can be traced back to the hierarchical characterization of the *overlap model with a shift based on probability* taken up in Section 2. The decision hierarchy is in alignment with the results based on the increase in the error as a function of the type of opinion evaluated; for the present problem, the order of priority reflects the increasing order of Mean Best Price Error on the test set: (i) Probability (0.86 MBPE, 0.61% of normalized error); (ii) Credibility (2.34 MBPE, 1.79% of normalized error); (iii) Plausibility (2.48 MBPE, 0.90% of normalized error); (iv) Possibility (2.87 MBPE, 2.21% of normalized error).

Table 2. Performance in predicting the best price by considering different sets of opinions and the most important features.

Model Evaluation Considering the Most Important Features		
Chosen Opinions	**Mean Best Price Error (mln)**	**Normalized Error (%)**
Probability, Plausibility, Credibility, Possibility	1.01	0.72
Plausibility, Credibility, Possibility	2.25	1.72
Probability, Plausibility	0.91	0.64
Credibility, Possibility	2.41	1.85
Probability	0.86	0.61
Plausibility	2.48	1.90
Credibility	2.34	1.79
Possibility	2.87	2.21

Following the logic in the approach of *overlap with shift based on probability*, for each player, the most promising final decision is taken based on the following steps:

1. The final decision is made based on the Probability opinion, if available, with an expected error equal to 0.86 MBPE (0.61% of normalized error), proportional to the weight α_1 estimated by the model;
2. If the Probability opinion is not available, the final decision is made based on the Credibility opinion, if available, with an expected error of 2.34 MBPE (1.79% of normalized error), in proportion to the α_3 weight estimated by the model;
3. If the Probability and Credibility opinions are not available, the final decision is taken based on the Plausibility opinion, if available, with an expected error of 2.48 MBPE (0.90% of normalized error), in proportion to the α_2 weight estimated by the model;
4. If the Probability, Credibility, and Plausibility opinions are not available, the final decision is taken based on the Possibility opinion, if available, with an expected error of 2.87 MBPE (2.21% of normalized error), in proportion to the α_4 weight estimated by the model.

To optimize the performance in conditions of info-incompleteness, an extension can be added to the aforementioned steps that involve the simultaneous contribution of multiple opinions. In particular, in the case the opinions of Plausibility, Credibility, and Possibility are available, it is advisable to make the decision considering the weighted sum of the related contributions, as this solution provides a lower Mean Best Price Error than what would be obtained by considering only the opinion concerning Credibility.

By extending the logic of the *overlap with shift based on probability* approach to the joint evaluation of several opinions, the most promising final decision is optimized based on the following steps:

1. The final decision is made based on the Probability opinion, if available, with an expected error equal to 0.86 MBPE (0.61% of normalized error), proportional to the weight α_1 estimated by the model;
2. If the Probability opinion is not available, the final decision is made based on the opinions of Plausibility, Credibility, and Possibility, if available, with an expected error equal to 2.25 MBPE (1.72% of normalized error), based on the sum of the opinions weighted by the values α_2, α_3, and α_4 identified by the model;
3. If the Probability and Plausibility opinions are not available, the final decision is made based on the Credibility opinion, if available, with an expected error of 2.48 MBPE (1.90% of normalized error), in proportion to the α_3 weight estimated by the model;
4. If the Probability, Plausibility, and Credibility opinions are not available, the final decision is made based on the Possibility opinion, if available, with an expected error of 2.87 MBPE (2.21% of normalized error), in proportion to the α_4 weight estimated by the model.

It is further interesting to note that, unlike the starting theoretical model, in which Plausibility was characterized by a higher priority than Credibility, in this case, the opposite is true. For the problem of football players' evaluation, the experts' opinion is weaker.

To provide a visual example of the prediction performances, Figure 4 shows a comparison between the ground truth and the best price predicted by the model trained with the subset of the most important features (1.01 MBPE, 0.72% of normalized error).

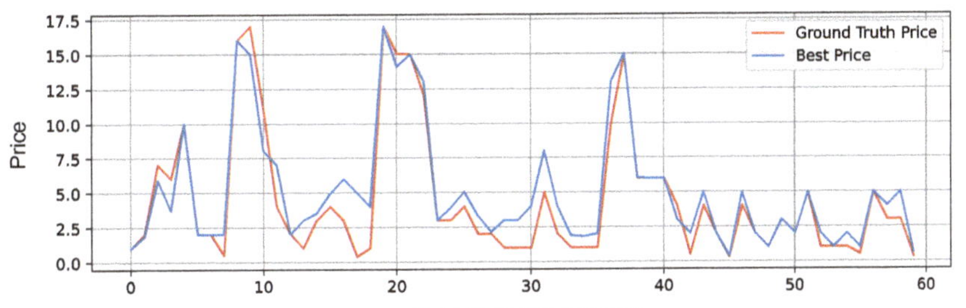

Figure 4. Comparison between the ground truth and the best price prediction obtained through the proposed model, after feature selection, based on Probability, Plausibility, Credibility, and Possibility opinions.

To validate the statistical evidence of the above results, it was decided to perform hypothesis testing concerning the differences between the models trained through different sets of opinions. To compute p-values, two-thousand bootstrapping sets have been generated from the predictions obtained by evaluating the models on the test set. Each generated set is compared with the ground truth to compute an MBPE score, which in turn is subtracted with the MBPE score obtained by evaluating another model. For each pair of models, a distribution of the differences is generated and the p-value is computed. Table 3 shows the results of the analysis by considering 0.05 as the threshold for significance.

The results show that, for the problem addressed in the present study, there is no statistical significance in adopting all the four opinions instead of considering the Probability only; to obtain optimal performances, there is strong evidence for considering the Probability, both exclusively and in conjunction with Plausibility, Credibility, and Possibility. Furthermore, the choice of considering both Credibility and Possibility is significantly better than that of considering Possibility only. Regarding other couples of hypotheses, no further statistical significance was found.

Table 3. Significance testing results regarding the differences in predicting the best price between pairs of opinions' subsets.

	p-Values of the Pair Differences							
	P_1, P_2, P_3, P_4	P_2, P_3, P_4	P_1, P_2	P_3, P_4	P_1	P_2	P_3	P_4
P_1, P_2, P_3, P_4	>0.05	0.002	0.29	<0.001	0.08	<0.001	<0.001	<0.001
P_2, P_3, P_4	0.002	>0.05	<0.001	0.512	< 0.001	0.1	0.67	0.067
P_1, P_2	0.29	<0.001	>0.05	<0.001	0.620	<0.001	<0.001	<0.001
P_3, P_4	<0.001	0.512	<0.001	>0.05	<0.001	0.8	1.204	0.04
P_1	0.08	<0.001	0.620	<0.001	>0.05	<0.001	<0.001	<0.001
P_2	<0.001	0.1	<0.001	0.8	<0.001	>0.05	0.50	0.23
P_3	<0.001	0.67	<0.001	1.204	<0.001	0.50	>0.05	0.109
P_4	<0.001	0.067	<0.001	0.04	<0.001	0.23	0.109	>0.05

6. Discussion

The results obtained through the experimentation of the proposed model applied to the prediction of evaluation of football players show that the opinions related to Plausibility, Credibility, and Possibility are not useful in conditions of info-completeness. Strong evidence was found in employing the Probability only to make the final decision. This result is coherent with [10,11], as the crowdsourcing, from the introduction of Transfermarkt, has become the main source for evaluation of football players. Conversely, the opinions concerning Plausibility, Credibility, and Possibility are essential in conditions of info-incompleteness; in this context, the optimal decision can be achieved by adopting an arbitrary set of opinions, except for the couple Credibility–Possibility, which was found to be more promising than the Possibility considered exclusively.

In conclusion, the performance in estimating values of football players is 0.86 MBPE (0.61% of normalized error) in info-completeness conditions, while it is 2.47 ± 0.188 MBPE (1.89 ± 0.15% of normalized error) in info-incompleteness.

The above results allow the research concerning Artificial Intelligence methodologies and decision support systems to be extended, as they provide an approach that potentially improves the hypotheses for the solution of predictive tasks. The approach tested in this study is simple and applicable to any decision context: if, in the prediction phase, one or more opinions regarding the output are available, it could be useful to consider these inputs to improve the performance of the model. For instance, in the case we want to consider, accuracy could be improved by a case study different from the one chosen in this work, such as the classification of images, the use of probable, plausible, credible, and possible opinions, if available. The conditional was used as the aforementioned field of application has yet to be tested; the results of this experiment could be different from those obtained in the present study (e.g., the Credibility opinion could provide better performances than the Probability).

The model proposed in the present study can be conceived as a sort of *human-in-the-loop* model, i.e., a predictor that requires human interaction. Instead of a human, our model proposes the Probability, Plausibility, Credibility, and Possibility opinions, provided by certain sets of humans, which can be extracted automatically by certain information sources. Having a human being available for supporting the model in real-time inference is very cost-effective; instead, in the case the same support is found in accessible information sources, such as the web, the external support to the model becomes cheaper. The potential for improvement is high, as human-in-the-loop models were recently proved to be effective, especially in medicine and cybersecurity [22].

7. Conclusions and Future Work

In the present study, the problem of assessing the best opinion in conditions of uncertainty and info-incompleteness was addressed. To achieve this objective, we proposed a solution that provides a learning model that, starting from the observations related to Probability, Plausibility, Credibility, and Possibility, together with other relevant characteristics, provides the weights associated with the considered set of opinions. The proposed model minimizes the error of the results provided by the best price function, defined as the weighted sum of the considered opinions. The experiment was performed on a real case study concerning the market evaluation of soccer players by building a dataset based on the information sources of Transfermarkt, Football Manager, and FIFA21. The input space concerns features acquired one year before the subsequent transfer, while the ground truth is represented by the cost of the actual transfer. The experiments were carried out for a total of 398 occurrences by varying the set of opinions acquired for taking decisions, both in conditions of info-completeness and info-incompleteness; in the first case, the Probability was considered, while, in the second case, we limited the hypotheses to the set of Plausibility, Credibility, and Possibility only.

The results prove the consistency to the prodromal study taken as reference and that it is possible to reach an error for the price prediction of 0.86 MBPE (0.61% of normalized error) and 2.47 ± 0.188 MBPE (1.89 ± 0.15% of normalized error) on the test set in conditions of info-completeness and info-incompleteness, respectively. Furthermore, from the analysis of statistical significance, it was found that the Probability opinion is fundamental in conditions of info-completeness; instead, in conditions of info-incompleteness, it is possible to adopt any set that considers Plausibility, Credibility, and Possibility. Finally, it was found that the employment of the Credibility–Possibility pair represents a better choice compared to the assumption involving the Possibility opinion only.

A possible future work regards the extension of the present study for the assessment of opinions' relevance by minimizing the occurrence and reliability functions in conditions of uncertainty and info-incompleteness. Regarding a real-world application, it is possible to define a decision support system to assist the Atmosphere Arc model [23], which brings the real economy into the digital economy through a Decentralized Content Management System (DCMS) and an Oracle. The DCMS contains the documentation of human work, while the Oracle analyzes the documentation and distributes a blockchain token. The proposed model can be employed to support token production and the documentation of the activities concerning a soccer society, providing a better estimation of the value.

Author Contributions: All authors contributed equally to the present work. All authors have read and agreed to the published version of the manuscript.

Funding: This research received no external funding.

Institutional Review Board Statement: Not applicable.

Informed Consent Statement: Not applicable.

Data Availability Statement: Not applicable.

Conflicts of Interest: The authors declare no conflict of interest.

References

1. Iovane, G.; Gironimo, P.D.; Chinnici, M.; Rapuano, A. Decision and Reasoning in Incompleteness or Uncertainty Conditions. *IEEE Access* **2020**, *8*, 115109–115122. [CrossRef]
2. Polya, G. *Mathematics and Plausible Reasoning: Induction and Analogy in Mathematics*; Princeton University Press: Princeton, NJ, USA, 1954.
3. Polya, G. *Mathematics and Plausible Reasoning: Patterns of Plausible Inference*; Princeton University Press: Princeton, NJ, USA, 1990.
4. Shafer, G. *A Mathematical Theory of Evidence*; Princeton University Press: Princeton, NJ, USA, 1976.

5. Dempster, A.P. Upper and Lower Probabilities Induced by a Multivalued Mapping. *Ann. Math. Stat.* **1967**, *38*, 325–339. [CrossRef]
6. Smarandache, F.; Dezert, J. An introduction to DSMT. In *Advances and Applications of DSMT for Information Fusion (Collected Works)*; American Research Press (ARP): Rehoboth, NM, USA, 2009; Volume 3.
7. Dobson, S.; Goddard, J.A.; Dobson, S. *The Economics of Football*; Cambridge University Press: Cambridge, UK, 2017.
8. Behravan, I.; Mohammad, M.R.S. A novel machine learning method for estimating football players' value in the transfer market. *Soft Comput.* **2021**, *25*, 2499–2511. [CrossRef]
9. Müller, O.; Simons, A.; Weinmann, M. Beyond Crowd Judgments: Data-Driven Estimation of Market Value in Association Football. *Eur. J. Oper. Res.* **2017**, *263*, 611–624. [CrossRef]
10. Singh, P.; Lamba, P. Influence of crowdsourcing, popularity and previous year statistics in market value estimation of football players. *J. Discret. Math. Sci. Cryptogr.* **2019**, *22*, 113–126. [CrossRef]
11. Steffen, H.; Callsen-Bracker, H.; Kreis, H. When the crowd evaluates soccer players' market values: Accuracy and evaluation attributes of an online community. *Sport Manag. Rev.* **2014**, *17*, 484–492.
12. Pérez-González, B.; Fernández-Luna, A.; Vega, P.; Burillo, P. The relative age effect: Does it also affect perceived market value? The case of the Spanish LFP (Professional Football League). *J. Phys. Educ. Sport* **2018**, *18*, 1408–1411.
13. Majewski, S. Identification of Factors Determining Market Value of the Most Valuable Football Players. *Cent. Eur. Manag. J.* **2016**, *24*, 91–104. [CrossRef]
14. Yigit, A.T.; Samak, B.; Kaya, T. Football player value assessment using machine learning techniques. In *International Conference on Intelligent and Fuzzy Systems*; Springer: Cham, Switzerland, 2019; pp. 289–297.
15. Felipe, J.L.; Fernandez-Luna, A.; Burillo, P.; Riva, L.E.D.L.; Sanchez, J.; Garcia-Unanue, J. Money talks: Team variables and player positions that most influence the market value of professional male footballers in Europe. *Sustainability* **2020**, *12*, 3709. [CrossRef]
16. Franck, E.; Nüesch, S. Talent and/or popularity: What does it take to be a superstar? *Econ. Inq.* **2012**, *50*, 202–216. [CrossRef]
17. Kiefer, S. The Impact of the Euro 2012 on Popularity and Market Value of Football Players. *Int. J. Sport Financ.* **2014**, *9*, 95–110.
18. Yaldo, L.; Shamir, L. Computational Estimation of Football Player Wages. *Int. J. Comput. Sci. Sport* **2017**, *16*, 18–38. [CrossRef]
19. Devlin, J.; Chang, M.; Lee, K.; Toutanova, K. BERT: Pre-training of Deep Bidirectional Transformers for Language Understanding. *arXiv* **2019**, arXiv:1810.04805.
20. GitHub. IBM Developer Model Asset Exchange: Text Sentiment Classifier. Available online: https://github.com/IBM/MAX-Text-Sentiment-Classifier (accessed on 23 November 2021).
21. IBM Research. IBM Project Debater. Available online: https://research.ibm.com/haifa/dept/vst/debating_data.shtml (accessed on 23 November 2021).
22. Zagalsky, A.; Te'eni, D.; Yahav, I.; Schwartz, D.G.; Silverman, G.; Cohen, D.; Mann, Y.; Lewinsky, D. The design of reciprocal learning between human and artificial intelligence. *Proc. ACM-Hum.-Comput. Interact.* **2021**, *5*, 443. [CrossRef]
23. Atmosphere Arc Whitepaper (2018). Available online: https://atmospherearc.com/atmospherearc.pdf (accessed on 24 November 2021).

applied sciences

MDPI

Article

Development of a Stock Price Prediction Framework for Intelligent Media and Technical Analysis

Sibusiso T. Mndawe [1,*], Babu Sena Paul [2] and Wesley Doorsamy [2]

1 Department of Electrical and Electronic Engineering Science, University of Johannesburg, Johannesburg 2006, South Africa
2 Instituted for Intelligent Systems, University of Johannesburg, Johannesburg 2006, South Africa; bspaul@uj.ac.za (B.S.P.); wdoorsamy@uj.ac.za (W.D.)
* Correspondence: 201144173@student.uj.ac.za

Abstract: Equity traders are always looking for tools that will help them maximise returns and minimise risk, be it fundamental or technical analysis techniques. This research integrates tools used by equity traders and uses them together with machine learning and deep learning techniques. The presented work introduces a South African-based sentiment classifier to extract sentiment from new headlines and tweets. The experimental work uses four machine learning models for fundamental analysis and six long short-term memory model architectures, including a developed encoder-decoder long short-term memory model for technical analysis. Data used in the experiments is mined and collected from news sites, tweets from Twitter and Yahoo Finance. The results from 2 experiments show an accuracy of 96% in predicting one of the major telecommunication companies listed on the JSE closing price movement while using the linear discriminant analysis model and an RMSE of 0.023 in predicting a significant telecommunication company closing price using encoder-decoder long short-term memory. These findings reveal that the sentiment feature contains an essential fundamental value, and technical indicators also help move closer to predicting the closing price.

Keywords: sentiment analysis; LSTM; machine learning; deep learning; stock market; forecasting; fundamental analysis; technical analysis

Citation: Mndawe, S.T.; Paul, B.S.; Doorsamy, W. Development of a Stock Price Prediction Framework for Intelligent Media and Technical Analysis. *Appl. Sci.* **2022**, *12*, 719. https://doi.org/10.3390/app12020719

Academic Editor: João M. F. Rodrigues

Received: 3 December 2021
Accepted: 3 January 2022
Published: 12 January 2022

1. Introduction

The problem of minimising risk and maximising returns bundled with predicting future price movements is what stock market traders have been trying to solve for years. Many have provided tools and solutions to solve this massive problem of predicting the increase and decrease of selected companies' stock prices. These equity traders have depended on news headlines (fundamental analysis) and technical indicators (technical analysis) as tools for prediction. Stock price prediction and stock price movement prediction predict what the future prices will look like from observing past and present price data. Researchers have also realised that stock price prediction depends not only on historical data but also on social media data. In 2018, a social media influencer expressed her unhappiness with Snapchat on Twitter; her tweet caused the Snapchat share price to drop by 6%, wiping out $1.3 billion [1]. Tesla CEO Elon Musk caused Telsa's share price to decrease by 10% after sending a negative tweet concerning Tesla's share price [2]. The age of mobile devices has seen a vast increase in social media and news data. This social and news data is filled with essential facts and opinions that may be harvested to create a stock price movement prediction and closing price prediction tool. In developing this framework, the research aims to answer the following questions: Can a framework on social media and news headlines be used to forecast a company's stock price movement? Additionally, can that same framework use technical indicators as features to increase the accuracy of predicting a company's stock closing price?

This paper presents a framework for forecasting stock price movements and closing prices using machine learning and deep learning models. The framework uses a text classifier model for sentiment analysis on social media and news data. Using machine learning, the research also investigates predicting the increase and decrease of a selected company's stock based on sentiment score. Technical indicators are applied together with times series analysis and deep learning to predict the closing price of a telecommunication company listed on the Johannesburg Stock Exchange.

The presented research design is based on a quantitative research methodology, and the experimental method is used, whereby actual social media and stock price data is collected and analysed. This paper discusses using social media and technical indicators to predict stock price movements and closing prices; the article compares different machine learning models for stock price movement prediction (fundamental analysis) and deep learning models for closing price prediction (technical analysis). The remainder of the paper is arranged as follows. Section 2 looks at a summary of relevant literature. Section 3 discusses data collection and methodology. Results from the experiments are discussed in Section 4. The conclusion is provided in the final quarter of the paper.

2. Materials and Methods

2.1. Theoretical Fundamentals

2.1.1. Forecasting Using Sentiment Analysis

Sentiment analysis studies people's attitudes, opinions, emotions, and assessment towards concerning topics, issues, and current affairs. Shah et al. [3] developed a sentiment analysis dictionary for the financial sector to better understand the effects of news sentiments on the stock market. Their model considered the pharmaceutical market and how the news affected the stock. Their model also suggested buying, selling or holding based on the sentiment score. Wu et al. [4] proposed a sentence-based sentiment analyser based on the Chinese Sentiment Analysis Ontology Base and Hownet. This paper integrated sentiment analysis into a support vector machine using the rolling window method to explore the relationship between stock price movements and stock forum sentiment. The paper regards the sentiment feature as one of the leading indicators due to the valuable information it carries.

2.1.2. Forecasting Using Recurrent Neural Networks

Li et al. [5] proposed a multi-input LSTM model for stock market prediction that mines relevant information from low relational features and removes irrelevant information using additional input gates. The paper included data from the Chinese stock market and prices related to the stocks. Using a two-stage attention mechanism and related stock prices improved the efficiency and accuracy of the model. D. Lein Minh et al. [6] proposed a sentiment word-embedding Stock2Vec trained on Hardvard IV-4 and a two-way gated recurrent neural unit for stock price direction prediction. The paper used historical S&P 500 prices and articles from Bloomberg and Reuter to predict the S&P 500 stock price movement. The paper showed that Stock2Vec can handle financial datasets more efficiently and that the two-way gated recurrent unit outperforms advanced models, including the gated recurrent unit and the long short-term memory.

2.1.3. Forecasting Using Hybrid CNNxLSTM and ConvLSTM Network

Xingjian et al. [7] introduced ConvLSTM as a short intensity prediction method. Lee & Kim [8] developed NuNet, a framework constructed using the ConvLSTM network. This study's framework successfully learned high-level features from KOSPI200, FTSE100, and S&P500 stock market data. Livieris et al. [9] proposed a CNNxLSTM model for accurate gold price movement and gold price prediction. The model increased its performance by combining LSTM layers with convolutional layers. Chen et al. [10] proposed a framework constructed using the ConvLSTM model for short-term traffic flow prediction. The model performed better than vanilla LSTM, stacked LSTM, and bidirectional LSTM.

2.1.4. Stock Market Prediction

Khan et al. [11] used algorithms to evaluate the effects of data from financial news and social media on the stock market. This paper used spam tweet reduction and feature selection to increase the quality and performance of predictions. Khan et al. [12] developed a framework that checks whether public and political domain sentiments affect company market trends. This paper showed an improvement of about 3% due to the sentiment feature. Vargas et al. [13] used two technical indicators and financial news articles to input a deep learning model for stock price prediction. This study compared two models for financial news and technical indicators and showed that the addition of technical indicators and financial news stabilises and improves the output. Chen and Shih [14] proposed a stock movement prediction framework based on Chinese news and technical indicators. The paper also proposed the use of the GATSP algorithm. Both experiments show the effectiveness of the two methods.

2.2. *Background*

In this section of the paper, a brief background is given on the machine and deep learning models considered by the paper to achieve a better understanding. The section is split into two; the first portion covers machine learning models used in the fundamental analysis experiment, and lastly, deep learning models in the technical analysis experiment are discussed.

2.2.1. Machine Learning Models

A support vector machine (SVM) is a supervised machine learning model used in classification. An SVM finds an optimal way to divide a dataset into two categories and determines the hyperplane from any point in the training dataset [15].

Fisher formulated linear discriminant analysis in 1936, and Welch in 1939. This classification and discrimination model is constructed with labelled observations from a dataset and a set of a new unlabeled dataset used to predict the dataset [16].

The decision tree is a classification and covers regression machine learning algorithms influenced by real-life analogy. Each leaf node on the tree is assigned a class label, and the root and other non-terminal nodes split the records with different test conditions [17].

Random forest was introduced in 2001 by Leo Breiman; this classification or regression machine learning model also gets its name from real-life analogy. A random forest comprises many decision trees that work together to produce a class prediction, and the trees with the most predictions become the final prediction [18].

2.2.2. Deep Learning Models

Long short-term memory (LSTM) was introduced by Hochreiter and Schimidhuber. The LSTM stems from recurrent neural networks but differs in architecture. Recurrent neural networks suffer from the exploding and vanishing gradient problem, and the LSTM solves that problem, as it can learn long-term dependencies because of its feedback loop. The specialty of LSTM is the cell state, also known as the memory bank; this is a horizontal line running through the LSTM block that carries information from the previous timestamps [19].

2.3. *Methodology*

This paper develops two frameworks: one to predict stock price movements using fundamental analysis and the second to predict the stock closing price using technical analysis. The paper determines a sentiment feature from tweets and news headlines related to South African companies in the telecommunication industry. A comparison is employed between four machine learning models and a sentiment classifier to predict stock price movement and use five LSTM architectures and technical indicators to predict the closing price. It should be highlighted that the framework does not include a recommender system on which stock to buy or sell or validate social media postings. The presented research focuses on the closing prices on both experiments due to the occurrence of news headlines being posted either at the end or beginning of the day.

2.3.1. Sentiment Classifier

Data from Twitter was first collected using an API called GetOldTweets [20] and then taken through the pre-processing stage, including labelling, to prepare for the sentiment classification.

Step 1, Twitter scraper: In this step, the GetOldTweets python API gathered South African-based tweets to introduce South African grammar into the sentiment classifier. Tweets were collected from famous South African television shows, the Datamustfall movement, telecommunication companies on the JSE, and finally, the Stanford Twitter Sentiment Corpus. The paper uses hashtags "#" to collect relevant tweets. These hashtags are considered due to the amount of pertinent sentiment they provide to the classifier. The popular South African television dataset uses the following hashtags: Date-My-Family, South African Idols, and Uyajola99, which amounts to a total of 3000 tweets. The next 1000 come from the hashtag Datamustfall movement, and the last 3000 tweets are from telecommunication companies. The paper considered the following hashtags: Vodacom, Mtn, and Telkom, and excluded tweets from their companies' Twitter accounts for this data, as this only has relevant marketing tweets.

Step 2, Data cleaning: To prepare the dataset for the training of the sentiment classifier, the dataset needed to go through a pre-processing step. As a result, a python function was written using regular expressions and a few other techniques to clean the dataset. The function first drops all empty rows in the dataset and turns all the letters into lower case. Next, the function removes all URLs, usernames, dates, whitespace, and the hashtag sign (#). Finally, columns are renamed to text, and all the collected tweets are targeted and labeled such that 0 = negative, 2 = neutral, and 4 = positive. This was conducted using the same method used to label the text in the Stanford Twitter Sentiment Corpus [21]. The final clean dataset consists of 14,000 tweets.

Step 3, Sentiment analyser training: To train the sentiment classifier, a pre-trained language model, BERT (Bidirectional Encoder Representation from Transformers), was chosen [22]. The paper can train its sentiment analysis classifier using the Hugging Face Python library to fine-tune the BERT model [23].

2.3.2. Fundamental Analysis

Fundamental analysis studies the stock market by analysing new headlines, economic and social reports, and political forces that may affect the price movement [24]. This paper employs a similar strategy through the use of fundamental analysis by looking at news headlines and social media reports in its research. The fundamental analysis experiment is split into three different sub-experiments using four different models to evaluate performance: linear discriminant analysis, support vector machine, decision tree, and random forest. To assess the performance of the models, the paper uses accuracy, precision, recall, the F-measure, and a confusion matrix. The data is split into 80/20 for training and testing; three datasets for fundamental analysis were constructed, considering dates from 2012 to 2019. The first dataset was the news headlines dataset, scraped off the Money web site. This dataset consisted of news headlines linked to the Vodacom Group Limited Company. The second dataset was composed of tweets collected using the GetOldTweets API with the hashtag #vodacom. The total length of the combined dataset was 893 rows. The combined dataset was then labelled according to price movement on the closing price for that particular day, with +1 indicating an increase and −1 indicating a decrease.

The data collected from the scraped Moneyweb site and the #vodacom tweets were processed using the same python function mentioned above. This function removes all URLs, usernames, dates, whitespace, and the hashtag sign (#) to prepare the dataset for the experiments. The last stage to take the dataset through is the transformation stage. Here, categorical encoding and scaling were applied to the data to transform the dataset. These two techniques convert text into numerical representations and scale the data into a specific range, respectively [25]. Categorical encoding on the fundamental analysis experiment was applied to the target column. This column contains the variables 'UP' and 'Down', which categorises whether Vodacom's stock price went up or down on a given day. These up-down variables were replaced with 1 and −1. The dataset was also scaled using the sklearn MinMaxScaler Python library [26]; this technique scales all the numerical data between 0 and 1 but does not scale the target and sentiment column.

Feature engineering is the process of extracting new features by transforming the current features [25]. The sentiment was extracted from all news headlines and tweets in the collected dataset, and this sentiment feature determined whether the polarity of the text was negative, positive, or neutral.

2.3.3. Technical Analysis

Technical analysis studies market trends gathered from volume and price movement. This paper looks at Vodacom Group Limited, one of South Africa's biggest telecommunication companies. Like the fundamental analysis experiment, this technical analysis experiment is also split into three sub-experiments. Let us now introduce two datasets to this experiment, the closing price dataset (univariate) and the closing price with technical indicators dataset (multivariate). An 80/20 split was applied to the datasets for training and testing. Six different long short-term memory (LSTM) architectures are introduced: ordinary LSTM, bidirectional LSTM, stacked (deep) LSTM, convLSTM, CNNxLSTM. Encoder-decoder LSTM was applied in the last experiment. The ordinary LSTM architecture is comprised of 2 layers: 1 LSTM layer with 200 units and a dense layer at the output. Stacked LSTM has 3 layers: 1 LSTM layer with 200 units, a second with 100 units and a dense layer with 1 output unit. Bidirectional LSTM is comprised of 3 layers; the first layer is a bidirectional LSTM with 200 units, while the second LSTM layer has 100 units. ConvLSTM has three layers. First is a ConvLSTM layer with 64 filters; the second is a flatten layer, and this is followed by a dense layer with 1 output. To evaluate the models, the present paper considers using the mean squared error (MSE) and root mean squared error (RMSE). The MSE and RMSE are the sum of the variance estimator and the squared basis of the estimator. These two are used to determine the model's performance, and the result closest to zero shows which model performed better.

Mean Squared Error

$$\mathrm{MSE}(y,\hat{y}) = \frac{1}{n_{\mathrm{samples}}} \sum_{i=0}^{n_{\mathrm{samples}}-1} (y_i - \hat{y}_i)^2 \quad (1)$$

MSE = mean squared error
n = number of data points
y_i = observed values
$\hat{y}_i$ = predicted values

The present paper collected stock price data from 1 January 2012 to 1 December 2019 on Vodacom Group Limited. Pandas-DataReader, a Pandas Python library, pulls all this stock data from Yahoo finance with the tag VOD.JO [27]. The DataReader pulled Vodacom's opening price, high price, low price, volume, and closing price. The closing price was extracted from this data to make up the univariate dataset. The second dataset is comprised of three technical indicators: the three moving average, MACD, and Bollinger Bands. These indicators were considered due to their popularity in the equity trading space, as most traders use them. This is an excellent way to mimic what traders would use to enhance their decision making. This collected a total of 2040 rows and 17 columns. The data collected using the Python Pandas DataReader consisted of 2040 rows. The DataReader does most of the hard work in making sure the data is imported in a friendly format, ready for the modelling stage.

Before modelling, the last stage was data transformation to both the univariate and multivariate datasets. The present paper scaled the data using the sklearn MinMaxScaler Python library, which scales the data between zero and one.

The experiment on technical analysis extracted features from Vodacom using three technical indicators: the three moving average, MACD, and Bollinger Bands. In total, the 3 technical indicators yielded a total of 11 features.

3. The Developed Model

3.1. Data Summary

The data collected for Experiment 1 were split into three different datasets to test whether the addition of new data points and features extracted from the data helps in the quest to predict stock price movements. First, the news headlines dataset was introduced; this was purely comprised of news headlines collected from the Moneyweb site. The second dataset was the news headlines with sentiment analysis; this dataset had the addition of the sentiment feature extracted using the sentiment analysis model mentioned above. The last dataset was the news headlines and tweets with sentiment analysis; like the second dataset, this had sentiment extracted using the sentiment analysis model, but with an addition of tweets concatenated with the news headlines. The dataset mentioned above was used in the three experiments for the fundamental analysis. An 80/20 split was applied to the dataset for training and testing. Experiment 2 has two primary datasets used in all three sub-experiments: the univariate and multivariate datasets. The first dataset was comprised of the Vodacom Group Limited closing price pulled from Yahoo finance using Python Pandas Datareader library. The second dataset was the Vodacom Group Limited stock data, including the opening price, high price, low price, volume, closing price, and features extracted from three technical indicators, the three moving average, MACD, and Bollinger Bands. An 80/20 split was applied to the dataset for training and testing.

3.2. Experiments

The first experiment used a count vectoriser with ngram_range set to (1, 2). The count vectorizer model from the Scikit-learn Python library converts the text corpus into a matrix of token counts. The target column, together with the matrix of token counts, were used to train the models. Three forms of machine learning were considered in this experiment: the decision tree, random forest, and support vector machine.

In the second experiment, sentiment analysis was introduced to the dataset in Experiment 1.1. A sentiment feature was extracted from Vodacom's news headlines, whether negative, positive, or neutral. Other additional features were introduced, including Vodacom's opening, price, high price, low price, volume, and closing price. The scaled financial data and the sentiment feature's polarity are used to train the four machine learning models: the decision tree, random forest, support vector machine, and linear discriminant analysis.

The last experiment introduced the last dataset, which was composed of the news headlines, tweets, and sentiment feature. The paper follows the same method as Experiment 1.2. A new sentiment feature was extracted from the news headlines and tweets. Vodacom's opening, price, high price, low price, volume, and closing price are again included as features together with the extracted sentiment feature. The scaled financial data and the sentiment feature's polarity are used to train the four machine learning models: the decision tree, random forest, support vector machine, and linear discriminant analysis.

In the first experiment, the LSTM sequence model and univariate time series forecasting were introduced. After transforming the univariate dataset, the dataset must be reshaped before presenting it to the models. LSTM models expect a three-dimensional data shape at the input in this order; these dimensions are samples (batch size), time steps (a point of observation in the samples), and features (an observation at a time step) [19]. The two hybrid architectures, the CNNxLSTM and ConvLSTM input shape, are four-dimensional; the dimensions are samples, subsequences, time steps, and features. The present study applied the correct type of reshaping to all six LSTM architectures.

Next, the multivariate dataset, consisting of Vodacom's stock data and additional technical indicators as features, was introduced. The same transformation and reshaping were applied to the dataset with one change. The feature variable was altered, as this dataset now has more than one feature. The same six LSTM architectures are used in this experiment.

The last experiment introduces a different objective to that of the first two experiments. The same multivariate dataset from the previous experiment was used, but here, the encoder-decoder model with an altered multi-step dataset was presented. The data sequence was altered, as the model expects a multi-parallel time series.

3.3. Results from Fundamental and Technical Analysis Experiments

This section discusses results from the two main experiments, the fundamental and technical analysis experiments. We first begin with experiment one, which is split into three sub-experiments, whereby data is divided into three datasets for fundamental analysis: the news headlines dataset, the news headline and sentiment analysis dataset and lastly, the news headlines, tweets and sentiment analysis dataset. Experiment two is also split into three sub-experiments using three datasets, namely the univariate dataset, the multivariate dataset and, finally, the multi-step dataset for technical analysis.

3.3.1. Fundamental Analysis: Experiment 1

The main objective for Experiment 1 was to predict whether Vodacom's stock would go up or down for 14 days by analysing and extracting sentiments from news headlines and tweets linked to Vodacom. The experiment used a comparison of four machine learning models in its quest to predict the stock movement. To evaluate the performance of all three experiments, the experiment used accuracy, precision, recall, the f1-score, and a confusion matrix.

Table 1 shows the results from Experiment 1.1; this experiment yielded the lowest accuracy in the three sub experiments. It can also be seen that the random forest model was the best performing model on the news headlines dataset in Experiment 1.1. Experiment 1.2 introduced the same objectives as Experiment 1.1, and a sentiment feature and Vodacom's stock financial data were also introduced to the dataset. Experiment 1.2 returns a far better accuracy in all the models due to the abovementioned features. Linear discriminant analysis was the best performing model, with an accuracy of 94%, as seen in Table 2. The last experiment followed the same objective as the first two; in this experiment, tweets related to Vodacom were added onto the news headlines, the sentiment feature from this text, and Vodacom's stock financial data. The experiment outperformed both experiments, and the linear discriminant analysis was also the best performing model, with an accuracy of 96%, as seen in Table 3.

Table 1. Experiment 1.1 results.

Models	Accuracy
Support vector machine	44%
Decision tree	46%
Random forest	49%

Table 2. Experiment 1.2 results.

Models	Accuracy
Support vector machine	54%
Decision tree	75%
Random forest	66%
Linear discriminant analysis	94%

Table 3. Experiment 1.3 results.

Models	Accuracy
Support vector machine	49%
Decision tree	82%
Random forest	74%
Linear discriminant analysis	96%

3.3.2. Technical Analysis: Experiment 2

The objective of Experiment 2 was to predict Vodacom's closing price using 20 days of Vodacom data, thereby predicting day 21. The experiment introduced two datasets; the first was the univariate (closing price) dataset, and the second was the multivariate (technical indicators) dataset. Five LSTM architectures were presented in the first two experiments, and in the third, the encoder-decoder LSTM was introduced. All LSTM architectures applied the mean square error as their loss function and the Adam optimiser to update the weights during training. To evaluate the performance of the experiments, the mean squared error and root mean squared error were used.

Table 4 shows a summary of the results from Experiment 2.1. First, ordinary LSTM is introduced; this model has 2 layers with 200 units and a dense layer with 1 output. The model is trained for 100 epochs with a learning rate of 1 e^{-4}. The prediction results are displayed against the test data in Figure 1.

Table 4. Experiment 2.1 results.

Models	Ordinary LSTM	Stacked LSTM	Bidirect LSTM	CNN LSTM	Conv LSTM
MSE	48,991.89	48,766.52	48,946.1	53,773.47	54,060.72
RMSE	221.34	220.83	221.23	231.8	232.51

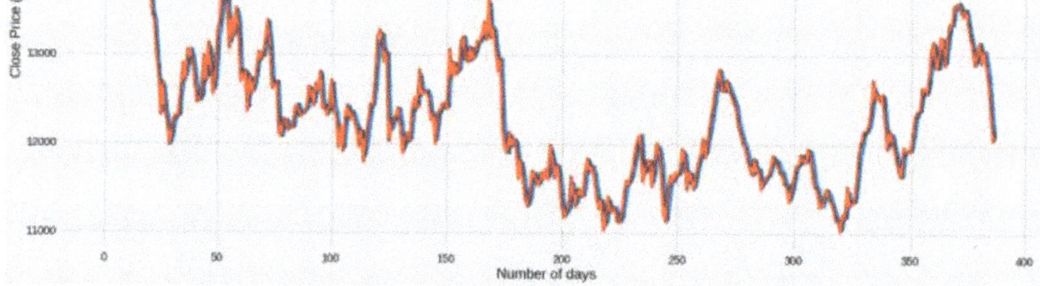

Figure 1. Ordinary LSTM Experiment 2.1.

Stacked LSTM, which was the best performing model in this section, was comprised of 2 LSTM layers with 200 and 100 units each and a dense layer at the output. The model was trained for 100 epochs with a learning rate of 1 e^{-4}. The prediction results are displayed against the test data in Figure 2.

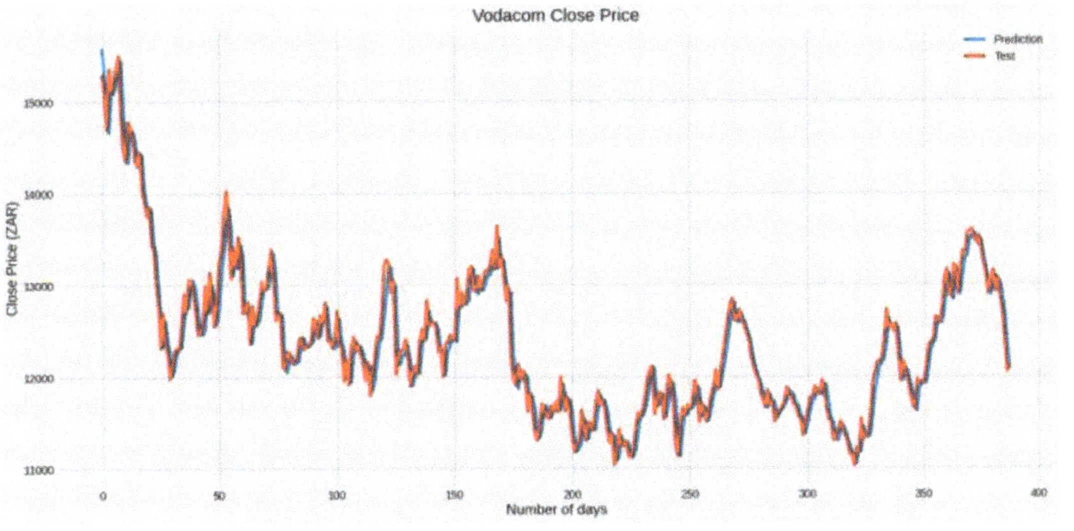

Figure 2. Stacked LSTM Experiment 2.1.

The third model was the bidirectional LSTM; this model was made up of three layers. The first layer had 200 units, the second had 100 units, and the last was a dense layer at the output. The model was trained for 100 epochs with a learning rate of 1 e^{-4}. The prediction results are displayed against the test data in Figure 3.

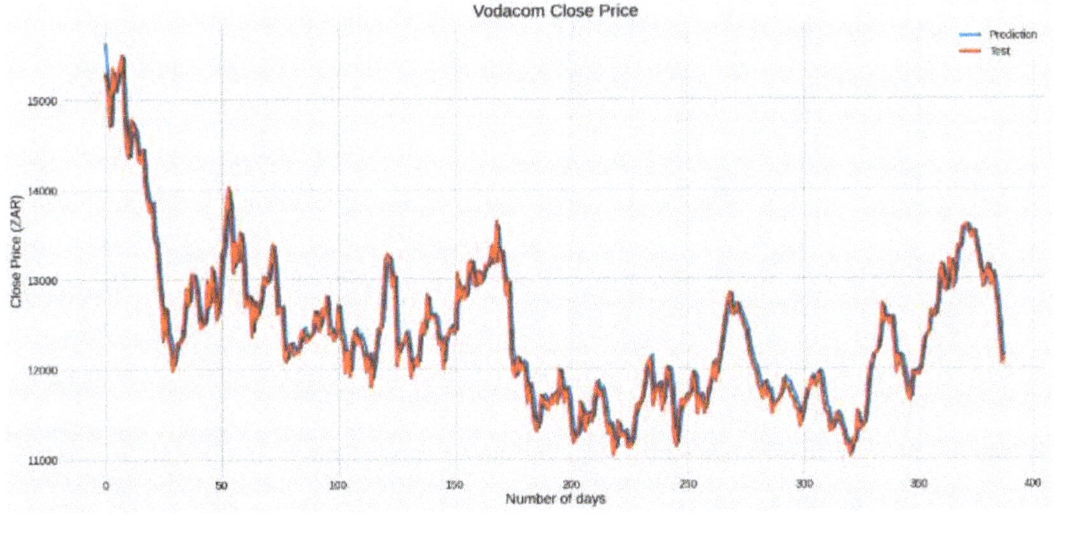

Figure 3. Bidirectional LSTM Experiment 2.1.

The convolutional neural network LSTM model has a total of eight layers, as seen in Figure 4. This model was trained for 100 epochs with a learning rate of 2 e^{-4}. The prediction results are displayed against the test data in Figure 5.

```
Model: "sequential_17"
_________________________________________________________________
Layer (type)                 Output Shape              Param #
=================================================================
time_distributed_33 (TimeDis (None, None, 10, 32)      96
_________________________________________________________________
time_distributed_34 (TimeDis (None, None, 5, 32)       0
_________________________________________________________________
time_distributed_35 (TimeDis (None, None, 160)         0
_________________________________________________________________
lstm_28 (LSTM)               (None, None, 200)         288800
_________________________________________________________________
lstm_29 (LSTM)               (None, 100)               120400
_________________________________________________________________
dense_27 (Dense)             (None, 30)                3030
_________________________________________________________________
dense_28 (Dense)             (None, 10)                310
_________________________________________________________________
dense_29 (Dense)             (None, 1)                 11
=================================================================
Total params: 412,647
Trainable params: 412,647
Non-trainable params: 0
_________________________________________________________________
```

Figure 4. CNNxLSTM.

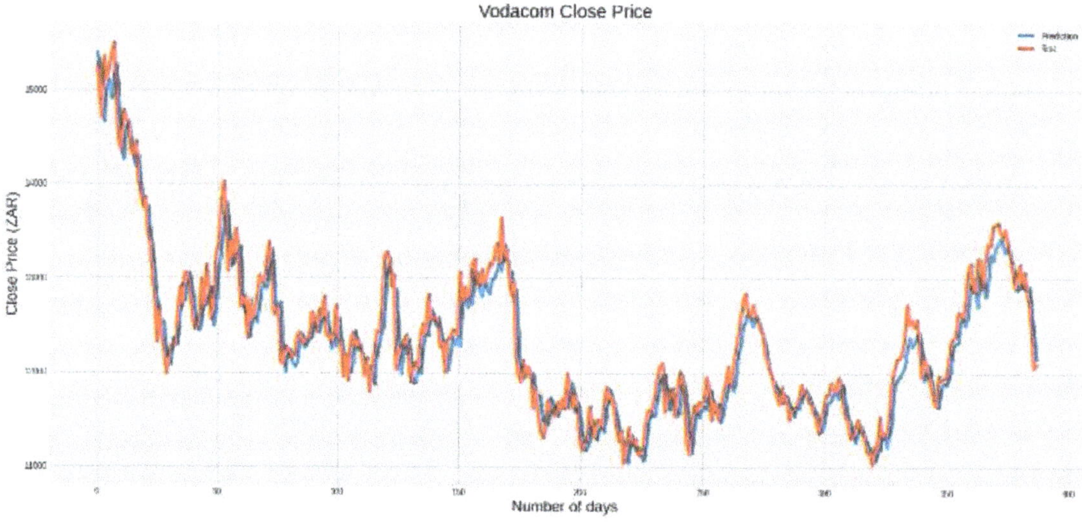

Figure 5. CNNxLSTM Experiment 2.1.

The last hybrid experiment is ConvLSTM, shown in Figure 6 with three layers. The first layer is a ConvLSTM with 64 filters; the second is a flatten layer and the third and final layer is a dense layer at the output. The model is trained for 100 epochs with a learning rate of $2\,e^{-4}$.

The second part of Experiment 2 introduces the multivariate dataset applied to the same five LSTM models as Experiment 2.1. With the addition of three technical indicators that generate 16 features to the dataset, there is a significant improvement in this experiment, as seen in Table 5. Experiment 2.2 follows the same process used in Experiment 2.1, and all models have the same architecture, including the same amount of layers, amount of epochs and the same learning rate as used in Experiment 2.1 to train the models.

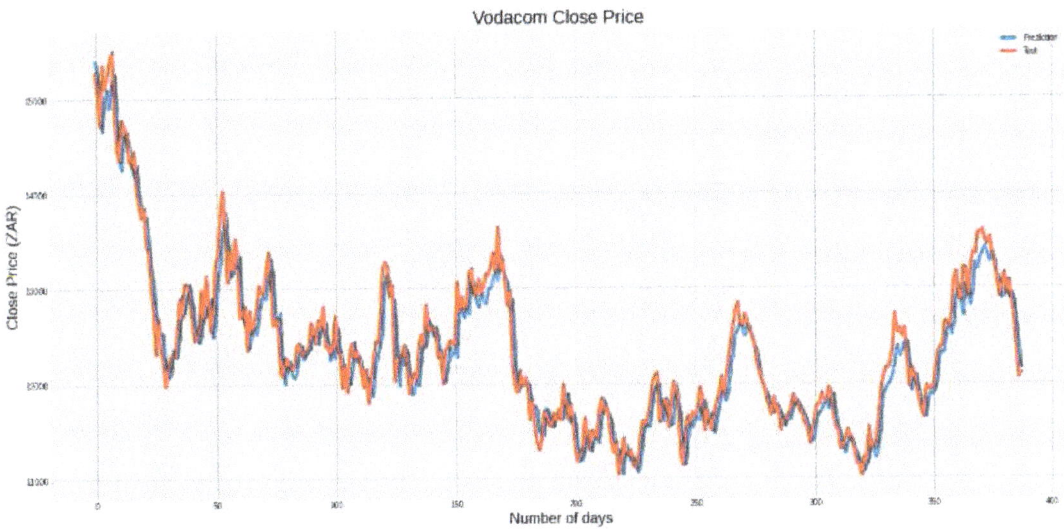

Figure 6. ConvLSTM Experiment 2.1.

Table 5. Experiment 2.2 Results.

Models	Ordinary LSTM	Stacked LSTM	Bidirect LSTM	CNN LSTM	Conv LSTM
MSE	5706.57	17,627.8	28,181	62,873.73	9295.55
RMSE	75.54	132.77	167.87	250.75	96.41

The ordinary LSTM is the best-performing model in Experiment 2.2, with an RMSE of 75.54, as shown in Figure 7.

Figure 7. Ordinary LSTM Experiment 2.2.

The stacked and bidirectional LSTM shows a significant improvement in RMSE compared to Experiment 2.1, with a final RMSE of 132.77 and 167.87. On the other hand, the hybrid models returned mixed results. The CNNxLSTM model Figure 8 saw an increase in RMSE, which resulted in impaired performance compared to experiment 2.1 in predicting the closing price. The ConvLSTM shows an improvement in prediction performance compared to the first experiment.

Figure 8. CNNxLSTM LSTM Experiment 2.2.

Overall, the performance of the models in Experiment 2.2 showed an improvement compared to the first experiment. This may be due to the addition of three technical indicators to the dataset.

The last experiment introduced the encoder-decoder LSTM model for the univariate and multivariate datasets. The MSE was used for the loss function to reduce the error in the prediction, and the Adam optimiser was used to update the weights during training. Like the first two experiments, MSE and RMSE were used to evaluate the model's performance.

The experiment applied multi-step to the dataset and predicted five days of closing price stock data from an input of 20 days. The model was comprised of four layers: an LSTM with 200 units, a repeat vector and another LSTM layer with 200 units, and finally, a dense layer at the output. The model was trained for 100 epochs.

Regarding the univariate dataset, the encoder-decoder LSTM returned the best performance in prediction with an RMSE of 0.023. The model was also trained and tested on the multivariate dataset and returned an RMSE of 507.49; Tables 6 and 7 show the five-day prediction results on the univariate and multivariate datasets.

Table 6. Experiment 2.3 results.

Test Data (ZAR)	Predictions (ZAR)
12,901	12,901.025
12,700	12,700.019
12,376	12,376.02
12,051	12,051.023
12,112	12,112.027

Table 7. Experiment 2.3 results.

Test Data (ZAR)	Predictions (ZAR)
12,901	12,959.947
12,700	12,939.186
12,376	12,875.924
12,051	12,830.503
12,112	12,801.53

4. Conclusions

The scientific novelty of the present research is that it aimed to construct a framework for intelligent media and technical analysis by forecasting stock price movement and closing prices using news headlines, tweets, and technical indicators. The paper first constructed a sentiment analysis classifier using BERT. A South African dataset was constructed by scraping South African-related tweets on Twitter; these were then concatenated with the Stanford Twitter Sentiment Corpus to create the training dataset. Two experiments were constructed: the fundamental analysis experiment to predict Vodacom's closing price movement and the technical analysis experiment to predict Vodacom's closing price. The first experiment was split into three sub-experiments which were run on three datasets. The first dataset was the news headlines dataset, the second was the news headlines and sentiment dataset, and the last was the news headlines, tweets, and sentiment dataset. These datasets were constructed by scraping news headlines related to Vodacom Group Limited on the Moneyweb site and tweets with the hashtag vodacom on Twitter.

The BERT-trained sentiment classifier extracted the sentiment feature from the collected text. The experiment included four machine learning models: support vector machine, linear discriminant analysis, decision tree, and random forest. The fundamental analysis experiment returned positive results, with the linear discriminant analysis being the best performing model in the experiment. The model achieved an accuracy of 96% in predicting Vodacom's closing price movement. The second experiment, the technical analysis experiment, presented a different objective than the first. In this experiment, Vodacom's closing price was predicted based on two datasets. The first dataset was constructed using Vodacom's closing price and named the univariate dataset. The second dataset utilized Vodacom's stock data, namely the opening price, high price, low price, volume, closing price, and features extracted from three technical indicators, the three moving average, MACD, and Bollinger Bands. This second dataset was called the multivariate dataset. Three sub-experiments were performed; the first two included five different LSTM architectures, namely ordinary LSTM, stacked (deep) LSTM, convLSTM, and CNNxLSTM. These LSTM architectures were chosen to compare and assess which model would perform the best on the data; as seen in the results, the most complex model was not always the best performing one.

The final experiment introduced the encoder-decoder LSTM architecture and yielded different outcomes between the univariate and multivariate datasets. For the first experiment, stacked LSTM archived the best results in predicting the closing price, with an RMSE of 220.83, and ordinary LSTM performed best with an RMSE of 75.54. The last experiment in Experiment 2 introduced encoder-decoder LSTM; this model achieved the best overall performance on the univariate dataset, with an RMSE of 0.023. In Experiment 1, the addition of the sentiment feature returned a significant improvement in predicting Vodacom's stock price movement due to the introduction of a South African-based sentiment classifier. The second experiment yielded mixed results. Though the developed encoder-decoder LSTM achieved the best performance on the univariate dataset, the multivariate dataset achieved the best overall performance in predicting Vodacom's closing price via the LSTM architectures.

The presented outcomes of the experiments provide beneficial results in answering the research questions by predicting price movements for fundamental analysis and predicting the closing price using technical indicators. The present research has shown that including a sentiment feature taken from the sentiment analysis classifier improves the prediction accuracy by a tremendous amount. The study has also seen the same results in the technical analysis experiment; the addition of technical indicators improves closing price predictions.

Future work shall consider extending the demographics of the dataset collected for the sentiment classification. The collection of more news headlines and tweets ensures that the training dataset is sufficiently large for the machine learning models to learn different underlying patterns in the data. The research could also include intraday prices to ensure the study has a larger dataset, as some of the models perform better with larger datasets. Other LSTM architectures could also be explored, such as deep LSTM with attention, encoder-decoder with attention, and sequence-to-sequence LSTM to improve overall performance. Recently, generative adversarial networks (GANs) have also been applied to time series data.

Author Contributions: Conceptualization, S.T.M.; Study design; Performed experiments and data analysis; Manuscript writing; review and editing, B.S.P.; Results review and critique; Manuscript review and editing W.D.; Study design evaluation; Results review and critique; Manuscript review and editing. All authors have read and agreed to the published version of the manuscript.

Funding: This research received no external funding.

Institutional Review Board Statement: Not applicable.

Informed Consent Statement: Not applicable.

Data Availability Statement: Not applicable.

Acknowledgments: The first author would like to thank Nedbank for its financial support.

Conflicts of Interest: The authors declare no conflict of interest.

References

1. Yurieff, K. Snapchat Stock Loses $1.3 Billion after Kylie Jenner Tweet. CNN. 23 February 2018. Available online: https://money.cnn.com/2018/02/22/technology/snapchat-update-kylie-jenner/index.html (accessed on 1 May 2020).
2. Bursztynsky, J. Tesla Shares Tank after Elon Musk Tweets the Stock Price Is 'too high'. CNBC. 1 May 2020. Available online: https://www.cnbc.com/2020/05/01/tesla-ceo-elon-musk-says-stock-price-is-too-high-shares-fall.html (accessed on 18 February 2012).
3. Dev, S.; Haruna, I.; Farhana, Z. Predicting the Effects of News Sentiments on the Stock Market. In Proceedings of the IEEE International Conference on Big Data (Big Data), Seattle, WA, USA, 10–13 December 2018.
4. Wu, D.D.; Ren, R.; Liu, T. Forecasting Stock Market Movement Direction Using Sentiment Analysis and Support Vector Machine. *IEEE Syst. J.* **2019**, *13*, 760–770.
5. Li, H.; Shen, Y.; Zhu, Y. Stock Price Prediction Using Attention-based Multi-Input LSTM. In Proceedings of the Machine Learning Research, Beijing, China, 14–16 November 2018; pp. 454–469.
6. Minh, D.L.; Sadeghi-Niaraki, A.; Huy, H.D.; Min, K.; Moon, H. Deep Learning Approach for Short-Term Stock Trends Prediction Based on Two-Stream Gated Recurrent Unit Network. *IEEE Access* **2018**, *6*, 55392–55404. [CrossRef]
7. Shi, X.; Chen, Z.; Wang, H.; Yeung, D.-Y.; Wong, W.-K.; Woo, W.-C. Convolutional LSTM network: A machine learning approach for precipitation nowcasting. In Proceedings of the 29th Annual Conference on Neural Information Processing Systems, Montreal, QC, Canada, December 2015.
8. Lee, S.W.; Kim, H.Y. Stock market forecasting with super-high dimensional time-series data. *Expert Syst. Appl.* **2020**, *161*, 113704. [CrossRef]
9. Livieris, I.E.; Pintelas, E.; Pintelas, P. A CNN–LSTM model for gold price time-series forecasting. *Neural Comput. Appl.* **2020**, *32*, 17351–17360. [CrossRef]
10. Chen, X.; Xie, X.; Teng, D. Short-term Traffic Flow Prediction Based on ConvLSTM Model. In Proceedings of the 2020 IEEE 5th Information Technology and Mechatronics Engineering Conference (ITOEC 2020), Chongqing, China, 12–14 June 2020.
11. Khan, W.; Ghazanfar, M.A.; Azam, M.A.; Karami, A.; Alyoubi, K.H.; Alfakeeh, A.S. Stock market prediction using machine learning classifiers and social media, news. *J. Ambient Intell. Humaniz. Comput.* **2020**, *11*, 1–24. [CrossRef]
12. Khan, W.; Malik, U.; Ghazanfar, M.A.; Azam, M.A.; Alyoubi, K.H.; Alfakeeh, A.S. Predicting stock market trends using machine learning algorithms via public sentiment and political situation analysis. *Soft Comput.* **2020**, *8*, 11019–11043. [CrossRef]
13. Vargas, M.R.; Anjos, C.E.M.d.; Bichara, G.L.G.; Evsukoff, A.G. Deep Learning for Stock Market Prediction Using Technical Indicators and Financial News Articles. In Proceedings of the 2018 International Joint Conference on Neural Networks (IJCNN), Rio de Janeiro, Brazil, 8–13 July 2018.
14. Chen, C.-H.; Shih, P. A Stock Trend Prediction Approach based on Chinese News and Technical Indicator Using Genetic Algorithms. In Proceedings of the 2019 IEEE Congress on Evolutionary Computation, Wellington, New Zealand, 10–13 June 2019.

15. Manning, C.D.; Raghavan, P.; Schütze, H. *Introduction to Modern Information Retrieval*, 2nd ed.; Cambridge University Press: Cambridge, UK, 2009.
16. Izenman, A.J. Discriminant Analysis and Other Linear. In *Modern Multivariate*; Springer: New York, NY, USA, 2013; pp. 237–238.
17. Tan, P.-N.; Steinbach, M.; Kumar, V. Classification: Basic Concepts, Decision Trees, and Model Evaluation. In *Introduction to Data Mining*; University of Minnesota: Minnesota, MN, USA, 2006; pp. 145–205.
18. Cutler, A.; Cutler, D.R.; Stevens, J.R. Random Forests. *Mach. Learn.* **2011**, *45*, 157–176.
19. Olah, C. Understanding LSTM Networks. Github. 27 August 2015. Available online: http://colah.github.io/posts/2015-08-Understanding-LSTMs/ (accessed on 2 June 2020).
20. Henrique, J. Get Old Tweets Python. 21 November 2018. Available online: https://pypi.org/project/GetOldTweets3/ (accessed on 28 January 2020).
21. Go, A.; Bhayani, R.; Huang, L. Twitter Sentiment Classification using Distant Supervision. *CS224N Proj. Rep. Stanf.* **2009**, *1*, 2009.
22. Devlin, J.; Chang, M.-W.; Lee, K.; Toutanova, K. BERT: Pre-training of Deep Bidirectional Transformers for Language Understanding. *arXiv* **2018**, preprint. arXiv:1810.04805.
23. Face, H. BERT. Hugging Face. 2020. Available online: https://huggingface.co/transformers/model_doc/bert.html (accessed on 10 September 2020).
24. Markets, A. An Introduction to Fundamental Analysis in Forex. Admiral Markets. 2 September 2020. Available online: https://admiralmarkets.com/education/articles/forex-analysis/introduction-to-forex-fundamental-analysis (accessed on 14 September 2020).
25. Ronaghan, S. Data Preparation for Machine Learning: Cleansing, Transformation & Feature Engineering. Towards Data Science. 20 September 2019. Available online: https://towardsdatascience.com/data-preparation-for-machine-learning-cleansing-transformation-feature-engineering-d2334079b06d (accessed on 26 August 2020).
26. Scikit Learn. Sklearn. Preprocessing. MinMaxScaler. Scikit Learn. 4 August 2007. Available online: https://scikit-learn.org/stable/modules/generated/sklearn.preprocessing.MinMaxScaler.html (accessed on 26 August 2020).
27. The PyData Development Team. Pandas-datareader. PyData. 21 August 2020. Available online: https://pandas-datareader.readthedocs.io/en/latest/ (accessed on 21 August 2020).

applied sciences

Article

Deep Data Assimilation: Integrating Deep Learning with Data Assimilation

Rossella Arcucci [1,*], Jiangcheng Zhu [2], Shuang Hu [3] and Yi-Ke Guo [1,4]

1 Data Science Institute, Imperial College London, London SW72AZ, UK; y.guo@imperial.ac.uk
2 State Key Laboratory of Industrial Control Technology, Zhejiang University, Hangzhou 310027, China; jiangchengzhu@zju.edu.cn
3 Ningbo Joynext Technology Inc., Ningbo 315000, China; shuang.hu@joynext.com
4 Department of Computer Science, Hong Kong Baptist University, Hong Kong
* Correspondence: r.arcucci@imperial.ac.uk

Abstract: In this paper, we propose Deep Data Assimilation (DDA), an integration of Data Assimilation (DA) with Machine Learning (ML). DA is the Bayesian approximation of the true state of some physical system at a given time by combining time-distributed observations with a dynamic model in an optimal way. We use a ML model in order to learn the assimilation process. In particular, a recurrent neural network, trained with the state of the dynamical system and the results of the DA process, is applied for this purpose. At each iteration, we learn a function that accumulates the misfit between the results of the forecasting model and the results of the DA. Subsequently, we compose this function with the dynamic model. This resulting composition is a dynamic model that includes the features of the DA process and that can be used for future prediction without the necessity of the DA. In fact, we prove that the DDA approach implies a reduction of the model error, which decreases at each iteration; this is achieved thanks to the use of DA in the training process. DDA is very useful in that cases when observations are not available for some time steps and DA cannot be applied to reduce the model error. The effectiveness of this method is validated by examples and a sensitivity study. In this paper, the DDA technology is applied to two different applications: the Double integral mass dot system and the Lorenz system. However, the algorithm and numerical methods that are proposed in this work can be applied to other physics problems that involve other equations and/or state variables.

Keywords: data assimilation; deep learning; neural network

Citation: Arcucci, R.; Zhu, J.; Hu, S.; Guo, Y.-K. Deep Data Assimilation: Integrating Deep Learning with Data Assimilation. *Appl. Sci.* **2021**, *11*, 1114. https://doi.org/10.3390/app11031114

Academic Editor: João M. F. Rodrigues
Received: 8 December 2020
Accepted: 21 January 2021
Published: 26 January 2021

1. Introduction and Motivations

The present work is placed in the context of the design of reliable algorithms for solving Data Assimilation (DA) for dynamic systems applications. DA is an uncertainty quantification technique by which measurements and model predictions are combined in order to obtain an accurate estimation of the state of the modelled system [1]. In the past 20 years, DA methodologies, used, in principle, mainly in atmospheric models, has become a main component in the development and validation of mathematical models in meteorology, climatology, geophysics, geology, and hydrology (often these models are referred to with the term predictive to underline that these are dynamical systems). Recently, DA has also been applied to numerical simulations of geophysical applications [2], medicine, and biological science [3,4] for improving the reliability of the numerical simulations. In practice, DA provides the answers to questions, such as: how well does the model under consideration represent the physical phenomena? What confidence can one have that the numerical results produced by the model are correct? How far can the calculated results be extrapolated? How can the predictability and/or extrapolation limits be extended and/or improved?

Accuracy in numerical models of dynamic systems does not come easily. The simulation errors come from three main sources: inaccurate input data, inaccurate physics models, and limited accuracy of the numerical solutions of the governing equations. Our ability to predict any physical phenomenon is determined by the accuracy of our input data and our modelling approach [5]. Errors arise at the different stages of the solution process, namely, the uncertainty in the mathematical model, in the model's solution, and in the measurements. These are the errors intrinsic to the DA problem. Moreover, there are the approximation errors that are introduced by the linearization, the discretization, and the model reduction. These errors incur when infinite-dimensional equations are replaced by a finite-dimensional system (that is, the process of discretization), or when simpler approximations to the equations are developed (e.g., by model reduction). Finally, given the numerical problem, an algorithm is developed and implemented as a mathematical software. At this stage, the inevitable rounding errors introduced by working in finite-precision arithmetic occur. These approaches are unable to fully overcome their unrealistic assumptions, particularly linearity, normality, and zero error covariances, despite the improvements in the complexity of the DA models to better match their application requirements and circumvent their implementation problems [6].

With the rapid developments in recent years, DL shows great capability in approximating nonlinear systems, and extracting high dimensional features. As the foundation and main driving force of deep learning, DNN is concerned less about numerical modelling. DNN took a data driven approach, where the models are built by learning from data. Instead of being deterministic [7,8], DL models such as NN are stochastic. Thus, they can well succeed when applied to deterministic systems, but without ever learning the relationship between the variables. For e.g., ML algorithms do not know when they violate the physics laws of [9,10] in weather forecasting. Before they begin to produce accurate results, machine learning methods often need large quantities of data, and, the bigger the architecture, the more data are required. In a wide variety of applications, DL is successfully used when the conditions allow. However, in cases where the dimensions are either very large, the data are noisy or the data do not cover the entire domain adequately; these approaches still fail. Furthermore, the network will not perform well if the available information is too noisy, too scarce, or if there is a lack of prominent features to reflect the problem. This can occur either if there is a lack of good characteristics or a lack of data on good ground reality. For this reason, a good quality of the data (smaller errors) can help to generate a more reliable DNN. DA is the Bayesian approximation of the true state of some physical system at a given time by combining time-distributed observations with a dynamic model in an optimal way. In some sense, DA increases the meaningfulness of the data and reduces the forecasting errors. Therefore, it is interesting to investigate the mechanism where two data driven paradigms, DA and DNN, can be integrated.

In this context, we developed the Deep Data Assimilation (DDA) model. DDA is a new concept that combines the DNN and DA. It faces the problem to reduce both input data error (by DA) and modelling error (by ML).

Related Works and Contribution of the Present Work

This paper is placed in the imperfect models and sensitivity analysis context for data assimilation and deep neural network. Sensitivity Analysis (SA) refers to the determination of the contributions of individual uncertainty on data to the uncertainty in the solution [11]. The sensitivity of the variational DA models has been studied in [12], where an Adjoint modeling is used in order to obtain first and second-order derivative information. In [13], sensitivity analysis is based on the Backward Error Analysis (B.E.A.), which figure out how much the errors propagation in DA process depend on the condition number of the background error covariance matrices. Reduced-order approaches are formulated in [12] in order to alleviate the computational cost that is associated with the sensitivity estimation. This method makes rerunning less expensive, the parameters must still be selected a priori and, consequently, important sensitivities may be missed [14]. For imperfect models, weak constraint DA (WCDA) can be employed [15], in which the model error term in the covariance evolution equation acts to reduce the dependence of the estimate on observations and prior states that are well separated in time. However, the time complexity of the WCDA models is a big limit that makes these models often unusable.

ML algorithms are capable of assisting or replacing, without the assumptions of traditional methods, the aforementioned conventional methods in assimilating data, and producing forecasts. Because any linear or nonlinear functions can be approximated by neural networks, DA has been integrated as a supplement in various applications. The technology and the model presented in this paper are very general and they can be applied to any kind of dynamical systems. The use of ML in correcting model forecasts is promising in several geophysics applications [16,17]. Babovic [18–20] applies neural network for error correction in forecasting. However, in this literature, the error correction neural network has not a direct relation with the system update model in each step and it does not train the results of the assimilation process. In [21], DA and ML are also integrated and a surrogate model is used in order to replace the dynamical system. In this paper we do not replace the the dynamical system with a surrogate models, as this would add approximation errors. In fact, in real case scenarios (i.e., in some operational centres), data driven models are welcome to support Computational Fluid Dynamics simulations [9,10,22], but the systems of Partial Differential Equations that are stably implemented to predict (weather, climate, ocean, air pollution, et al.) dynamics are not replaced due to the big approximations that this replacement would introduce. The technology that we propose in this paper integrates DA with ML in a way that can be used in real case scenarios for real world applications without changing the already implemented dynamical systems.

Other works have already explored the relationship between the learning, analysis, and implementation of data in ML and DA. These studies have concentrated on fusing methods from the two fields instead of taking a modular approach like the one implemented in this paper, thereby developing methods that are unique to those approaches. The integration of the DA and ML frameworks from a Bayesian perspective illustrates the interface between probabilistic ML methods and differential equations. The equivalence is formally presented in [23] and it illustrates the parallels between the two fields. Here, the equivalence of four-dimensional VarDA (4D-Var) and Recurrent NN, and how approximate Bayesian inverse methods (i.e., VarDA in DA and back-propagation in ML) can be used to merge the two fields. These methods are also especially suited to systems involving Gaussian process, as presented in [24–26]. These are developed data-driven algorithms that are capable, under a unified approach, of learning nonlinear, space-dependent cross-correlations, and of estimating model statistics. A DA method can be used via the DL framework to help the dynamic model (in this case, the ML model) to find optimal initial conditions. Although DA can be computationally costly, the recent dimensional reduction developments using ML substantially reduce this expense, while retaining much of the variance of the original [27] model. Therefore, DA helps to develop the ML model, while [28,29] also benefits from the ML techniques. Methods that merge DA models, such as Kalman filters and ML, exist to overcome noise or to integrate time series knowledge. The authors account for temporal details in human motion analysis in [30] by incorporating a Kalman filter into a neural network of LSTM. In [31], the authors suggest a methodology that combines NN and DA for model error correction. Instead, in [32], the authors use DA, in particular Kalman tracking, to speed up any learning-based motion tracking method to real-time and to correct some common inconsistencies in motion tracking methods that are based on the camera. Finally, in [33], the authors introduce a new neural network for speed and replace the whole DA process.

In this paper, we proposed a new concept, called DDA, which combines the DL into the conventional DA. In recent years, DL gained great success both in academic and industrial areas. In function approximation, which has unknown model and high nonlinearity, it shows great benefit. Here, we use DNN to describe the model uncertainty and the assimilation process. In an end-to-end approach, the neural networks are introduced and their parameters are iteratively modified by applying the DA methods with upcoming observations. The resulting DNN model includes the features of the DA process. We prove that this implies a reduction of the model error, which decreases at each iteration.

We also prove that the DDA approach introduces an improvement in both DA and DNN. We introduce a sensitivity analysis that is based on the backward error analysis to compare the error in DDA result and the error in DA. We prove that the error in the results obtained by DDA is reduced with respect the DA. We also prove that the use of the results of DA to train the DL model introduces a novelty in the SA techniques used to evaluate which inputs are important in NN. The implemented approach can be compared to the 'Weights' Method [34]. The distinction from the classical 'Weights' method is that the weights between the nodes are given in our approach by the error covariance matrices that are determined in the DA phase.

In summary, we prove that the DDA approach

- includes the features of the DA process into the DNN model with a consequent reduction of the model error;
- introduces a reduction in the overall error in the assimilation solution with respect to the DA solution; and,
- increases the reliability of the DNN model, including the error covariance matrices as weight in the loss function.

This paper is structured, as follows. Section 2 provides preliminaries on DA. We proposed the DDA methodology in Section 3, where we introduce the integration of deep learning with data assimilation. In Section 4, numerical examples are provided in order to implement and verity the DDA algorithm and, in Section 5, conclusions and future work are summarized.

2. Data Assimilation Process

Let

$$\dot{u} = \mathcal{M}(u, t, \theta) \tag{1}$$

be a forecasting model, where u denotes the state, $\mathcal{M}$ is a nonlinear map, θ is a parameter of interest, and $t \in [0, T]$ denotes the time. Let

$$v = \mathcal{H}(u) + \epsilon \tag{2}$$

be an observation of the state u, where $\mathcal{H}$ is an observation function and ϵ denotes the measurement error.

Data Assimilation is concerned with how to use the observations in (2) and the model in (1) in order to obtain the best possible knowledge of the system as a function of time.The Bayes theorem conducts the estimation of a function u^{DA}, which maximizes a probability density function, given the observation v and a prior from u [1,35].

For a fixed a time step $t_k \in [0, T]$, let u_k denote the estimated system state at time t_k:

$$u_k = M\, u_{k-1} \tag{3}$$

where the operator M represents a discretization of a first order approximation of $\mathcal{M}$ in (1). Let v_k be an observation of the state at time t_k and let

$$H : u_k \mapsto v_k. \tag{4}$$

be the discretization of a first order approximation of the linearization of $\mathcal{H}$ in (2). The DA process consists in finding u^{DA} (called analysis) as an optimal tradeoff between the prediction made based on the estimated system state (called background) and the available observation. The state u is then replaced by the function u^{DA}, which includes the information from the observation v in a prediction-correction cycle that is shown in Figure 1.

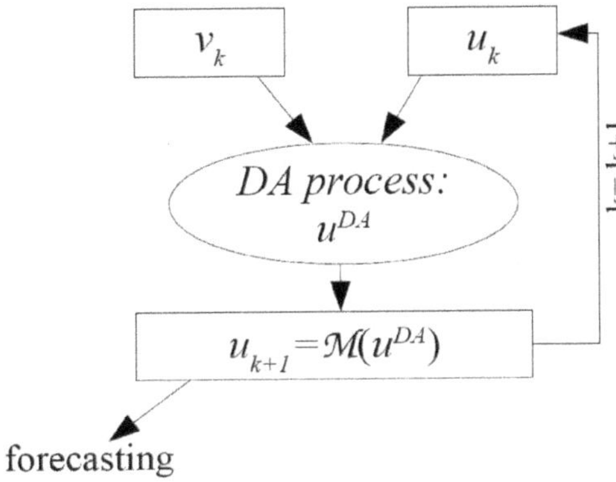

Figure 1. The prediction-correction cycle.

The analysis u^{DA} is computed as inverse solution of

$$v_k = Hu^{DA} + e_{R_k}, \tag{5}$$

subject to the constraint that

$$u^{DA} = u_k + e_{B_k}. \tag{6}$$

where $e_{R_k} = \mathcal{N}(\mu_R, R_k)$ and $e_{B_k} = \mathcal{N}(\mu_B, B_k)$ denote the observation error and model error, respectively, $\mathcal{N}(\cdot)$ denotes Gaussian distribution [1], and e_{R_k} includes the measurement error ϵ introduced in (2). Both of the errors include the discretization, approximation and the other numerical errors.

Because H is typically rank deficient, the (5) is an ill posed inverse problem [36,37]. The Tikhonov formulation [38] leads to an unconstrained least square problem, where the term in (6) ensures the existence of a unique solution of the (5). The DA process can be then described in Algorithm 1, where:

$$u^{DA} = argmin_u \left\{ \|u - u_k\|^2_{B_k^{-1}} + \|v_k - Hu\|^2_{R_k^{-1}} \right\} \tag{7}$$

where R_k and B_k are the observation and model error covariance matrices, respectively:

$$R_k := \sigma_0^2 I, \tag{8}$$

with $0 \le \sigma_0^2 \le 1$ and I be the identity matrix,

$$B_k = V_k V_k^T \tag{9}$$

with V_k background error deviation matrix [13] being computed by a set of n_k historical data $S = \{u_j\}_{j=1,\dots,n_k}$ available at time t_k and such that

$$V_k = \{V_{jk}\}_{j=1,\dots,n_k},$$

where

$$V_{jk} = u_j - \bar{u}$$

and

$$\bar{u} = mean_{j=1,\dots,n_k}\{u_j\}.$$

Algorithm 1 DA

Input: for $k = 0, \ldots, m$ temporal steps: observations v_k, matrices R_k, model M, H, background u_0, historical data $S = \{u_j\}_{j=1,\ldots,n_k}$,

Compute B_0 from u_0 and S

Initialize iteration $k = 1$

while $k < m$ **do**

Compute $u_k = M\, u_{k-1}$

Compute B_k

Compute

$$u_k^{DA} = argmin_u \left\{ \|u - u_k\|_{B_k^{-1}} + \|v_k - Hu\|_{R_k^{-1}} \right\} \tag{10}$$

Count up k for the next iteration

end

Output: u^{DA}

The DA process, as defined in (7), can be solved by several methods. Two main approaches that have gained acceptance as powerful methods for data assimilation on the last years are the variational approach and Kalman Filter (KF). The variational approach [39] is based on the minimization of a functional, which estimates the discrepancy between numerical results and measures. The Kalman Filter [40] is a recursive filtering instead. In both cases we seek an optimal solution. Statistically, KF seeks a solution with minimum variance. Variational methods seek a solution that minimizes a suitable cost function. In certain cases, the two approaches are identical and they provide exactly the same solution [1]. However, the statistical approach, though often complex and time-consuming, can provide a richer information structure, i.e., an average and some characteristics of its variability (probability distribution).

Assuming the acquisition of the observations in a fixed time steps, the variational approach is named 3DVar [1,6], and it consists in computing the minimum of the cost function:

$$J(u) = (u - u_k)^T B_k^{-1} (u - u_k) + (v_k - Hu)^T R_k^{-1} (v_k - Hu) \tag{11}$$

where

$$u^{DA} = argmin_u\, J(u).$$

Potential problems of a variational method are mainly given by three main points: it heavily relies on correct B_k matrices, it does not take system evolution into account, and it can end up in local minima.

Kalman Filter consists in computing the explicit solution of the normal equations:

$$u^{DA} = u_k + K_k(v_k - Hu_k) \tag{12}$$

where the matrix

$$K_k = B_k H^T (H B_k H^T + R_k)^{-1} \tag{13}$$

is called Kalman gain matrix and where B_k is updated at each time step [6]:

$$B_k = M((1 - K_{k-1}H)B_{k-1})M^T. \tag{14}$$

A potential problem of Kalman Filter is that K_k is too large to store for large-dimensional problems.

A common point of 3DVar and KF is that the observations that are defined in the window $[0, T]$ (circle points in Figure 2) are assimilated in the temporal point, where the state is defined (grey rhombus in Figure 2 at time t_k). After the assimilation, a new temporal window $[T, 2T]$ is considered and new observations are assimilated. Each process starts at the end point of the previous process and the two assimilation processes are defined in almost independent temporal windows. The result of the assimilation process u^{DA} (green pentagons in Figure 2), as computed by 3DVar or KF, is called analysis. The explicit expression for the analysis error covariance matrix, denoted as A_k, is [1]:

$$A_k = (I - K_k H) B_k. \quad (15)$$

where I denotes the identity matrix.

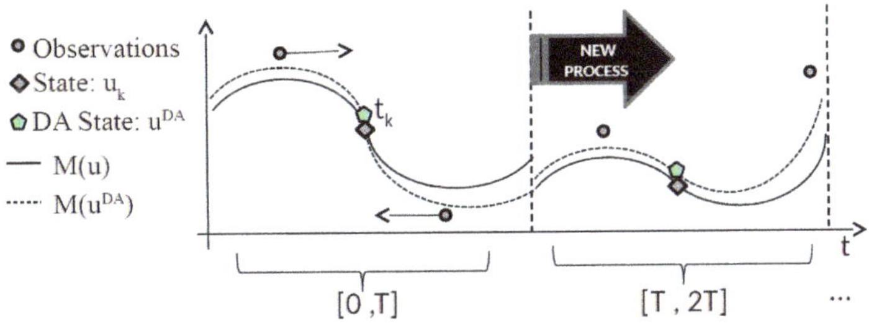

Figure 2. The Data Assimilation (DA) process.

Even if the DA process takes information in a temporal window into account, it does not really take the past into account. This is due to the complexity of the real application forecasting models for which a long time window would be too computationally expensive. In next section, we introduce the Deep Data Assimilation (DDA) model, in which we replace the forecasting model M (black line in Figure 3) with a new Machine Learning model $\mathcal{M}$ (red line in Figure 3) trained while using DA results and taking the past into account.

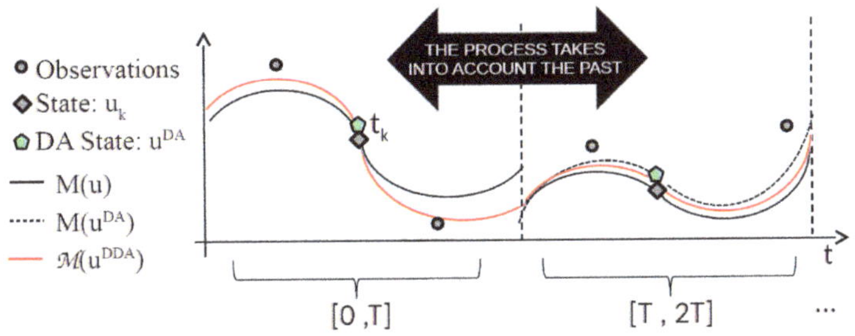

Figure 3. The Deep Data Assimilation (DDA) process.

3. The Deep Data Assimilation (DDA) Model

Data assimilation (DA) methodologies improve the levels of confidence in computational predictions (i.e., improve numerical forecasted results) by incorporating observational data into a prediction model. However, the error propagation into the forecating model is not improved by DA, so that, at each step, correction have to be based from scratch without learning from previous experience of error correction. The strongly nonlinear character of many physical processes of interest can result in the dramatic amplification of even small uncertainties in the input, so that they produce large uncertainties in the system behavior [5]. Because this instability, as many observations are assimilated as possible to the point where a strong requirement to DA is to enable real-time utilization of data to improve predictions. Therefore, it is desirable to use machine learning method in order to learn a function which accumulates the process of previous assimilation process. We use NN to model this process. The key concept is in recording each step of state correction during an assimilation period and then learn a function in order to capture this updating mechanism. The system model is then revised by composing this learned updating function with the current system model. Such a process continues by further learning the assimilation process with the updated system model.

The resulting NN-forecasting model is then a forecasting model with an intrinsic assimilation process.

We introduce a neural network $\mathcal{G}_\omega$, starting by a composition with the model M, as described in Figure 4. The training target is:

$$u_k = \mathcal{G}_\omega(Mu_{k-1}). \tag{16}$$

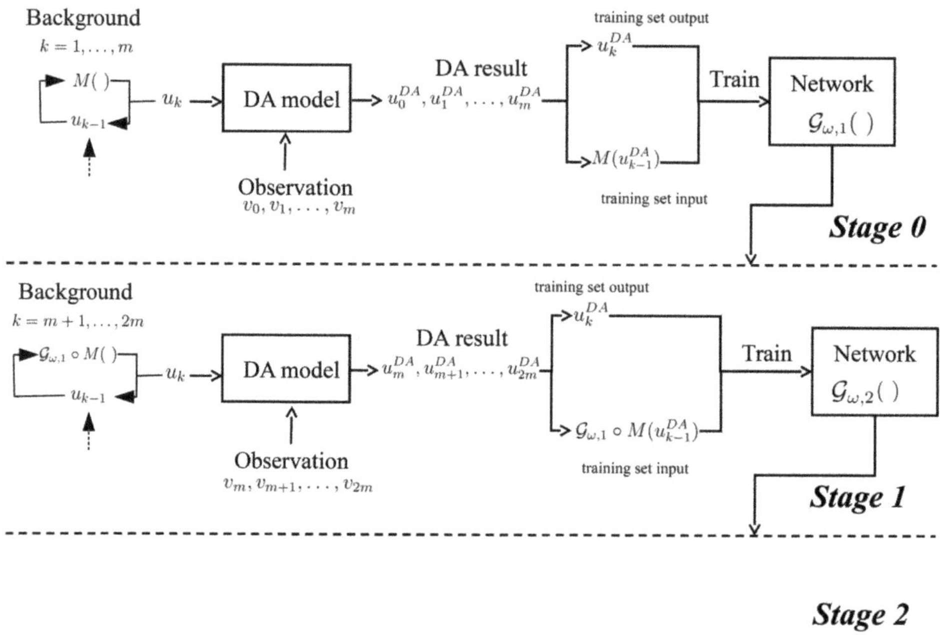

Figure 4. The schematic diagram of DDA.

We train $\mathcal{G}_\omega$ by exploiting the results of DA, as described in Algorithm 2. The $\mathcal{G}_\omega$ model can be the iterative form $\mathcal{G}_{\omega,i} \circ \mathcal{G}_{\omega,i-1} \circ \ldots \circ \mathcal{G}_{\omega,1} \circ M$, where $\mathcal{G}_{\omega,i}$ is defined as the network that was trained in ith iteration and $\circ$ denotes the composition function. Let $\mathcal{M}_i$ be the model that was implemented in the ith iteration, it is:

$$\mathcal{M}_i = \mathcal{G}_{\omega,i} \circ \mathcal{G}_{\omega,i-1} \circ \ldots \circ \mathcal{G}_{\omega,1} \circ M, \tag{17}$$

from which $\mathcal{M}_i = \mathcal{G}_{\omega,i} \circ \mathcal{M}_{i-1}$, and $\mathcal{M}_0 = M$.

For iteration $i > 1$, the training set for $\mathcal{G}_{\omega,i}$ is $\{(u_k, u_k^{DDA})\}$, which exploits the $\{(u_{k-1}^{DDA}, u_k^{DDA})\}$ from the last updated model $\mathcal{M}_i$. In this paper, we implement a Recurrent Neural Network (RNN) based on Elman networks [41]. It is a basic RNN structure that loops through the hidden layer output as an internal variable (that corresponds to the Jordan network [42] that outputs $u_{t+\ldots}$ as a variable). As shown in Figure 5, the RNN is specifically expressed as

$$\begin{aligned} h_t &= tanh(w_1[u_k, h_{t-1}] + b_1) \\ u_k^{DDA} &= w_2(tanh(w_1[u_k, h_{t-1}] + b_1)) + b_2 \end{aligned} \tag{18}$$

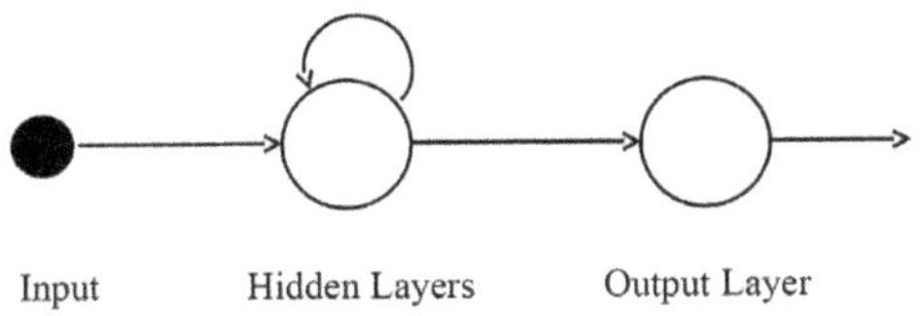

Figure 5. Recurrent Neural Network.

Algorithm 2 $\mathcal{A}(DDA)$

Input: for $k = 0, \ldots, m$ temporal steps: observations v_k, matrices R_k, background u_0, historical data $S = \{u_j\}_{j=1,\ldots,n_k}$, model M, H
Compute B_0 from u_0 and S
Initialize the $u_0^{DDA} = u_0$
Initialize the $\mathcal{M}_1 = M$
Initialize iteration $k = 1$
while $k < m$ **do**
  Compute $u_k = \mathcal{M}_k u_{k-1}^{DDA}$

$$u_k^{DDA} = argmin_u \Big\{ \|u - u_k\|_{B_k^{-1}} + \|Hu - v_k\|_{R_k^{-1}} \Big\} \tag{19}$$

  Generate the training set $\mathcal{D}_{k_i} = \{u_{k_i}, u_{k_i}^{DDA}\}$ with $k_i \in [k \cdot m, (k+1) \cdot m]$
  Train a neural networks $\mathcal{G}_{\omega,k}$ with $\mathcal{D}_{k_i}$
  Put trained $\mathcal{G}_{\omega,k}$ in the model: $\mathcal{M}_{k+1} \leftarrow \mathcal{G}_{\omega,k} \circ \mathcal{M}_k$
  Put $B_{k+1} \leftarrow A_k$ (where A_k is defined in (15))
  Count up k for the next iteration
end
Output: u^{DDA}

This choice to train the NN using data from DA introduces a reduction of the model's error at each time step that can be quantified as reduction of the background error covariance matrix, as proved in next theorem.

Theorem 1. *Let B_{k-1} and B_k be the error covariance matrices at step $k-1$ and step k in Algorithm 2, respectively. We prove that*

$$\|B_k\|_\infty \leq \|B_{k-1}\|_\infty \tag{20}$$

i.e., the error in background data is reduced at each iteration.

Proof. We prove the thesis by a reductio ad absurdum. We assume

$$\|B_k\|_\infty > \|B_{k-1}\|_\infty \tag{21}$$

At step $k-1$, from Step 10 of Algorithm 2 we have $B_k = A_{k-1}$, i.e., from (15) and (13), the (21) gives

$$\left\| \left(I - \frac{H_{k-1}B_{k-1}H_{k-1}^T}{H_{k-1}B_{k-1}H_{k-1}^T + R_{k-1}} \right) B_{k-1} \right\|_\infty > \|B_{k-1}\|_\infty \tag{22}$$

i.e., from the property of $\|\cdot\|_\infty$: $\|a\|_\infty \|b\|_\infty > \|a\, b\|_\infty$, we have

$$\left\| I - \frac{H_{k-1}B_{k-1}H_{k-1}^T}{H_{k-1}B_{k-1}H_{k-1}^T + R_{k-1}} \right\|_\infty \|B_{k-1}\|_\infty > \|B_{k-1}\|_\infty \tag{23}$$

which means that

$$\left\| I - \frac{H_{k-1}B_{k-1}H_{k-1}^T}{H_{k-1}B_{k-1}H_{k-1}^T + R_{k-1}} \right\|_\infty > 1 \tag{24}$$

Becaue of the definition of $\|\cdot\|_\infty$, we can then assume that $\exists\, ij$, such that

$$1-\frac{b_{ij}}{b_{ij}+r_{ij}}>1 \tag{25}$$

i.e., such that

$$\frac{b_{ij}}{b_{ij}+r_{ij}}<0. \tag{26}$$

which is absurd. In fact, if we consider the two possible conditions $i=j$ and $i\neq j$. In case $i=j$, b_{ii} and r_{ii} are diagonal elements of the errors covariance matrices. In other words, they are values of variances that mean that they are positive values and the (26) is not satisfied. In case $i\neq j$, $r_{ij}=0$ as the error covariance matrix R_k is diagonal, as defined in (8). Subsequently, we have that $\frac{b_{ij}}{b_{ij}+r_{ij}}=\frac{b_{ij}}{b_{ij}}=1$ and the (26) is also not satisfied. Subsequently, the (20) is proven. □

Another important aspect is that the solution of the DDA Algorithm 2 is more accurate than the solution of a standard DA Algorithm 1.

In fact, even if some sources of errors cannot be ignored (i.e., the round off error), then the introduction of the DDA method reduces the error in the numerical forecasting model (below denoted as ξ_k). The following result held.

Theorem 2. *Let δ_k denote the error in the DA solution:*

$$u_k^{DA}=u_k^{true}+\delta_k\,,$$

and let $\hat{\delta}_k$ be the error in DDA solution:

$$u_k^{DDA}=u_k^{true}+\hat{\delta}_k\,,$$

it is:

$$\|\hat{\delta}_k\|_\infty\leq\|\delta_k\|_\infty \tag{27}$$

Proof. Let ξ_k be error in the solution of the dynamic system M in (3), i.e., $u_k=M(u_{k-1})+\xi_k$. From the backward error analysis that was applied to the DA algorithm [13], we have:

$$\|\delta_k\|_\infty=\mu_{DA}\,\|\xi_k\|_\infty \tag{28}$$

and

$$\|\hat{\delta}_k\|_\infty=\mu_{DDA}\,\|\xi_k\|_\infty \tag{29}$$

where μ_{DA} and μ_{DDA} denote the condition numbers of the DA numerical model and the DDA numerical model, respectively.

It has been proved, in [37], that the condition number of the DA and DDA numerical models are directly proportional to the condition numbers of the related background error covariance matrices B_k. Subsequently, thanks to the (20), we have $\mu_{DDA}\leq\mu_{DA}$, from which:

$$\|\hat{\delta}_k\|_\infty=\mu_{DDA}\,\|\xi_k\|_\infty\leq\mu_{DA}\,\|\xi_k\|_\infty=\|\delta_k\|_\infty \tag{30}$$

□

The DDA framework in this article implement a Recurrent neural network (RNN) structure as $\mathcal{G}_\omega$. The loss function is defined, as described in Theorem 3 and the network is updated at each step by the gradient loss function on its weights and parameters:

$$\omega_{i+1}\leftarrow\omega_{i+1}+\alpha\frac{\partial\mathcal{L}_i}{\partial\omega}. \tag{31}$$

Theorem 3. *Let* $\{(u_k, u_k^{DDA})\}$ *be the training set for* $\mathcal{G}_{\omega,i}$*, which exploits the data* $\{(u_{k-1}^{DDA}, u_k^{DDA})\}$ *from the past updated model* $\mathcal{M}_i$ *in (17). The loss function for the Back-propagation, being defined as mean square error (MSE) for the DA training set, is such that*

$$\mathcal{L}_k(\omega) = \left\| B_k H^T O_k^{-1} d_k - \mathcal{G}_{\omega,k}(B_{k-1} H^T O_{k-1}^{-1} d_{k-1}) \right\|_2 \tag{32}$$

where $d_k = v_k - H_k u_k$ *is the misfit between observed data and background data, and* $O_k = H_k B_k H_k^T + R_k$ *is the Hessian of the DA system.*

Proof. The mean square error (MSE)-based loss function for the Back-propagation is such that

$$\mathcal{L}_i = ||u_k^{DDA} - \mathcal{G}_{\omega,i} \circ \mathcal{M}_i(u_{k-1}^{DDA})||_2, \tag{33}$$

The solution u_k^{DDA} of the DDA system, which is obtained by a DA function, can be expresses as [1]:

$$u_k^{DDA} = B_k H^T (H B_k H^T + R_k)^{-1} (v_k - H u_k) \tag{34}$$

Let posed $d_k = v_k - H u_k$ and $O_k = H B_k H^T + R_k$, (34) gives:

$$u_k^{DDA} = B_k H^T O_k^{-1} d_k \tag{35}$$

Substituting, in (33), the expression of u_k^{DDA} given by (35), the (32) is proven. □

4. Experiments

In this section, we apply the DDA technology to two different applications: the Double integral mass dot system and the Lorenz system:

- Double integral mass dot system:
  For the physical representation of a generalized motion system, the double-integral particle system is commonly used. The method, under the influence of an acceleration regulation quantity, can be seen as the motion of a particle. The position and velocity are the variables that affect the state at each time step. The status as a controlled state gradually converges to a relatively stable value, due to the presence of a PID controller in the feedback control system. The performance of the controller is the system-acting acceleration, or it can be interpreted as a force that linearly relates to the acceleration. Double integral system is a mathematic abstraction of Newton's Second Law that is often used as a simplified model of several controlled systems. It can be represented as a continuous model, as:

$$\begin{aligned} \dot{u} &= Au + Bz + \xi, \\ v &= Cu + r \end{aligned} \tag{36}$$

  where the state $u = [u_1, u_2]^T$ is a two-dimensional vector containing position u_1 and velocity u_2 and where $z = \dot{u}_2$ is the controlled input. The coefficients matrices A, B, and C are time-invariant system and observations matrices. r is the noise of the observations, which is Gaussian and two dimensional. The system disruption ξ comprises two aspects: random system noise and instability of the structural model.
- Lorenz system:
  Lorenz developed a simplified mathematical model for atmospheric convection in 1963. It is a common test case for DA algorithms. The model is a system of three ordinary differential equations, named Lorenz equations. For some parameter values and initial conditions, it is noteworthy for having a chaotic behavior. The Lorenz equations are given by the nonlinear system:

$$\begin{aligned} \frac{dp}{dt} &= -\sigma(p-q), \\ \frac{dq}{dt} &= \rho p - q - pr, \\ \frac{dr}{dt} &= pq - \beta r. \end{aligned} \tag{37}$$

where p, q and r are coordinates, and σ, ρ, and β are parameters. In our implementation, the model has been discretized by second order Runge–Kutta method [43] and, for this test case, we posed $\sigma = 10, \rho = 8/3$ and $\beta = 28$.

We mainly provide testing in order to validate the results that we proved in the previous section. Afterwards, we mainly focus on:

4.1 DDA accuracy: we prove that the accuracy of the forecasting result increases as the number of time steps increases, as proved in Theorem 1;

4.2 DDA vs. DA: in order to address the true models, we face the circumstances where DA is not sufficient to reduce the modeling uncertainty and we show that the DDA algorithm, we have proposed is applicable to solve this issue. Such findings validate what we showed in Theorem 2; and,

4.3 R-DDA vs. F-DDA: we are demonstrating that the accuracy of the R-DDA is better than that of the F-DDA. It is also due to the way the weight is calculated, i.e., by including the DA covariance matrices, as shown in Theorem 3.

We implemented the algorithms on the Simulink (Simulation and Model-Based Design) in Matlab 2016a.

4.1. DDA Accuracy

4.1.1. Double Integral Mass Dot System

When considering model uncertainty, a system disturbance in (36) is introduced so that:

$$\xi_k = \xi_{smu}(u_k) + \xi_{iid,k}, \tag{38}$$

where $\xi_{smu}(\cdot)$ denotes the structural model uncertainty and ξ_{iid} denotes the i.i.d random disturbance.

One typical model error is from the parallel composition as:

$$\xi_{smu}(u_k) = D\frac{e^{u_k+1}}{1+e^{u_k+1}}. \tag{39}$$

where the amplitude vector $D = [d_1, d_2]$ is adjustable, according to the reality.

A cascade controller is designed to prevent divergence from the device in order to accomplish the model simulation. As a sinusoidal function, the tracking signal is set. 10,000 samples are collected from a 100 s simulation with a 100 Hz sample frequency. The training set for the DDA Algorithm 2 is made of the time series $\{u_k, u_k^{DDA}\}$ of m couples. First of all, we run the framework for the dynamic system Equation (36) and record the observation v, control input z. A corresponding DA runs alongside program updates, and outputs a u^{DA} prediction sequence. The network $\mathcal{G}_{\omega,1}$ is trained on the dataset of m steps.

Subsequently, the system and the DA update the data from the $m+1$ step to $2m$ step. Although the device upgrade process in DA now incorporates the qualified neural network $\mathcal{G}_{\omega,1}$, as:

$$u_k = \mathcal{G}_{\omega,1}(Mu_{k-1} + Gu_k) \tag{40}$$

Figure 6 shows what we expected regarding the accuracy of DDA. In fact, as it is shown, the mean forecasting error decrease as the training window length increases. The Figure also shows a comparison of the mean forecasting error values with respect the results that were obtained from DA. Figure 7 is a convergence curve when matlab trains a network using the Levenberg–Marquardt backpropagation [44,45] method. It can be seen that, with just few epochs, the best value for the mean square error is achieved.

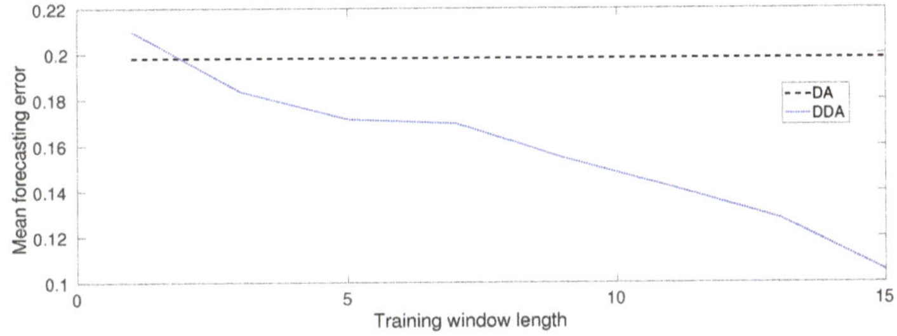

Figure 6. Mean forecasting error-first test case.

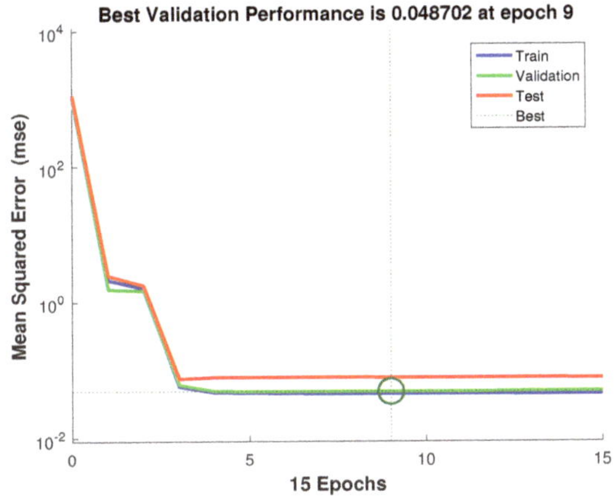

Figure 7. Training process-first test case.

4.1.2. Lorenz System

We implement a Lorenz system with structural model uncertain as:

$$u_k = M_k(u_{k-1}) + \xi_{smu,k} \tag{41}$$

where $u_k = [p_k, q_k, r_k]^T$ denotes the state and $M_k(\cdot)$ denotes the discrete function. The structural model uncertainty is denoted here as $\xi_{smu,k} = c\, u_{k-1}$, with $c = 0.01$, and $M_k(\cdot)$, such that:

$$\begin{aligned} p_{k+1} =& p_k + \sigma\frac{\Delta t}{2}[2(q_k - p_k) + \Delta t(\rho p_k - q_k - p_k r_k) - \sigma\Delta t(q_k - p_k)], \\ q_{k+1} =& q_k + \frac{\Delta t}{2}[\rho p_k - q_k - p_k r_k + \rho(p_k + \sigma\Delta t(q_k - p_k)) - q_k \\ & - \Delta t(\rho p_k - q_k - p_k r_k) - (p_k + \sigma\Delta t(q_k - p_k))(r_k + \Delta t(p_k q_k - \beta r_k))], \\ r_{k+1} =& r_k + \frac{\Delta t}{2}[p_k q_k - \beta r_k + (p_k + \Delta t\sigma(q_k - p_k))(q_k + \Delta t(\rho p_k - q_k - p_k r_k)) \\ & - \beta r_k - \Delta t(p_k q_k - \beta r_k)]. \end{aligned} \tag{42}$$

First of all, for the dynamic system, Equation (37), we run the system and record the true value of the system u. Subsequently, by adding Gaussian noise to the true value of u, v is created. The DA algorithm is then applied to produce a sequence with a DA result of u^{DA} and a length of m. The training set is made of the time series $\{M_k(u_{k-1}^{DA}), u_k^{DA}\}$. Afterwards, the neural network $\mathcal{G}_\omega$ is trained.

Figure 8 shows the accuracy results of DDA for Lorenz test case. Additionally, in this case, the results confirm our expectation. In fact, as it is shown, the mean forecasting error decrease as the training window length increases. Additionally, in this figure, a comparison of the mean forecasting error values with respect the results obtained from DA is shown. Figure 9 is a convergence curve when matlab trains a network using the Levenberg–Marquardt backpropagation method and, also in this case, the best value for the mean square error is achieved with just a few epochs.

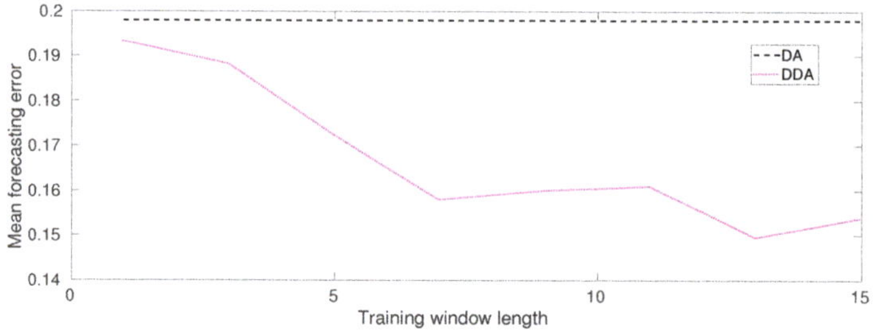

Figure 8. Mean forecasting error-second test case.

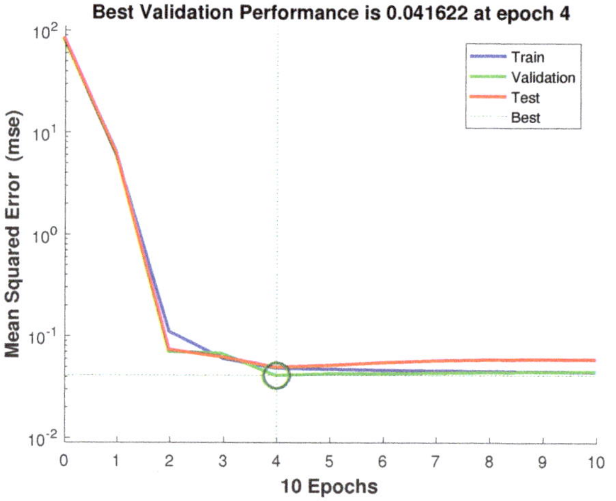

Figure 9. Training process-second test case.

4.2. DDA vs. DA

In this section, we compare the forecasting results based on DA model and the DDA model, respectively. For the DA, the system model update is formed as:

$$u_k = M_k(u_{k-1}). \tag{43}$$

where M_k is the discrete forecasting system in (3). For DDA, the system update model is as:

$$u_k = \mathcal{M}_k(u_{k-1}). \tag{44}$$

where $\mathcal{M}_k$ is defined in (17).

4.2.1. Double Integral Mass Dot System

Figure 10 shows the result of DA and DDA model on a double integral system that considers the model error. Two vertical dash lines are at which the neural network is trained. Obviously, the forecasting state value based on DDA is more closer to the true value than DA. The DA results in a larger residual. We also calculate the ratio of residual to true value as a forecasting error metric. The average forecasting error is significantly reduced from 16.7% to 1.5% by applying DDA model. Table 1 lists the run-time cost of every training window in Figure 10.

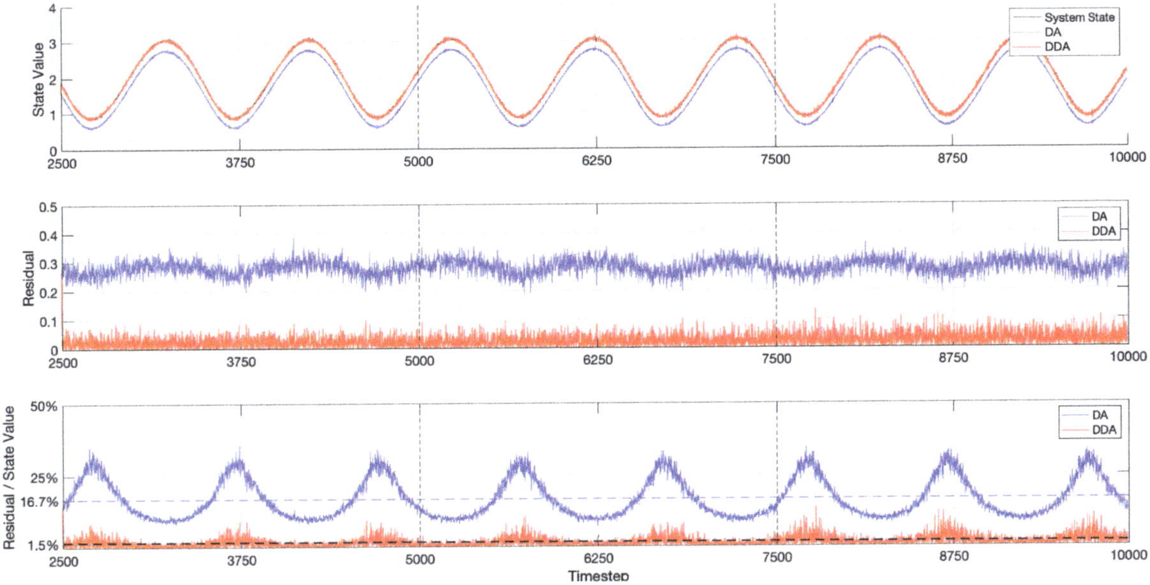

Figure 10. Simulation result of DDA on double integral system considering the model error. The vertical dash-lines refer to the training windows.

4.2.2. Lorenz System

We compare the forecasting results that are shown in Figure 11. The $k \in [0, 1000]$ generated training set is not plotted in full. The forecast starts at $k = 1000$. We can see that the DA model's forecasting errors in all three axes are large. DA model trajectories can not track the real state. Although the DDA model trajectories track the true value very well. Figure 12 is a 3D plot of this forecasting part of the Lorenz system. This demonstrates that the DA model fails to predict the right hand part of this butterfly. In this test, for the right wing trajectory of butterfly, the DDA model outperforms the DA model in forecasting.

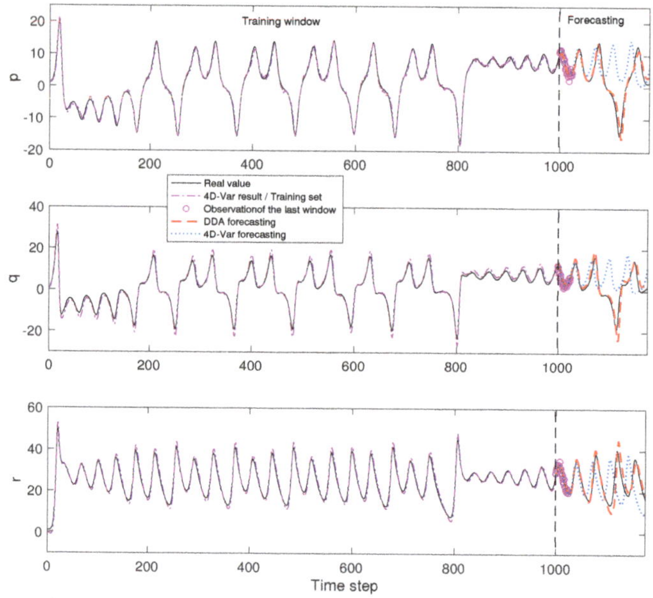

Figure 11. DA and DDA trajectories on 3 axis.

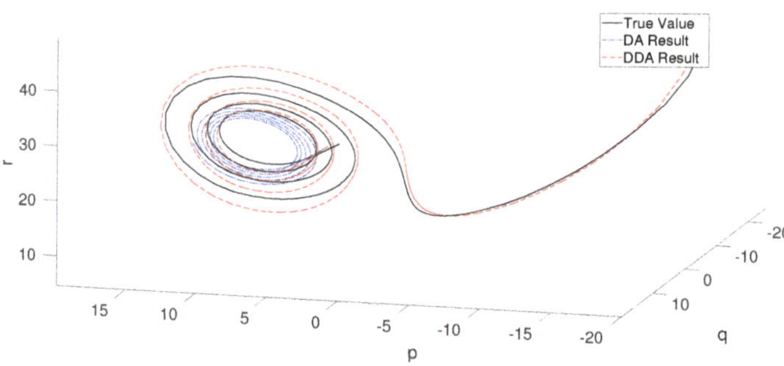

Figure 12. Forecasts of the DA and DDA trajectories. It is the three-dimensional (3D)-plot of Figure 11 for timestep $\in (1000, 1176)$.

4.3. F-DDA vs. R-DDA

In this section, we introduce the DDA using the feedfoward neural network (we named F-DDA) and compare the results with the R-DDA, i.e., the DDA we described in this paper that uses the recurrent neural network. The feed-foward (FC) network is the earliest and most basic neural network structure. This means that in the next layer, the output of each neuron in the upper layer becomes the input of each neuron, and it has a structure with corresponding values of weight. Inside the FC network, there is no cycle or loop. In the fitting of functions, fully conected neural networks are widely cited (especially the output of continuous value functions). The multilayer perceptron (MLP), which contains three hidden layers, is used in this paper.

In the training step of the DDA framework, the training set is the DDA result over a period of time. Take the first cycle ($i = 0$) as an example. The training set is $u_0^{DDA}, u_1^{DDA}, u_2^{DDA}, \ldots, u_m^{DDA}$ (the results of consecutive time series), and this data set generates the model. Because this network is a feedfoward neural network and it is unable to manage time series, we consider that the training set for the M category is independent of each other.

4.3.1. Double Integral Mass Dot System

In this section, we compared the performance of F-DDA and R-DDA in the double integral mass dot system. Because our data were previously generated by simulink, random noise, and observation noise are fixed and can be compared.

Figure 13 shows that F-DDA obtains a similar result when comparing to R-DDA in this case. It is mainly because the double integral mass dot system is not strongly dependent on the time historical states. Meanwhile, when considering the execution time cost that is shown in Table 1, F-DDA is more preferable, as it takes shorter training time than R-DDA.

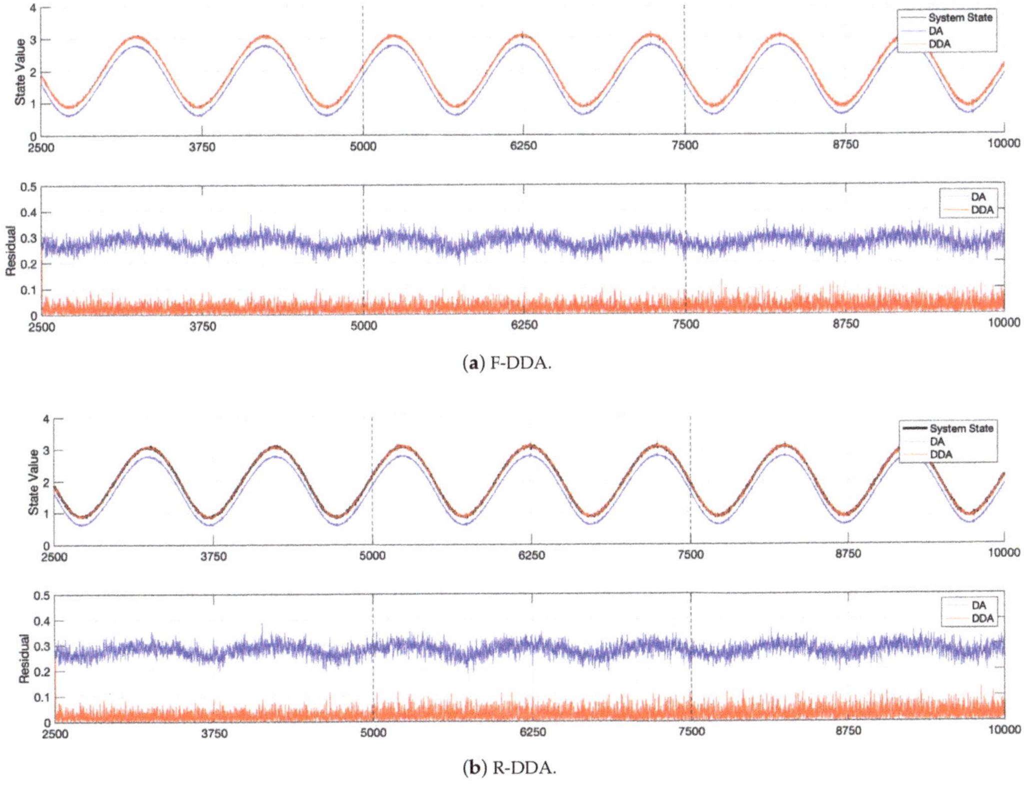

Figure 13. The simulation result of DDA on double integral system. The vertical dash-lines refer to the training windows.

Table 1. The execution time cost for both F-DDA and R-DDA in every training and forecasting window when considering the model error.

Train	**t1**	**t2**	**t3**
F-DDA	0.9605 s	0.5723 s	0.4181 s
R-DDA	102.2533 s	16.4835 s	15.0160 s
Forecast	**t1**	**t2**	**t3**
F-DDA	17.9610 s	36.2358 s	56.4286 s
R-DDA	18.3527 s	35.9934 s	53.8951 s

4.3.2. Lorenz System

Figure 14 is a comparison of the forecasting results of F-DDA and R-DDA for the Lorenz system. It can be seen that the R-DDA (black dashed line) has a more accurate prediction of the true value (blue solid line) and the curve fits better. In contrast, F-DDA (red dotted line) has a certain time lag and it is also relatively inaccurate in amplitude. This is mainly because the Lorenz system is strongly time-dependent. Table 2 provides the run-time cost for both F-DDA and R-DDA in training and forecasting. It is worth taking more training time for R-DDA to achieve a better prediction than F-DDA.

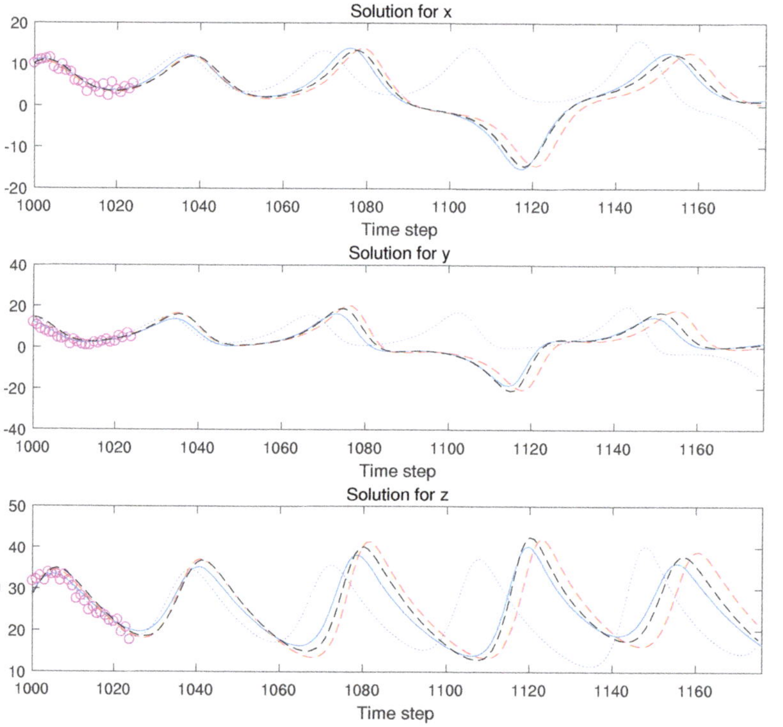

Figure 14. Trajectories of DA and DDA forecasting. It is the timesteps as Figure 11 for $\in (1000, 1176)$. R-DDA (black dashline) vs. F-DDA (red dashline).

Table 2. Run-time cost for F-DDA and R-DDA in training and forecasting of Lorenz system.

	Train Time	Forecast Time
F-DDA	3.7812 s	3.8581 s
R-DDA	24.2487 s	3.2901 s

5. Conclusions and Future Work

In this article, we discuss the DDA: the integration of DL and DA and we validate the algorithm by a numerical examples. The DA methods have increased strongly in complexity in order to better suit their application requirements and circumvent their implementation problems. However, the approaches to DA are unable to fully overcome their unrealistic assumptions, particularly of zero error covariances, linearity, and normality. DL shows great capability in approximating nonlinear systems, and extracting high-dimensional features. Together with the DA methods, DL is capable of helping traditional methods to make forecasts without the conventional methods' assumptions. On the other side, the training data provided to DL technologies include several numerical, approximation, and round off errors that are trained in the DL forecasting model. DA can increase the reliability of the DL models reducing errors by including information on physical meanings from the observed data.

This paper showed that the cohesion of DL and DA is blended in the future generation of technology that is used in support of predictive models.

So far, the DDA algorithm still remains to be implemented and verified in varies specific domains, which have huge state space and more complex internal and external mechanisms. Future work includes adding more data to the systems. A generalization of DDA could be developed if it is used for different dynamical systems.

The numerical solution of dynamic systems could be replaced by DL algorithms, such as Generative Adversarial Networks or Convolutional Neural Networks combined with LSTM, in order to make the runs faster. This will accelerate the forecast process towards a solution in real time. When combined with DDA, this future work has the potential to be very fast.

DDA is encouraging, as speed and accuracy are typically terms that are mutually exclusive. The results that are shown here are promising and demonstrate how, in computationally demanding physical models, the merger of DA and ML models, especially in computationally demanding physical models.

Author Contributions: R.A. and Y.-K.G. conceived and designed the experiments; J.Z. and S.H. performed the experiments; All the authors analyzed the data; All the authors contributed reagents/materials/analysis tools. All authors have read and agreed to the published version of the manuscript.

Funding: This research was funded by the Imperial College-Zhejiang University Joint Applied Data Science Lab. The work is supported by the EP/T000414/1 PREdictive Modelling with QuantIfication of UncERtainty for MultiphasE Systems (PREMIERE) and the EP/T003189/1 Health assessment across biological length scales for personal pollution exposure and its mitigation (INHALE).

Acknowledgments: This work is supported by the EP/T000414/1 PREdictive Modelling with QuantIfication of UncERtainty for MultiphasE Systems (PREMIERE) and the EP/T003189/1 Health assessment across biological length scales for personal pollution exposure and its mitigation (INHALE).

Conflicts of Interest: The authors declare no conflict of interest.

Acronyms

DA	Data Assimilation
DDA	Deep Data Assimilation
DL	Deep Learning
DNN	Deep Neural Network
LSTM	Long Short Term Memory
ML	Machine Learning
NN	Neural Network
VarDA	Variational Data Assimilation

References

1. Kalnay, E. *Atmospheric Modeling, Data Assimilation and Predictability*; Cambridge University Press: Cambridge, MA, USA, 2003.
2. Blum, J.; Le Dimet, F.X.; Navon, I.M. Data assimilation for geophysical fluids. In *Handbook of Numerical Analysis*; Elsevier: Amsterdam, The Netherlands, 2009; Volume 14, pp. 385–441.
3. D'Elia, M.; Perego, M.; Veneziani, A. A variational data assimilation procedure for the incompressible Navier-Stokes equations in hemodynamics. *J. Sci. Comput.* **2012**, *52*, 340–359. [CrossRef]
4. Potthast, R.; Graben, P.B. Inverse problems in neural field theory. *SIAM J. Appl. Dyn. Syst.* **2009**, *8*, 1405–1433. [CrossRef]
5. Christie, M.A.; Glimm, J.; Grove, J.W.; Higdon, D.M.; Sharp, D.H.; Wood-Schultz, M.M. Error analysis and simulations of complex phenomena. *Los Alamos Sci.* **2005**, *29, 6–25*.
6. Asch, M.; Bocquet, M.; Nodet, M. *Data Assimilation: Methods, Algorithms, and Applications*; SIAM: Philadelphia, PA, USA, 2016; Volume 11.
7. Weinan, E. A proposal on machine learning via dynamical systems. *Commun. Math. Stat.* **2017**, *5*, 1–11.
8. Li, Q.; Chen, L.; Tai, C.; Weinan, E. Maximum principle based algorithms for deep learning. *J. Mach. Learn. Res.* **2017**, *18*, 5998–6026.

9. Boukabara, S.A.; Krasnopolsky, V.; Stewart, J.Q.; Maddy, E.S.; Shahroudi, N.; Hoffman, R.N. Leveraging Modern Artificial Intelligence for Remote Sensing and NWP: Benefits and Challenges. *Bull. Am. Meteorol. Soc.* **2019**, *100*, ES473–ES491. [CrossRef]
10. Dueben, P.D.; Bauer, P. Challenges and design choices for global weather and climate models based on machine learning. *Geosci. Model Dev.* **2018**, *11*, 3999–4009. [CrossRef]
11. Cacuci, D. *Sensitivity and Uncertainty Analysis*; Chapman & Hall/CRC: New York, NY, USA, 2003.
12. Daescu, D.; Navon, I. Sensitivity analysis in nonlinear variational data assimilation: theoretical aspects and applications. In *Advanced Numerical Methods for Complex Environmental Models: Needs and Availability*; Farago, I., Zlatev, Z., Eds.; Bentham Science Publishers: Sharjah, UAE, 2013.
13. Arcucci, R.; D'Amore, L.; Pistoia, J.; Toumi, R.; Murli, A. On the variational data assimilation problem solving and sensitivity analysis. *J. Comput. Phys.* **2017**, *335*, 311–326. [CrossRef]
14. Cacuci, D.; Navon, I.; Ionescu-Bujor, M. *Computational Methods for Data Evaluation and Assimilation*; CRC Press: Boca Raton, FL, USA, 2013.
15. Fisher, M.; Leutbecher, M.; Kelly, G.A. On the equivalence between Kalman smoothing and weak-constraint four-dimensional variational data assimilation. *Q. J. R. Meteorol. Soc.* **2005**, *131*, 3235–3246. [CrossRef]
16. Gagne, D.J.; McGovern, A.; Haupt, S.E.; Sobash, R.A.; Williams, J.K.; Xue, M. Storm-based probabilistic hail forecasting with machine learning applied to convection-allowing ensembles. *Weather Forecast.* **2017**, *32*, 1819–1840. [CrossRef]
17. Campos, R.M.; Krasnopolsky, V.; Alves, J.H.G.; Penny, S.G. Nonlinear wave ensemble averaging in the Gulf of Mexico using neural networks. *J. Atmos. Ocean. Technol.* **2019**, *36*, 113–127. [CrossRef]
18. Babovic, V.; Keijzer, M.; Bundzel, M. From global to local modelling: A case study in error correction of deterministic models. In Proceedings of the Fourth International Conference on Hydro Informatics, Cedar Rapids, IA, USA, 23–27 July 2000.
19. Babovic, V.; Cañizares, R.; Jensen, H.R.; Klinting, A. Neural networks as routine for error updating of numerical models. *J. Hydraul. Eng.* **2001**, *127*, 181–193. [CrossRef]
20. Babovic, V.; Fuhrman, D.R. Data assimilation of local model error forecasts in a deterministic model. *Int. J. Numer. Methods Fluids* **2002**, *39*, 887–918. [CrossRef]
21. Brajard, J.; Carassi, A.; Bocquet, M.; Bertino, L. Combining data assimilation and machine learning to emulate a dynamical model from sparse and noisy observations: A case study with the Lorenz 96 model. *arXiv* **2020**, arXiv:2001.01520.
22. Rasp, S.; Dueben, P.D.; Scher, S.; Weyn, J.A.; Mouatadid, S.; Thuerey, N. WeatherBench: A benchmark dataset for data-driven weather forecasting. *arXiv* **2020**, arXiv:2002.00469.
23. Geer, A.J. *Learning Earth System Models from Observations: Machine Learning or Data Assimilation?* Technical Report 863; ECMWF: Reading, UK, 2020. [CrossRef]
24. Bocquet, M.; Brajard, J.; Carrassi, A.; Bertino, L. Bayesian inference of chaotic dynamics by merging data assimilation, machine learning and expectation-maximization. *Found. Data Sci.* **2020**, *2*, 55–80. [CrossRef]
25. Raissi, M.; Perdikaris, P.; Karniadakis, G. Inferring solutions of differential equations using noisy multi-fidelity data. *J. Comput. Phys.* **2016**, *335*. [CrossRef]
26. Perdikaris, P.; Raissi, M.; Damianou, A.; Lawrence, N.; Karniadakis, G. Nonlinear information fusion algorithms for data-efficient multi-fidelity modelling. *Proc. R. Soc. A Math. Phys. Eng. Sci.* **2017**, *473*, 20160751. [CrossRef]
27. Chao, G.; Luo, Y.; Ding, W. Recent advances in supervised dimension reduction: A survey. *Mach. Learn. Knowl. Extr.* **2019**, *1*, 341–358. [CrossRef]
28. Quilodrán Casas, C.; Arcucci, R.; Wu, P.; Pain, C.; Guo, Y.K. A Reduced Order Deep Data Assimilation model. *Phys. D Nonlinear Phenom.* **2020**, *412*, 132615. [CrossRef]
29. Makarynskyy, O. Improving wave predictions with artificial neural networks. *Ocean Eng.* **2004**, *31*, 709–724. [CrossRef]
30. Coskun, H.; Achilles, F.; Dipietro, R.; Navab, N.; Tombari, F. Long Short-Term Memory Kalman Filters: Recurrent Neural Estimators for Pose Regularization. In Proceedings of the IEEE International Conference on Computer Vision, Venice, Italy, 22–29 October 2017; pp. 5525–5533. arXiv:1708.01885v1. [CrossRef]
31. Zhu, J.; Hu, S.; Arcucci, R.; Xu, C.; Zhu, J.; Guo, Y.k. Model error correction in data assimilation by integrating neural networks. *Big Data Min. Anal.* **2019**, *2*, 83–91. [CrossRef]
32. Buizza, C.; Fischer, T.; Demiris, Y. Real-Time Multi-Person Pose Tracking using Data Assimilation. In Proceedings of the IEEE Winter Conference on Applications of Computer Vision, Snowmass Village, CO, USA, 1–5 March 2020; pp. 1–8.
33. Arcucci, R.; Moutiq, L.; Guo, Y.K. Neural assimilation. In *International Conference on Computational Science*; Springer: Berlin/Heidelberg, Germany, 2020; pp. 155–168.
34. Foo, Y.W.; Goh, C.; Li, Y. Machine learning with sensitivity analysis to determine key factors contributing to energy consumption in cloud data centers. In Proceedings of the 2016 International Conference on Cloud Computing Research and Innovations (ICCCRI), Singapore, 4–5 May 2016; IEEE: Piscataway, NJ, USA, 2016; pp. 107–113.
35. Purser, R. A new approach to the optimal assimilation of meteorological data by iterative Bayesian analysis. In Proceedings of the Preprint of the 10th AMS Conference on Weather Forecasting and Analysis, Clearwater Beach, FL, USA, 25–29 June 1984; American Meteorological Society: Boston, MA, USA, 1984; pp. 102–105.
36. Engl, H.W.; Hanke, M.; Neubauer, A. *Regularization of Inverse Problems*; Springer Science & Business Media: Berlin/Heidelberg, Germany,1996; Volume 375.

37. Nichols, N. Mathematical concepts in data assimilation. In *Data Assimilation*; Lahoz, W., Khattatov, B., Menard, M., Eds.; Springer: Berlin/Heidelberg, Germany, 2010.
38. Hansen, P. *Rank Deficient and Discrete Ill-Posed Problems*; SIAM: Philadelphia, PA, USA, 1998.
39. Le Dimet, F.; Talagrand, O. Variational algorithms for analysis and assimilation of meteorological observations: Theoretical aspects. *Tellus* **1986**, *38A*, 97–110. [CrossRef]
40. Kalman, R.E. A new approach to linear filtering and prediction problems. *J. Basic Eng.* **1960**, *82*, 35–45. [CrossRef]
41. Elman, J.L. Finding structure in time. *Cogn. Sci.* **1990**, *14*, 179–211. [CrossRef]
42. Jordan, M.I. Serial order: A parallel distributed processing approach. In *Advances in Psychology*; Elsevier: Amsterdam, The Netherlands, 1997; Volume 121, pp. 471–495.
43. Lawless, A.S. *Data Assimiliation with the Lorenz Equations*; University of Reading: Reading, UK, 2002.
44. Levenberg, K. A method for the solution of certain non-linear problems in least squares. *Q. Appl. Math.* **1944**, 2, 164–168. [CrossRef]
45. Marquardt, D.W. An algorithm for least-squares estimation of nonlinear parameters. *J. Soc. Ind. Appl. Math.* **1963**, *11*, 431–441. [CrossRef]

applied sciences

MDPI

Article

Deep Learning and Internet of Things for Beach Monitoring: An Experimental Study of Beach Attendance Prediction at Castelldefels Beach

Mari Carmen Domingo

Department of Network Engineering, BarcelonaTech (UPC) University, 08860 Barcelona, Spain; cdomingo@entel.upc.edu

Abstract: Smart seaside cities can fully exploit the capabilities brought by Internet of Things (IoT) and artificial intelligence to improve the efficiency of city services in traditional smart city applications: smart home, smart healthcare, smart transportation, smart surveillance, smart environment, cyber security, etc. However, smart coastal cities are characterized by their specific application domain, namely, beach monitoring. Beach attendance prediction is a beach monitoring application of particular importance for coastal managers to successfully plan beach services in terms of security, rescue, health and environmental assistance. In this paper, an experimental study that uses IoT data and deep learning to predict the number of beach visitors at Castelldefels beach (Barcelona, Spain) was developed. Images of Castelldefels beach were captured by a video monitoring system. An image recognition software was used to estimate beach attendance. A deep learning algorithm (deep neural network) to predict beach attendance was developed. The experimental results prove the feasibility of Deep Neural Networks (DNNs) for beach attendance prediction. For each beach, a classification of occupancy was estimated, depending on the number of beach visitors. The proposed model outperforms other machine learning models (decision tree, k-nearest neighbors, and random forest) and can successfully classify seven beach occupancy levels with the Mean Absolute Error (MAE), accuracy, precision, recall and F1-score of 0.03, 92.7%, 92.9%, 92.7%, and 92.7%, respectively.

Keywords: Internet of Things; network architecture; deep learning; smart cities

Citation: Domingo, M.C. Deep Learning and Internet of Things for Beach Monitoring: An Experimental Study of Beach Attendance Prediction at Castelldefels Beach. *Appl. Sci.* **2021**, *11*, 10735. https://doi.org/10.3390/app112210735

Academic Editors: João M. F. Rodrigues, Pedro J. S. Cardoso and Marta Chinnici

Received: 2 October 2021
Accepted: 11 November 2021
Published: 14 November 2021

1. Introduction

The term smart city was first used in the 1990s for the use of Information and Communication Technologies (ICT) to develop modern infrastructures within cities [1]. Afterwards, the smart city concept evolved and is no longer limited to the diffusion of ICT, it is also related to people and community needs. Smart cities are cities that use ICT to develop new urban intelligence functions in such a way that community and quality of life are enhanced [2]. New decision-making paradigms have been developed for optimizing the continuous, real-time allocation of resources to satisfy demands in large urban environments [3].

Coastal areas are some of the most active and biologically diverse ecosystems in the world. Their population is constantly growing; at least 60% of the world's population lives within 100 km of the coast [4]. Furthermore, 80% of all tourism takes place in coastal areas, and beaches are among the most popular destinations [5].

The Internet of Things (IoT) and Artificial Intelligence (AI) are two cornerstone technologies enabling smart cities. The IoT refers to a world of networked smart devices, where every day interconnected objects transform into smart objects able to collect and share data, thanks to the combination of the Internet and powerful technologies such as Radio-Frequency Identification (RFID), real-time localization, and embedded sensors [6,7]. AI refers to machines working in an intelligent way. Machine learning is a subset of AI that provides machines with the ability to learn without being explicitly programmed.

A huge amount of information (Big Data) is extracted from IoT devices. The analysis of Big Data through Artificial Intelligence (AI) is very helpful to improve the performance of smart cities' services. Although AI has been popular since the early 1950s [8], its application has been slow. The popularity of AI rose from 2014 to 2017 as a result of the growth of Big Data. The concept of smart city became more popular in the same period [9]; therefore, there is a correlation between smart cites with Big Data and AI [7].

Smart seaside cities can fully exploit the capabilities brought by IoT and artificial intelligence to improve the efficiency of city services in traditional smart city applications [10–12]: smart home, smart healthcare, smart transportation, smart surveillance, smart environment, cyber-security, etc. However, smart coastal cities are characterized by their own specific application domain, namely, beach monitoring.

Beach attendance prediction is a beach monitoring application of particular importance for coastal managers to successfully plan beach services. In this paper, an experimental study that uses IoT data and deep learning to predict the number of beach visitors at Castelldefels beach (Barcelona, Spain) was developed. Its purpose is to predict the number of beach visitors that will go to the beach in the future (e.g., any day during the next prime swimming season), depending on the month, the weekday, the time, the weather data (forecast), and if it is a working day or holiday. The dataset of the deep learning algorithm contains data regarding all the previously mentioned attributes. The number of beachgoers was estimated using image recognition software from 2016 to 2018.

Econometrics and machine learning aim at building a predictive model for a variable of interest, using explanatory variables (or features). In econometrics, probabilistic models are built to describe economic phenomena, while machine learning uses algorithms capable of learning, usually for classification purposes [13]. In recent years, machine learning models have been found to be more effective than traditional econometric models [13].

Machine learning overcomes the limitations of econometric models in terms of prediction. With respect to predictions, econometric models are affected by forecasting errors due to overfitting. This modelling error happens in complex models when the higher the variance is, the lower the bias is. Machine learning can also be affected by overfitting. In machine learning overfitting refers to good performance on the training data, but poor generalization to other data. Nevertheless, in machine learning there exist several techniques to prevent overfitting: early stopping, training with more data, data augmentation, feature selection, regularization, etc. Machine learning models are able to obtain a high prediction accuracy with almost any type of data and perform classification efficiently [14]. An important advantage of machine learning models is that their hyperparameters can be highly tuned to improve the prediction. Furthermore, once the basic machine learning models are trained, the researcher can identify which is the best one for a particular dataset. In addition, advanced machine learning models like deep learning algorithms show very good results with unbalanced datasets and Big Data [14]. Deep learning is a promising class of machine learning models that has become a popular subject in the field of science. Deep learning has been used successfully for signal processing, pattern recognition, and statistical analysis.

Many researchers have shown that neural networks outperform econometrics models in forecasting accuracy. In [15], tourism data was used to forecast the arrival of tourists from USA to Durban (South Africa). It was shown that neural networks perform better than exponential smoothing, ARIMA, multiple regression, and genetic regression models. In [16], the application of three time-series forecasting techniques, namely exponential smoothing, univariate ARIMA, and Elman's Model of Artificial Neural Networks (ANN), was investigated to predict travel demand (i.e., the number of arrivals) from different countries to Hong Kong. The results show that neural networks are the best method for forecasting visitor arrivals, especially those series without obvious patterns.

We selected a machine learning algorithm to predict beach attendance due to the efficiency of these algorithms for classification purposes. Particularly, a deep learning algorithm (deep neural network) was selected due to the higher performance of neural networks for prediction compared to econometrics and its promising results in the field of science.

The contributions of this paper are summarized as follows:

- We propose to use IoT data and deep learning to estimate beach attendance.
- A deep learning (fully connected deep neural network) algorithm was developed for beach attendance prediction.
- Beach attendance is predicted based on the following attributes: the time of day, the day of the week, the season, and the weather conditions. We use these attributes as the seven input variables for training our proposed deep neural network (DNN) model.
- The last attribute (number of beach visitors), the output variable, is taken as the ground-truth of the target attribute for model training. Images from cameras are analyzed to estimate the number of beach visitors using an image recognition software. This output is used to train the deep neural network during the training phase (using the training set); it is also used later to test the model (using the testing set) in terms of accuracy and mean absolute error (MAE).
- Our proposed deep learning classifier outperforms other machine learning models (decision tree, *k*-nearest neighbors, and random forest) and can successfully differentiate between seven beach occupancy levels, with the Mean Absolute Error (MAE), accuracy, precision, recall and F1-score of 0.03, 92.7%, 92.9%, 92.7%, and 92.7%, respectively.

To the best of our knowledge, our proposal is the first deep learning algorithm for beach attendance prediction. The experimental results prove its feasibility.

The paper is structured as follows. Section 2 discusses the related work. Section 3 introduces the proposed deep neural network model and the dataset. In Section 4, we present the experimental settings and the evaluation metrics and show the results. In Section 5, the results are discussed. Finally, the paper is concluded in Section 6.

2. Related Work

It has always been extremely important to maintain the quality of beaches, since sand and water conditions can affect human health. Furthermore, most overcrowded beaches are seriously affected by pollution and traffic congestion. Therefore, the physical and social carrying capacities are essential factors to estimate the comfort of most crowded beaches [17]; the physical carrying capacity refers to the maximum number of individuals that can physically fit on a beach, whereas the social carrying capacity is associated with crowding perception in the presence of a large amount of beach visitors.

Several studies [18–22] recognize the importance of evaluating beach attendance to maintain the recreational capacity of the beaches. It is an essential factor to plan different beach services in terms of security, rescue, health and environmental assistance. These studies analyze how beach attendance is affected by weather conditions and other aspects such as time, season, etc. In [18], the authors quantify the number and location of beach visitors during 2012 using video images at the Lido of Sète beach, France. An automatic counting algorithm in Matlab is used to estimate beach attendance; they also study beach users' behavior. In [19], video images are analyzed using an algorithm developed in Matlab; the purpose is to quantify visitors to two city beaches of Barcelona from November 2001 to December 2005 and to predict beach occupation. They tried to find a mathematical expression to model beach attendance. The observed mean number of daily users is adjusted to a time-dependent Fourier polynomial for the two beaches in Barcelona. The occupation data for 2002–2004 are averaged to obtain an estimation of the occupation trend for a typical year. This averaged function is projected into a 14-term Fourier polynomial. The Fourier fit obtained and the original time series of the average occupation in a typical year adjusted 74% and 69% of the absolute value of the original time series for the two beaches, respectively. In [20], images from the Argus video monitoring system that belongs to ten beaches from the Spanish Mediterranean coast are selected from 1 July to 20 October 2009; beach occupancy is estimated based on certain density levels. In [21], web cameras are used to count the number of beach visitors of three well-known Australian beaches using a people counting computer program. The authors also analyzed how the number of beach visitors is affected by certain weather and ocean conditions. In [22], a monitoring system consisting of sensors, cameras, and smartphones is used to evaluate the occupational state of a beach in Cagliary (Italy) using the collected environmental and crowding data. The beach crowd density is evaluated based on beach images, with a support vector machine (SVM) classifier that distinguishes between three levels of crowd density: low, medium, and high.

All these studies [18–22] confirm that there is a relationship between beach attendance and certain attributes, such as time of day, day of week, season, and weather conditions.

Smart seaside cities can benefit from machine learning and Internet of Things (IoT) to fight against public health crises such as COVID-19. IoT devices (cameras, drones, etc.) on beaches can be used to control crowd density. It has been analyzed how COVID-19 transmission is affected by the beachgoer behavior [23]. UAV imagery and publicly available beachcam video data collected during the summer of 2020 at the recreational beach oceanfront in Virginia Beach, USA, were analyzed [23]. It was found that beach users are concentrated in approximately 43% of the total beach surface area, whereas approximately one-third of landward beach surface is left vacant. Static webcam images of the boardwalk also indicated relatively consistent use throughout the day, high use at beach access points and points of interest (i.e., King Neptune statue), and low use of face coverings for observed northbound boardwalk users (8.7%) [23]. These data are useful for authorities to supervise beach areas in real-time and make decisions regarding social distancing, use of masks, and other measures to contain the pandemic.

Beach attendance is also an essential factor to plan different beach services in terms of security, rescue, health, and environmental assistance. Beach access conditions, number and size of parking areas, and other services (restaurants, leisure activities, public restrooms, etc.) are affected as well. The emergence of COVID-19 has brought new implications for beach access. There is a need to implement and enforce additional mitigation strategies (physical distancing, limiting gatherings, supporting hygiene, etc.) so that there is no widespread community transmission of the virus. Beach attendance estimation is essential to plan the reopening of beaches and offer beach services safely. Real-time occupation was analyzed in 14 beaches of the coast of Guipuzcoa (Basque Country, Northern Spain) [24]. Video images from 12 stations located along 50 km of coastline were processed. A machine learning algorithm (AdaBoost, SVM and Quadratic Regression) was used to count beach users. The occupancy level (full, high, medium, or low) of every beach was sent to local authorities through a web/mobile app as well as special warnings under particular circumstances to allow them to take action in cases where carrying capacity limits were about to be reached [25].

In this paper we describe a deployment to predict beach attendance using machine learning at Castelldefels beach (Barcelona, Spain).

3. Materials and Methods

In this paper, experimental research that uses IoT data and deep learning to successfully predict the number of beach visitors at Castelldefels beach (Barcelona, Spain) was developed.

Cameras on beaches can be used to control crowd density. Images can also be analyzed together with other attributes (such as weather conditions determined by IoT sensors at the weather station) for beach attendance prediction.

Our dataset is based on the images from the video monitoring system of Castelldefels beach (Barcelona, Spain) and on the weather data from a weather station. The images and weather data were obtained using IoT devices.

One of the major obstacles to build a real intelligent system [26] is dealing with random disturbances, processing a huge amount of imprecise data, interacting with a dynamically changing environment, and coping with uncertainty. Neural networks make it possible to solve particular problems by using a customer developed algorithm with an intelligent behavior. Therefore, our proposed application for beach attendance prediction will be based on these.

3.1. Deep Neural Network Model

Deep learning has been used to predict beach attendance based on certain weather conditions and other attributes. Deep learning is a promising approach to extract data from IoT devices, even in complex environments where other machine learning techniques are confused [27]. Therefore, we propose to use a fully connected DNN to predict beach attendance. The IoT data about the required attributes is fed as inputs to the DNN algorithm so that the output layer can estimate the number of beach visitors (beach occupancy).

A DNN is a neural network with more than two layers (including just the hidden layers and the output layer). DNNs can model complex non-linear relationships.

The DNN takes the input data and extracts automatically appropriate representations for detection or classification purposes [28]. Each layer extracts and amplifies those features that are more relevant for decision making, whereas irrelevant features are suppressed. Each layer is connected to neighboring layers, with different weights attached to the connection.

Weights and biases are both learnable parameters inside neural networks. The weights determine how each feature affects the prediction. Bias represents how far off the predictions are from their intended value.

The weight for the connection from the kth neuron in the $(l-1)$th layer to the jth neuron in the lth layer is expressed as w^l_{jk}. The bias of the jth neuron in the lth layer is defined as b^l_j. Therefore, the activation of the jth neuron in the lth layer, a^l_j, is given by

$$a^l_j = g\left(\sum_k w^l_{jk} a^{l-1}_k + b^l_j\right) \tag{1}$$

where the sum is over all neurons k in the $(l-1)$th layer.

The rectified linear unit (ReLU) was used as the activation function

$$g(z) = \max(0, z) \tag{2}$$

The weights and bias are estimated by minimizing a loss function [29].

Next, we describe our dataset.

3.2. Videometry

Our dataset is based on the images from the video monitoring system of Castelldefels beach [30] (Barcelona, Spain) and on the weather data from a weather station of Meteocat (Meteorological Service of Catalonia) [31].

The Castelldefels beach is a 5 km long strip of sand located in Spain around 18 km away from the south of Barcelona, between the delta of Llobregat river and the Garraf Massif. People usually visit the beach to swim in the calm waters of the Mediterranean Sea, sunbathe, do water sports or activities with children, or just walk along the shoreline. In Castelldefels, the prime bathing season is from 1 June to 31 August. Castelldefels is a popular destination for Spanish holidaymakers; it has an excellent location, close to Barcelona City, which is considered within the world ranking among the 20 most visited cities by foreign tourists [32] and among the 10 most visited cities in Europe [33]. The Barcelona metropolitan area, with a population of over 5 million, is the most populous urban area on the Mediterranean coast and one of the largest in Europe. The beaches of the Barcelona metropolitan area are visited each year by 10 million people, who spend (directly or indirectly) almost 60 million euros during the prime summer season [34]. The most visited beaches during the summer for the north metropolitan area of Barcelona are Sant Adria, Badalona, and Montgat, and for the south metropolitan area are Gava and Castelldefels. We selected Castelldefels beach due to its popularity, the presence of a video monitoring system that has operated at Castelldefels beach since 5 October 2010, and the location of our campus, the Castelldefels School of Telecommunications and Aerospace Engineering (EETAC-UPC) (Barcelona-Tech University), 10 min away from Castelldefels beach.

Coastal ecosystems require continuous observation, which is achieved by means of coastal video monitoring. Remote sensing techniques offer cost-efficient, long-term data collection, with high resolution in time and space. Shore-based video systems enable automated data collection, encompassing a much greater range of time and spatial scales than previously possible. Video systems are very useful for automatic shoreline detection and data analysis [35] and intertidal [36] and subtidal bathymetry. Shore-based video monitoring systems are also very useful for breaking-wave height estimation from digital images [37]. In our case, the video monitoring system of Castelldefels beach consists of five video cameras located at the tower in Plaza de las Palmeres, 30 m away from Pineda beach in Castelldefels (see Figure 1). An example of the images captured by the five video cameras is displayed in Figure 2.

Figure 1. Video monitoring system of Castelldefels beach (Barcelona, Spain).

Figure 2. Example of the images captured by the five video cameras.

Images have been collected since April 2010 using SIRENA software developed at IMEDEA (CSIC), and they are publicly available on their website [30]. The scientific exploitation of the images is a joint agreement between the Coastal Ocean Observatory [38] of the Institut de Ciències del Mar [39] ICM-CSIC and the Coastal Morphodynamics group (UPC-Barcelona Tech University). Every daylight hour, the cameras take one picture per second for a ten minute period; in our work, a snapshot image was used to count the number of beach visitors per hour from 9:00 to 19:00 h during June, July, and August (prime bathing season) from 2016 to 2018 using an image recognition software named "CountThings" [40].

3.3. Image Recognition Software

The image recognition software "CountThings" [40] is being used professionally to count objects in many different fields, such as medicine and construction. The number of beachgoers derived from manual counting was compared with automated counts in the snapshots of five days in the summer from 2016 to 2018 (see Figure 3).

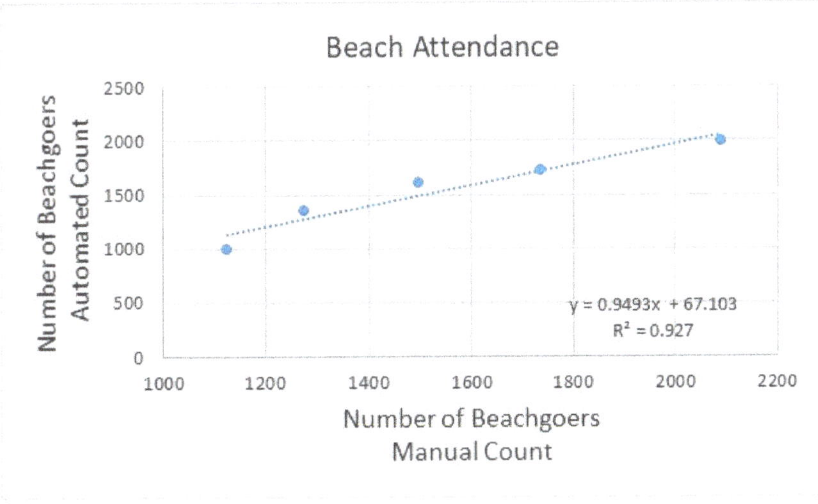

Figure 3. Comparison between manual and automated count of beach visitors.

We compared the counts using a linear regression analysis. The comparison showed an R^2 of 0.927 and a p-value < 0.001. Therefore, we verified that the automated count is not significantly different from the manual count, the error is acceptable, and this methodology is suitable for counting the number of beach visitors.

Figure 4 shows an example of the computer vision algorithm results at Castelldefels beach. The circles indicate the beach visitors that were detected by the algorithm. The algorithm correctly identified 106 beachgoers (true positives, *TP*), whereas it returned false negatives (*FN*) for cyclists, beachgoers covered by umbrellas, and beachgoers surrounded by a darker background.

The detection algorithm has a high level of accuracy (R^2 of 0.9270), but it also failed at times, with cloudy or rainy weather or producing blurry images, as shown in Figure 5. In these cases, the number of beachgoers was counted manually. The cases where the detection algorithm does not work well are uncommon (e.g., it rains only very seldom), and overall, the use of the detection algorithm saved time and provided good results.

Figure 4. Camera captured image (**left**) and image result of the detection algorithm (**right**) at Castelldefels beach, 16 June 2018.

Figure 5. Image result of the detection algorithm for images with issues (blurry, **left**; rainy, **right**) at Castelldefels beach.

Next, we present our results for beach attendance detection. The total average beach attendance per day for the summer from 2016 to 2018 is shown in Figure 6.

Beach attendance was calculated for June, July, and August for mornings and afternoons and was very similar for all the years studied. The year with the highest beach attendance was 2017, although the weather was similar when compared to 2016 and 2018, with 2 rainy days in 2016, 5 in 2017, and 6 in 2018. Beachgoers prefer going to the beach in the mornings for all the months, and in most cases beach attendance during the morning was double than that in the afternoon. June showed a lower beach attendance because the weather was worse for most of the month. The daily distribution of beach users is analyzed in Figure 7 for working days (Tuesdays) and weekends (Saturdays and Sundays) during the summer (June, July, and August) from 2016 to 2018. The number of beachgoers was already significant at 9:00 h and kept increasing to reach a maximum at 11:00 h; afterwards, it decreased progressively. Beach attendance was higher during the weekends (especially on Sundays), which suggests that, apart from tourists, a significant number of local people go to the beach on weekends when they are not working. The number of beach visitors was similar, independent of the day, from 17:00 h to 19:00 h.

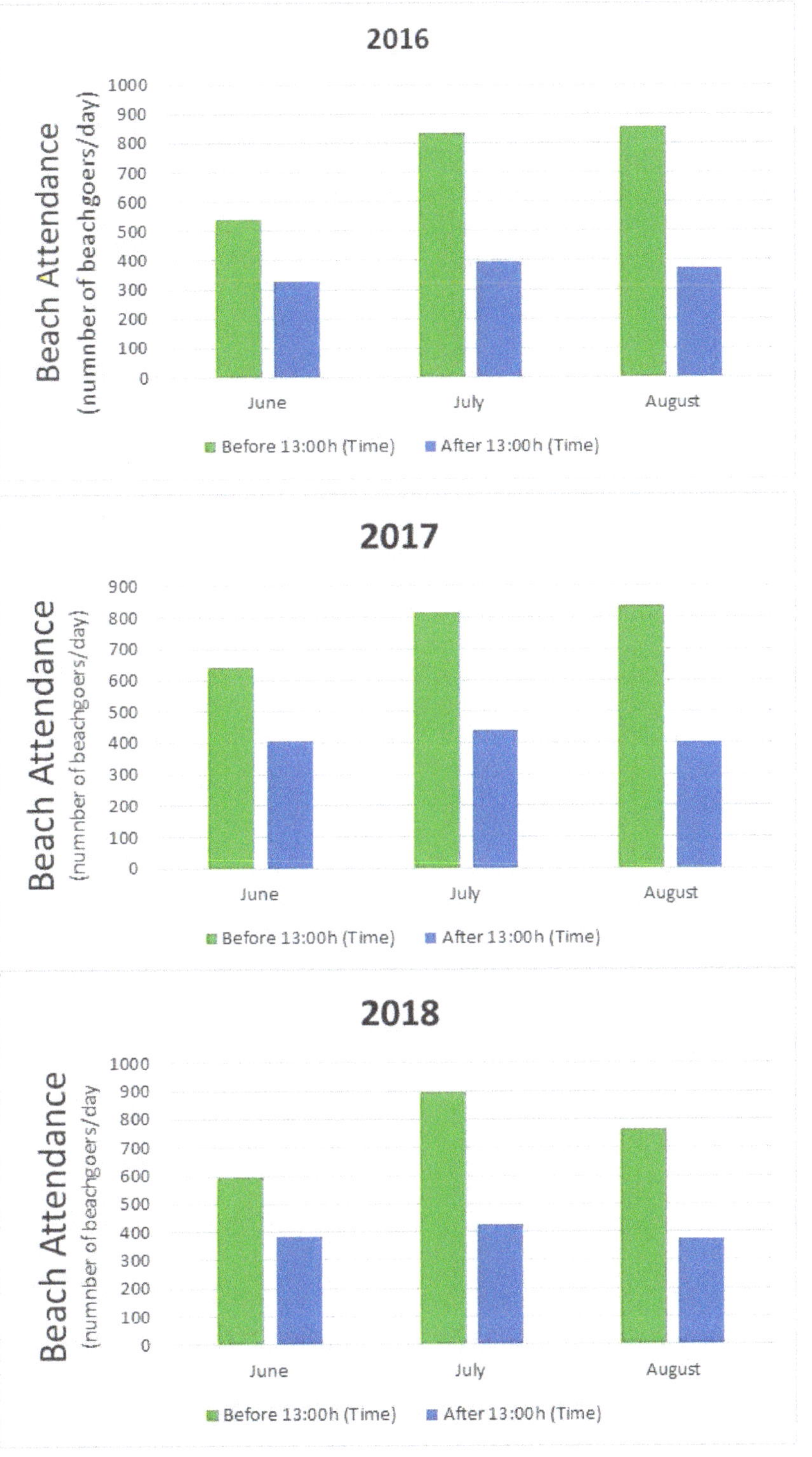

Figure 6. Morning and afternoon total average beach attendance per day.

Figure 7. Hourly distribution of beach users for Tuesdays, Saturdays, and Sundays for June, July, and August from 2016 to 2018.

3.4. Dataset

In total, there are 19,984 data samples in our dataset, and the attributes associated with the data samples are grouped into the following categories.

(1) Datetime Attributes: To facilitate the data processing, datetime is digitalized into three attributes, including the integer-based month, weekday, and time.
(2) Outdoor Attributes: The three outdoor attributes available in the dataset are the temperature (in Celsius), accumulated rainfall (in mm), and air velocity (in m/s).
(3) Calendar Attribute: We also consider the attribute of working day or holiday (Saturday, Sunday, or local holiday).
(4) Number of Beach Visitors Attribute: This attribute refers to the counted number of beach visitors (beach occupancy).

We use the datetime, outdoor, and calendar attributes as the 7 input variables for training our proposed DNN model. The last attribute (number of beach visitors), output variable, is taken as the ground-truth of the target attribute for model training. This output is used to train the deep neural network during the training phase (using the training set); it is also used later to test the model (using the testing set) in terms of accuracy and mean absolute error (MAE).

There are 2498 rows, and each row corresponds to a given record of the data set. Every column represents one variable. There are 7 input variables: month, weekday, time, temperature, accumulated rainfall, air velocity, working day/holiday, and an output variable: resulting number of beachgoers for a particular month, weekday, etc. Therefore, there are 2498 records $*$ (7 input + 1 output) variables = 19,984 data samples in our dataset. The number of beachgoers is obtained using an image recognition software named CountThings and not the proposed deep learning algorithm. The number of beachgoers is counted by the image recognition software, counting the number of beachgoers from each of the 5 snapshots (each one of the 5 video cameras captures a snapshot). The number of images analyzed by the image recognition software is 2498 $*$ 5 = 12,490 images.

In terms of spatial distribution, changes are observed between pre-COVID-19 and pandemic years, as shown in Figure 8. An example of this change can be observed in the heat maps of two high attendance days, 9 August 2018 (left) and 19 August 2020 (pandemic year) (right), at 11:00 h. Before COVID-19, there was a higher concentration of people near the shore (non-uniform distribution). During COVID-19, pandemic beach users are located following a uniform distribution, trying to respect social distance recommendations.

Figure 8. Different spatial distribution patterns with maximum beach occupancy levels in 2018 (**left**) and 2020 (**right**).

4. Experiments and Results

Next, we present the experimental settings and the evaluation metrics and show the results.

4.1. Experimental Settings

The proposed deep learning algorithm was implemented in Python 3.7.3.

A neural network can learn something different from what its trainer had in mind [41]. A case of useless learning is when the neural network memorizes the training examples without learning what they have in common. A trained network is able to generalize when it can classify data from the same class as the learning data but that it has never seen before. This ability requires that the network is tested with an independent dataset. Therefore, the data samples were divided as follows: 70% of the samples were used for the training dataset (70% of 2498 records = 1748 records for the training phase) and 30% for the testing dataset (30% of 2498 records = 750 for the testing phase).

The input attributes were normalized. The proposed DNN consists of one input layer with 7 neurons that match the 7 input attributes, 6 hidden layers each with 300 neurons, and one output layer with one neuron for the modeling target.

Our proposed DNN was trained for the beach visitors classification task. In the dataset, the number of beach visitors was assigned to one of the 7 classes (or beach occupancy levels): 0–49, 50–99, 100–149, 150–199, 200–249, 250–299, and 300+. The class distribution is shown in Table 1. The total number of records (1748) refers to the training phase.

Table 1. Class distribution of beach visitors.

Classes (Beach Visitors)	Number of Samples	Percentage of Total Samples
0–49	272	15.56
50–99	685	39.19
100–149	372	21.28
150–199	280	16.02
200–249	17	0.97
250–299	91	5.20
300+	31	1.77

Overfitting refers to a deep learning model that has a small loss function on the training data, and the prediction accuracy is high; however, on the test data, the loss function is relatively large and the prediction accuracy is low. To overcome the overfitting constraint, a grid search was conducted to find the best hyperparameters of the deep neural network using 10-fold cross validation. In a 10-fold cross validation scheme, the dataset is divided into 10 blocks of approximately equal size. In this case, 90% or 9 blocks of the data are used for training, and 10% or 1 block of the data is used for testing. This process is repeated 10 times, with a different data block used for testing each time. Table 2 shows the number of training and testing instances in each partition scheme. The total number of records for the training (1748) and testing (750) phases is specified. The resulting values for the hyperparameters of the neural network training are presented in Table 3. In order to solve the overfitting problem, we also introduced the dropout parameter (dropout value of 0.1), which achieved the regularization effect.

Table 2. Class distribution of beach visitors.

Training-Testing Partition (%)	Total Training Records	Total Testing Records
70–30	1748	750
10-fold cross validation	2248	250

Table 3. Hyperparameters.

Hyperparameter	Specification
Number of hidden layers	6
Number of neurons per hidden layer	300
Optimizer	Adam
Learning rate	0.001
Activation function	ReLU

We evaluated the SGD, RMSprop, Adam, Adagrad, Adamax, and Nadam optimizers using the grid search technique. The Adaptive Moment Estimation (Adam) optimizer was selected to minimize the loss function and speed up the training process because it obtains the best results.

The Adam optimizer is one of the most popular gradient descent optimization algorithms since it is computationally efficient and has very little memory requirement. This method calculates the individual adaptive learning rate for each parameter from estimates of first and second moments of the gradients.

The Adam algorithm (see Algorithm 1) first updates the exponential moving averages of the gradient (m_t) and the squared gradient (v_t), which is the estimate of the first and second moment. The hyperparameters $\beta_1, \beta_2 \in [0, 1)$ control the exponential decay rates of these moving averages, as shown in the following equations:

$$m_t = \beta_1 m_{t-1} + (1 - \beta_1) g_t \quad (3)$$

$$v_t = \beta_2 v_{t-1} + (1 - \beta_2) g_t^2 \quad (4)$$

where g is the current gradient value of error function for the neural network training.

Algorithm 1 Adam, our proposed algorithm for the training process.

1: Declare the parameters Objective function $f(\theta)$, hyperparameter learning rate α, exponential decay rates β_1, β_2 for moment estimates, tolerance parameter $\epsilon > 0$ for numerical stability
2: Initialize first moment vector $m_0 = 0$, second moment vector $v_0 = 0$ and timestep $t = 0$
3: **while** θ_t has not converged **do**
3.1 update timestep $t = t + 1$
3.2 compute gradient of objective using $g_t = \nabla_\theta f_t(\theta_t - 1)$
3.3 update first moment estimate and second moment estimate using Equations (3) and (4), respectively.
3.4 compute unbiased first and second moment estimate using Equations (5) and (6), respectively.
3.5 update objective parameters using Equation (7).
end while
4: return final parameter θ_t

Moving averages are initialized as 0. The moment estimates are biased around 0, especially during the initial timesteps. This initialization bias can be easily counteracted resulting in bias-corrected estimate.

$$\hat{m}_t = \frac{m_t}{1-\beta_1^t} \tag{5}$$

$$\hat{v}_t = \frac{v_t}{1-\beta_2^t} \tag{6}$$

Finally, we update the parameter θ_t as shown below:

$$\theta_t = \theta_{t-1} - \frac{\alpha \hat{m}_t}{\sqrt{\hat{v}_t} + \varepsilon} \tag{7}$$

We used in our experiments for the Adam optimizer a learning rate $\alpha = 10^{-3}$ and two decay parameters β_1= 0.9 and β_2 = 0.999 [42].

Experiments were carried out on a laptop running 64-bit Windows 10 Home on an Intel Core i5-8265U and using 8 GB of memory.

4.2. Evaluation Metrics

Five different metrics were used to evaluate the performance of the proposed scheme: Mean Absolute Error (MAE), accuracy, precision, recall, and F1-score. F1-score is the harmonic mean of precision and recall.

The selected metrics can be calculated as follows:

$$Accuracy = \frac{TP+TN}{TP+TN+FP+FN} \tag{8}$$

where TP represents true positives, TN represents true negatives, FP represents false positives, and FN represents false negatives.

$$MAE = \frac{1}{K}\sum_{k=1}^{K}|g_k - g\prime_k| \tag{9}$$

where g_k and $g\prime_k$ represent the real and predicted number of beach visitors, respectively, and K denotes the total number of testing samples.

$$Precision = \frac{TP}{TP+FP} \tag{10}$$

$$Recall = \frac{TP}{TP+FN} \tag{11}$$

$$\mathrm{F1-score} = \frac{2 \times Precision \times Recall}{Precision + Recall} \tag{12}$$

4.3. Performance Evaluation

Figures 9 and 10 illustrate the loss function convergence curve and accuracy curve of the neural network training and testing phases for beach occupancy prediction, respectively. The decreasing loss and the increasing accuracy curves in Figures 9 and 10, respectively, generally suggest that the neural network model is learning to generalize on the target problem.

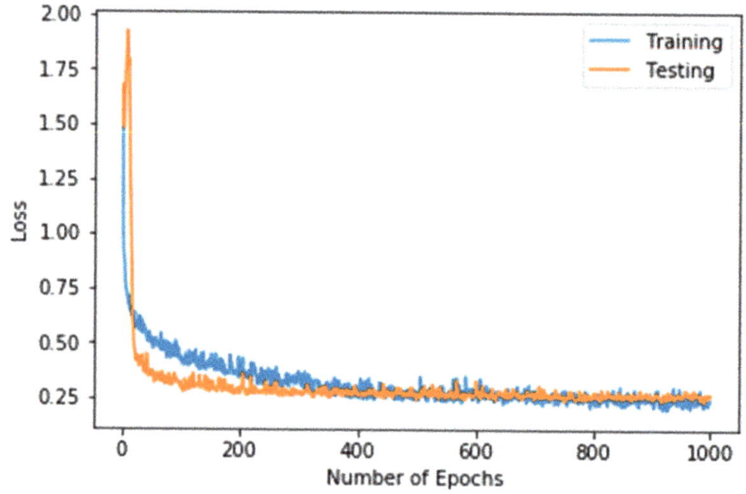

Figure 9. Loss curve for training and testing phases.

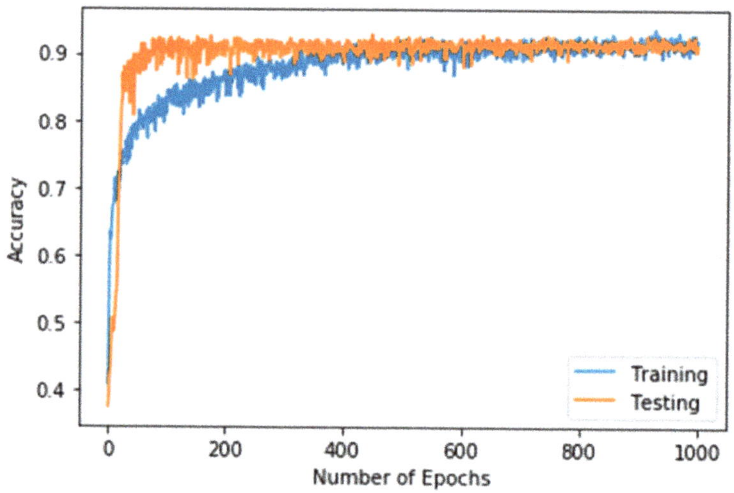

Figure 10. Accuracy for training and testing phases.

To figure out the performance of the developed DNN, Figure 11 shows the confusion matrices on the test set (1) only for June, (2) only for July, (3) only for August, and (4) for the whole dataset (June, July, and August). The classification error for June (see Figure 11a) came mainly from the prediction of classes 150–199, 200–249, 250–299, and 300+. The reason is that there are fewer training samples for these classes, since the prime bathing season starts in June, and during this month, fewer people come to the beach. In July (Figure 11b) and August (Figure 11c), beach visitors of all classes are well predicted, with the exception of class 200–249 for July, since there are very few training samples for this class. No sample of class 300+ can be predicted for August, because during this month, the residents in Spain have holidays; they tend to travel far away from the cities, and Castelldefels beach is located very close to the city of Barcelona. All classes are well predicted for the whole dataset (Figure 11d).

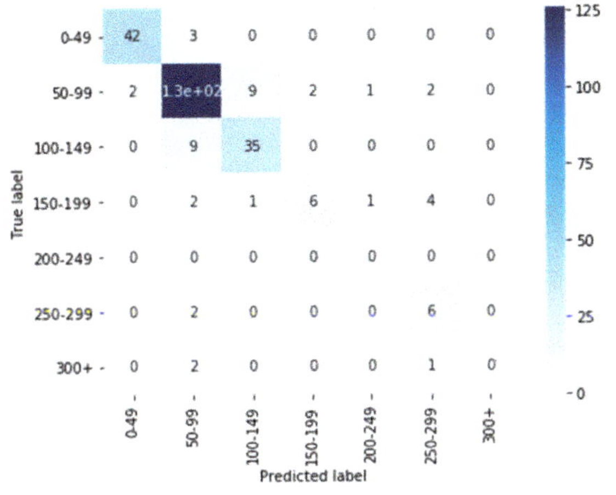

(a)

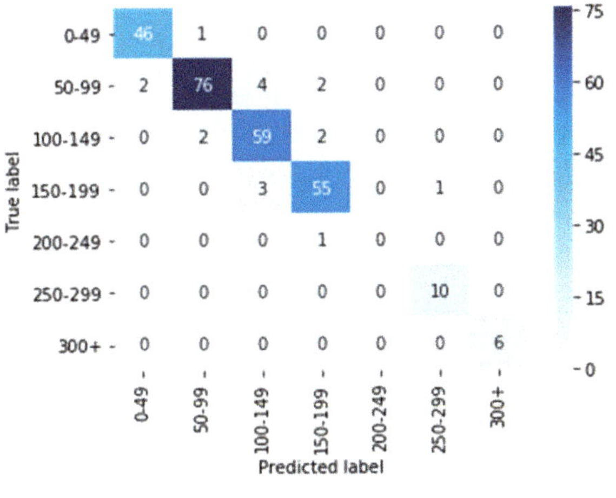

(b)

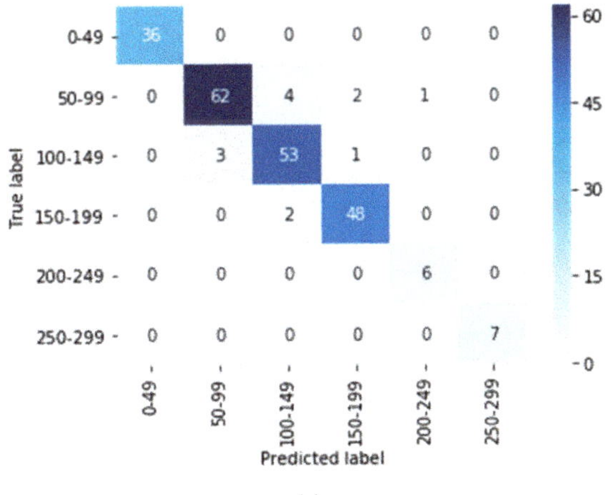

(c)

Figure 11. *Cont.*

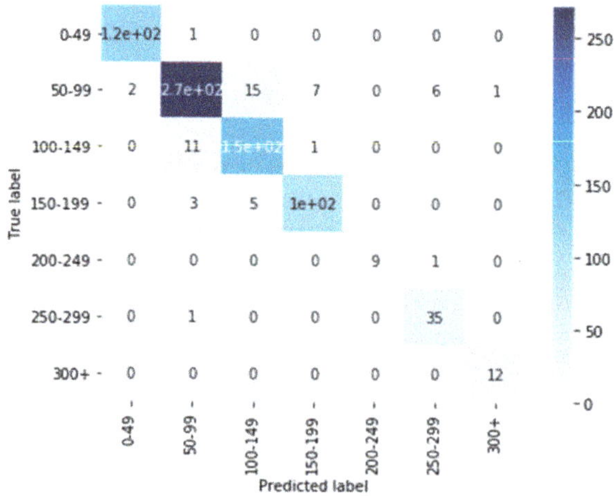

(**d**)

Figure 11. Confusion matrixes for (**a**) June, (**b**) July, (**c**) August, and (**d**) the whole dataset.

The accuracy, precision, recall, and F1-score for all the classes and the whole data set were computed using the confusion matrix. The results as well as the number of instances per class for the test set are shown in Table 4. Accuracy is not a very good measure of performance when dealing with unbalanced datasets (like our case) because it counts the number of correct predictions regardless of the type of class. For this reason, it is biased towards the majority classes.

Table 4. Comparison of accuracy and F1-score for all classes.

Classes (Beach Visitors)	Accuracy	Precision	Recall	F1-Score	Number of Samples per Class for the Test Set
0–49	0.99	0.98	0.99	0.99	121
50–99	0.94	0.94	0.90	0.92	302
100–149	0.96	0.88	0.92	0.90	159
150–199	0.98	0.93	0.93	0.93	110
200–249	1	1	0.90	0.95	10
250–299	0.99	0.83	0.97	0.90	36
300+	1	0.92	1	0.96	12

As we can see in Table 4, the accuracy is extremely high (0.99 or 1), even when the model fails to identify some samples of minority classes (such as classes 200–249, 250–299, and 300+). Since accuracy gives biased results with unbalanced data, it is not a good metric to use. Therefore, we selected precision and recall, together with F1-score, as more reliable measures.

Table 5 reports the results of our analysis for every category in the test set using the MAE, accuracy, precision, recall, and F1-score. The results of 10 independent runs are shown. We present the results (1) only for June, (2) only for July, (3) only for August, and (4) for the whole dataset.

The DNN achieves the highest F1-score, with a value of 94.3866%, for August and a high F1-score of 92.7114% for the whole data set. The lowest results in terms of F1-score are obtained when the DNN is trained just for June, since the video cameras were unavailable during several days, and these days were essential to train the DNN. This fact suggests that training with more images would improve the classification. Furthermore, from the confusion matrixes, it can be observed that in June (when the beach attendance is lower) there are fewer samples to train categories with a high number of beach visitors, and this causes the DNN to fail in the prediction. Nevertheless, the F1-scores for all months maintain a high level, which proves the feasibility of the DNN. Precision and recall have similar values. A higher F1-score is caused by higher precision and recall values, and vice versa. The accuracy values are also high and similar to the F1-score values. Finally, we notice that for the whole dataset, the best MAE of 0.03059 is obtained.

Table 5. MAE, accuracy, precision, recall, and F1-score.

Dataset Size	MAE	Accuracy	Precision	Recall	F1-Score
June	0.0553	0.84298	0.844526	0.836719	0.837623
July	0.03069	0.94219	0.935523	0.935185	0.934539
August	0.04055	0.94801	0.946325	0.944	0.943866
whole dataset	**0.03059**	**0.9269333**	**0.929422**	**0.926933**	**0.927114**

Table 6 presents the evaluation of the DNN in every category using the precision, recall, and F1-scores for 10 independent runs. The DNN performs well in the classification task for all categories.

Table 6. Per-class precision, recall, and F1-scores for the time period 2016–2018.

Classes (Beach Visitors)	Precision	Recall	F1-Score
0–49	0.981	0.981	**0.982**
50–99	0.943	0.898	0.919
100–149	0.865	0.925	0.892
150–199	0.92	0.956	**0.9323**
200–249	0.925	0.849	0.879
250–299	0.939	0.914	0.926
300+	0.93	0.97	**0.949**

The best per-class F1-scores are obtained for different types of categories: a very reduced number of beach visitors (0–49), a very high number (300+), and a medium number (150–199). The categories 50–99 and 200–249 show the worst results for the F1-score due to the increase of false negatives (recall). The category 100–149 has an F1-score of 89.2% due to the increase of the false positives (precision). Nevertheless, the precision rate, recall, and F1-score are relatively stable for all categories.

The performance of the proposed DNN for the whole dataset was investigated with different batch sizes. The results of 10 independent runs are shown in Table 7. With a batch size of 64, the best F1-score of 92.3412% is achieved.

Table 7. F1-score for different batch sizes.

Batch Size	F1-Score
16	0.88511
32	0.917197
64	**0.923412**
128	0.917907
256	0.91545

In order to check the robustness of the results, the model was revaluated for the time periods 2018–2020 and 2019–2021. Tables 8 and 9 present the evaluation of the DNN in every category using the precision, recall, and F1-scores for the selected time periods and 10 independent runs. The DNN performs well in the classification task for all categories. In the time period 2018–2020, the best per-class F1-scores are obtained for different types of categories: a reduced number of beach visitors (50–99) and a high/very high number (250–299)/(300+). The category 100–149 shows the worst results for the F1-score due to the increase of false negatives (recall). The category 200–249 has an F1-score of 90% due to the increase of the false positives (precision). In the time period 2019–2021, the best per-class F1-scores are obtained for categories with a high/very high number of beach visitors (200–249; 250–299)/(300+). The categories 100–149 and 150–199 show the worst results for the F1-score due to the increase of false negatives (recall). We observe that for both time periods, the precision rate, recall, and F1-score are relatively stable for all categories. Therefore, we can conclude that the robustness of the model is not affected by the time periods chosen.

Table 8. Per-class precision, recall, and F1-scores for the time period 2018–2020.

Classes (Beach Visitors)	Precision	Recall	F1-Score
0–49	0.891	0.927	0.91
50–99	0.935	0.912	**0.923**
100–149	0.931	0.856	0.892
150–199	0.941	0.905	0.922
200–249	0.893	0.903	0.90
250–299	0.931	0.981	**0.955**
300+	0.945	0.927	**0.936**

Table 9. Per-class precision, recall, and F1-scores for the time period 2019–2021.

Classes (Beach Visitors)	Precision	Recall	F1-Score
0–49	0.931	0.917	0.924
50–99	0.915	0.931	0.923
100–149	0.951	0.888	0.918
150–199	0.932	0.861	0.895
200–249	0.934	0.94	**0.937**
250–299	0.893	0.963	**0.927**
300+	0.951	0.936	**0.943**

4.4. Network Depth

The DNN topology was also investigated in detail to improve the modelling performance.

Two critical DNN topology parameters are the network depth (number of hidden layers) and width (number of neurons per hidden layer) [43]. We tested different DNN topologies, where the number of hidden layers varies between 1 and 6 and the number of neurons per hidden layer varies between 30 and 300.

The performance impact of the network depth is shown in Figure 12. The MAE of 10 independent runs is shown. The number of neurons per hidden layer is 300, and the number of layers varies between 1 and 6. It can be observed that when the number of hidden layers is increased, the modeling performance of the deep neural network is improved. The median MAE for the number of beach visitors improves from 0.04655 with one hidden layer to 0.03045 for six hidden layers, with a 34.6% improvement. However, the improvement slows down when more layers are added. The improvement becomes insignificant with more than four hidden layers. The median MAE for the number of beach visitors shows a 34.2% decrease from 0.04655 with one hidden layer to 0.03065 with four hidden layers; however, when the number of hidden layers is increased to six, only a 0.65% decreased is achieved, with 0.00036 modeling accuracy variation. We observe that the deep neural network can maintain a good modelling performance when the number of layers is increased and does not suffer overfitting. We conclude that the model is not biased with the training data and keeps improving when the number of neurons raises.

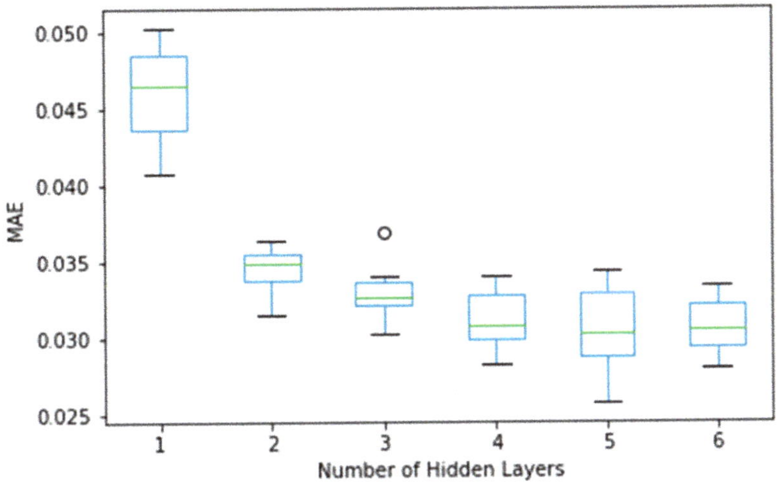

Figure 12. MAE vs. number of hidden layers.

4.5. Network Width

Next, we study the impact of network width, or the number of neurons per hidden layer. The performance impact of the network width is shown in Figure 13. The results are shown with six hidden layers, and the width varies between 25 and 300. We see that the modeling performance is better when the number of neurons is increased. It can also be observed that the modeling performance after 175 neurons per hidden layer improves very slowly when more neurons are added. However, the convergence speed in network width is faster than that of network depth. Doubling one hidden layer to two layers decreases the median MAE from 0.04655 to 0.03485, by 25.13%, and further doubling to four layers improves the MAE to 0.03045, by 12.62%.

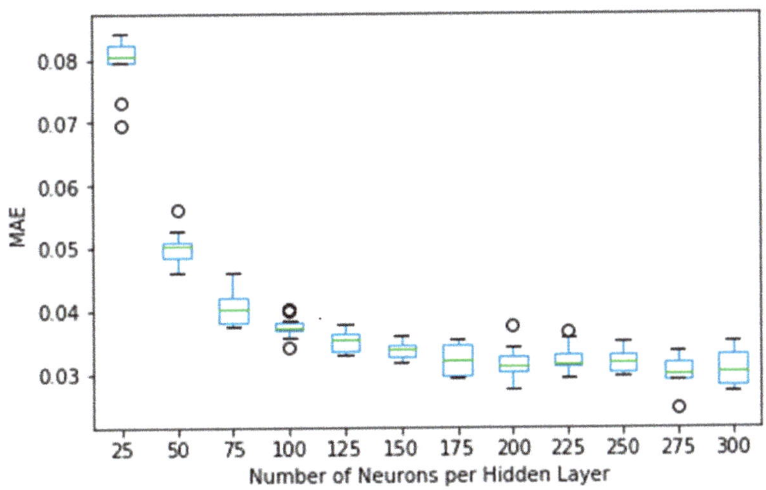

Figure 13. MAE vs. number of neurons per hidden layer.

Regarding the network width, doubling the number of neurons per layer from 25 neurons to 50 reduces the MAE from 0.08045 to 0.05025, by 37.54%, and further doubling to 100 neurons achieves a 0.0372 median MAE, with 25.97% improvement. These results reflect the fact that an increase in the number of hidden layers and the number of neurons per hidden layer has a different impact on the deep neural network. In a fully connected neural network, every neuron in each layer of the network is connected to every other neuron in the adjacent forward layer. If such a neural network has n neurons per layer and m hidden layers, the total number of neuron links is $O(mn^2)$. Since the number of neuron links scales linearly with network depth but exponentially with network width [43], the network performance improves more significantly as the network widens.

4.6. Optimal Network Topology

Next, we study the optimal network topology. We consider five different network topologies that maintain the total number of neurons (1800). The first topology, denoted as FIRST, consists of a neural network with six hidden layers and 300 neurons evenly distributed in each layer. The second topology (SECOND) considers fewer hidden layers, with more neurons per layer than the FIRST case. Specifically, the neural network consists of three hidden layers, with 600 neurons per layer. The third topology (THIRD) considers more hidden layers, with fewer neurons per layer than the FIRST case. It consists of 12 layers, with 150 neurons per layer. The fourth topology (FOURTH) assumes there is an increase in the number of neurons for deeper hidden layers. The number of neurons is increased for each hidden layer as follows: 200 (for the first hidden layer), 200, 300, 300, 400, and 400 (for the last hidden layer). The last topology (FIFTH) assumes there is a decrease in the number of neurons for deeper hidden layers (the reverse order of the proposed FOURTH topology). The modelling performance using the five topologies is shown in Figure 14. The fifth topology (FIFTH) achieves the best overall performance, with a median MAE of 0.02735, which is 9.59, 12.34, 23.82 and 1.26% more accurate than the models FIRST, SECOND, THIRD and FIFTH, respectively. The FOURTH and FIFTH topologies achieve a better modeling performance. Therefore, we conclude that a fatter and especially a thinner topology improves the modeling performance.

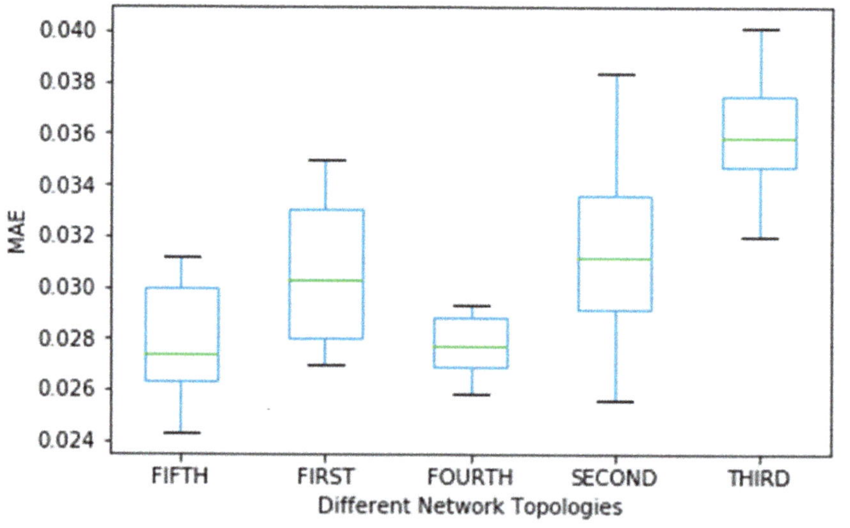

Figure 14. MAE vs. different network topologies.

4.7. Comparison to Other Models

Python libraries enable the deployment of many traditional models. Scikit-learn is an open-source machine learning library that supports supervised and unsupervised learning. It offers several tools for model fitting, data preprocessing, model selection and evaluation, and many other utilities.

We compared our proposal with three well-known models provided by the scikit-learn library, which support multiclass classification. These classifiers are decision tree, k-nearest neighbor (kNN), and random forest.

We also employed 10-fold cross validation for the evaluation and comparison of these machine learning algorithms. Table 10 shows the results achieved by our deep neural network model and these traditional models.

Table 10. Per-class precision, recall, and F1-scores. Comparison between our proposed DNN and other traditional machine learning models.

	Accuracy	Precision	Recall	F1-Score
Our DNN	**0.927**	**0.929**	**0.927**	**0.927**
Decision tree	0.892	0.87	0.88	0.875
K-nearest neighbor	0.593	0.65	0.68	0.664
Random forest	0.412	0.61	0.56	0.584

The analysis shows that our proposed DNN model achieves the highest accuracy, precision, recall, and F1-score of 92.7%, 92.9%, 92.7%, and 92.7%, respectively, compared to traditional machine learning models. We also observe that, of the traditional models, decision tree achieves the highest accuracy, precision, recall, and F1-score of 89.2%, 87%, 88%, and 87.5%, respectively, followed by k-nearest neighbor, with an accuracy, precision, recall, and F1-score of 59.3%, 65%, 68%, and 66.4%, respectively. We can conclude that our proposed algorithm outperforms other traditional machine learning algorithms and is able to perform beach attendance predictions adequately.

5. Discussion

Several authors [18–22,24] have analyzed how beach attendance is affected by weather conditions and other aspects such as time, season, etc. The results indicate that beach attendance is affected by the season, day, hour, and meteorological conditions. Our results confirm the same trend.

Our results regarding the daily temporal distribution of beachgoers are consistent with other beaches along the Mediterranean Sea [18–20].

Other authors have used classical econometric models to simulate the seasonality and cyclicality of the processes. In [18], the authors modelled the number of beachgoers as a function of temperature, for temperatures lower than 30 °C, using the regression coefficient $R^2 = 0.87$ at the Lido of Sète Beach, France. In [21], regression methods are used to determine how the number of beachgoers is affected by the season, day, weather, and ocean conditions (maximum significant wave height, water temperature) in Australian beaches. The authors use ordinal logistic regression to distinguish between three categories of beach visitor numbers: high, moderate, and low. The authors in [19] sought a mathematical expression to model beach attendance. The observed mean number of daily users was adjusted to a time-dependent Fourier polynomial for two beaches in Barcelona. The occupation data for 2002–2004 were averaged to obtain an estimation of the occupation trend for a typical year. This averaged function was projected into a 14-term Fourier polynomial. The Fourier fit obtained and the original time series of the average occupation in a typical year adjusted 74% and 69% of the absolute value of the original time series for the two beaches, respectively. In contrast, our proposed DNN improves these results. It can predict well the number of beach visitors assigned to any of the seven classes or beach occupancy levels and obtains a good performance, with an accuracy of 92.7%.

6. Conclusions

In this paper, experimental research that uses IoT data and deep learning to estimate beach attendance at Castelldefels beach (Barcelona, Spain) was developed, and beach attendance was predicted. Images of Castelldefels beach were captured by a video monitoring system.

An image recognition software was used to estimate the number of beachgoers per hour from 9:00 to 19:00 h during June, July, and August (prime swimming season) from 2016 to 2018. It was verified that the automated count was not significantly different from the manual count, and thus this methodology is suitable for the evaluation of beach attendance. The detection algorithm was not able to estimate the number of beachgoers with cloudy, rainy weather or blurry images. To solve this problem, the number of beach visitors was counted manually. However, the detection algorithm saved time and provided good results for a huge number of images. It would also be necessary to perform additional training of the detection algorithm to improve its accuracy in distinguishing special cases: beach goers riding their bicycles, beach goers covered partially by umbrellas at the beach, etc.

It was shown that weather, time, season, and working day/holiday have significant effects on the number of beachgoers. More resources (e.g., lifeguards, police officers that patrol the beaches, etc.) will be required during weekends or public holidays to protect beach visitors, especially during prime seasons.

Furthermore, a deep learning algorithm was trained for the first time for beach attendance prediction. The experimental results prove the feasibility of DNNs for beach attendance prediction. The confusion matrices on the test set were shown for June, July, and August. All classes are well predicted for the whole dataset. In our testing experiments, the proposed DNN yields a very good performance with an MAE, accuracy, precision, recall, and F1-score of 0.03, 92.7%, 92.9%, 92.7%, and 92.7%, respectively. Our proposed deep learning classifier outperforms other machine learning models (decision tree, k-nearest neighbor, and random forest) and can successfully differentiate between seven beach occupancy levels. The best F1-scores are obtained for a very reduced number of beach visitors (0–49), a very high number (300+), and a medium number (150–199), with values of 98.2%, 93.23%, and 94.9%, respectively. The impact of the DNN topology was also investigated. The results show that the DNN performance improves when the number of hidden layers is increased. It also improves with more neurons per hidden layer. The modeling accuracy benefits from an increase in the number of neurons for deeper hidden layers (fourth topology), and the best results are obtained when there is a decrease in the number of neurons for deeper hidden layers (fifth topology).

This research has two limitations. First, only one beach in Castelldefels is considered. This research could be extended to other beaches in Castelldefels. Second, the weather data are taken from a weather station of Meteocat (Meteorological Service of Catalonia) that is located in Viladecans, near but not at the respective beach. However, we expect that these registered weather conditions are reasonably accurate for our case study.

This work has shown that coastal videometry and image processing are very efficient tools for beach attendance detection. Furthermore, beach attendance prediction was successfully developed, and it is of particular importance for coastal managers to plan beach services in terms of security, rescue, health, and environmental assistance.

Funding: This work was funded by the Agencia Estatal de Investigación of Ministerio de Ciencia e Innovación of Spain under project PID2019–108713RB-C51 MCIN/AEI/10.13039/501100011033.

Institutional Review Board Statement: Not applicable.

Informed Consent Statement: Not applicable.

Data Availability Statement: Not applicable.

Acknowledgments: This work was supported by the Agencia Estatal de Investigación of Ministerio de Ciencia e Innovación of Spain under project PID2019–108713RB-C51 MCIN/AEI/10.13039/501100011033.

Conflicts of Interest: The authors declare no conflict of interest. The funders had no role in the design of the study; in the collection, analyses, or interpretation of data; in the writing of the manuscript, or in the decision to publish the results.

References

1. Albino, V.; Berardi, U.; Dangelico, R.M. Smart Cities: Definitions, Dimensions, Performance and Initiatives. *J. Urban Technol.* **2015**, *22*, 3–21. [CrossRef]
2. Batty, M.; Axhausen, K.W.; Giannotti, F.; Pozdnoukhov, A.; Bazzani, A.; Wachowicz, M.; Ouzounis, G.; Portugali, Y. Smart Cities of the Future. *Eur. Phys. J. Spec. Top.* **2012**, *214*, 481–518. [CrossRef]
3. Doboli, A.; Curiac, D.; Pescaru, D.; Doboli, S.; Tang, W.; Volosencu, C.; Gilberti, M.; Banias, O.; Istin, C. *Cities of the Future: Employing Wireless Sensor Networks for Efficient Decision Making in Complex Environments*; CEAS Technical Report Nr 831; University at Stony Brook: Stony Brook, NY, USA, 2008.
4. WWF. *Living Blue Planet Report; Species, Habitats and Human Well-Being*; WWF: Gland, Switzerland, 2015. Available online: http://assets.worldwildlife.org/publications/817/files/original/Living_Blue_Planet_Report_2015_Final_LR.pdf?1442242821&_ga=1.161602740.969507462.1487893751 (accessed on 31 August 2021).
5. Sánchez, L.; Gazo, M.; Sánchez, J. Nurturing Ocean Literacy through Responsible Tourism: Best Practices for Marine Wildlife Watching during Ecotourism Activities. Submon, Barcelona, Spain. 2016. Available online: http://www.wildsea.eu/ftp/library/00_WILDSEA_BEST_PRACTICE_MANUAL.pdf (accessed on 31 August 2021).
6. Domingo, M.C. An Overview of the Internet of Things for People with Disabilities. *J. Netw. Comput. Appl.* **2012**, *35*, 584–596. [CrossRef]
7. Volosencu, C. Identification in Sensor Networks. In Proceedings of the 9th WSEAS International Conference on Automation and Information (ICAI'08), Bucharest, Romania, 24–26 June 2008; pp. 175–183.
8. Allam, Z.; Dhunny, Z.A. On Big Data, Artificial Intelligence and Smart Cities. *Cities* **2019**, *89*, 80–91. [CrossRef]
9. Allam, Z.; Newman, P. Redefining the Smart City: Culture, Metabolism and Governance. *Smart Cities* **2018**, *1*, 4–25. [CrossRef]
10. Mohammadi, M.; Al-Fuqaha, A. Enabling Cognitive Smart Cities Using Big Data and Machine Learning: Approaches and Challenges. *IEEE Commun. Mag.* **2018**, *56*, 94–101. [CrossRef]
11. Ullah, Z.; Al-Turjman, F.; Mostarda, L.; Gagliardi, R. Applications of Artificial Intelligence and Machine Learning in Smart Cities. *Comput. Commun.* **2020**, *154*, 313–323. [CrossRef]
12. Atitallah, S.B.; Driss, M.; Boulila, W.; Ghézala, H.B. Leveraging Deep Learning and IoT Big Data Analytics to Support the Smart Cities Development: Review and Future Directions. *Comput. Sci. Rev.* **2020**, *38*, 100303. [CrossRef]
13. Charpentier, A.; Flachaire, E.; Ly, A. Econometrics and machine learning. *Econ. Stat.* **2018**, *505*, 147–169. [CrossRef]
14. Shobana, G.; Umamaheswari, K. Forecasting by Machine Learning Techniques and Econometrics: A Review. In Proceedings of the 2021 6th International Conference on Inventive Computation Technologies (ICICT), Coimbatore, India, 20–22 January 2021; pp. 1010–1016. [CrossRef]
15. Burger, C.J.S.C.; Dohnal, M.; Kathrada, M.; Law, R. A Practitioners Guide to a Time-series Methods for Tourism Demand Forecasting—A Case Study of Durban, South Africa. *Tour. Manag.* **2001**, *2*, 402–409. [CrossRef]
16. Cho, V. A Comparison of Three Different Approaches to Tourist Arrival Forecasting. *Tour. Manag.* **2003**, *24*, 323–330. [CrossRef]
17. Pereira, C. Beach Carrying Capacity Assessment: How Important is it? *J. Coast. Res.* **2002**, *36*, 190–197.
18. Balouin, Y.; Rey-Valette, H.; Picand, P.A. Automatic Assessment and Analysis of Beach Attendance Using Video Images at the Lido of Sète Beach, France. *Ocean Coast. Manag.* **2014**, *102*, 114–122. [CrossRef]
19. Guillen, J.; Garcia-Olivares, A.; Ojeda, E.; Osorio, A.; Gonzalez, R. Long-term Quantification of Beach Users Using Video Monitoring. *J. Coast. Res.* **2008**, *24*, 1612–1619. [CrossRef]
20. Martínez, E.; Gómez, M.B. Weather, Climate and Tourist Behavior: The Beach Tourism of the Spanish Mediterranean Coast as a Case Study. *Eur. J. Tour. Hosp. Recreat.* **2012**, *3*, 77–96.
21. Zhang, F.; Wang, X.H. Assessing Preferences of Beach Users for Certain Aspects of Weather and Ocean Conditions: Case Studies from Australia. *Int. J. Biometeorol.* **2013**, *57*, 337–347. [CrossRef]
22. Girau, R.; Anedda, M.; Fadda, M.; Farina, M.; Floris, A.; Sole, M.; Giusto, D. Coastal Monitoring System Based on Social Internet of Things Platform. *IEEE Int. Things J.* **2020**, *7*, 1260–1272. [CrossRef]
23. Kane, B.; Zajchowski, C.A.B.; Allen, T.R.; McLeod, G.; Allen, N.H. Is it Safer at the Beach? Spatial and Temporal Analyses of Beachgoer Behaviors during the COVID-19 Pandemic. *Ocean Coast. Manag.* **2021**, *205*, 105533. [CrossRef]
24. Epelde, I.; Liria, P.; de Santiago, I.; Garnier, R.; Uriarte, A.; Picón, A.; Galdrán, A.; Arteche, J.A.; Lago, A.; Corera, Z.; et al. Beach Carrying Capacity Management under COVID-19 Era on the Basque Coast by Means of Automated Coastal Videometry. *Ocean Coast. Manag.* **2021**, *208*, 105588. [CrossRef]
25. Perillo, G.M.E.; Botero, C.M.; Milanes, C.B.; Elliff, C.I.; Cervantes, O.; Zielinski, S.; Bombana, B.; Glavovic, B.C. Integrated Coastal Zone Management in the Context of COVID-19. *Ocean Coast. Manag.* **2021**, *210*, 105687. [CrossRef]
26. Li, H.; Gupta, M. *Fuzzy Logic and Intelligent Systems*; Kluwer: Boston, MA, USA, 1995.
27. Li, H.; Ota, K.; Dong, M. Learning IoT in Edge: Deep Learning for the Internet of Things with Edge Computing. *IEEE Netw.* **2018**, *32*, 96–101. [CrossRef]
28. Mao, Q.; Hu, F.; Hao, Q. Deep Learning for Intelligent Wireless Networks: A Comprehensive Survey. *IEEE Commun. Surv. Tutor.* **2018**, *20*, 2595–2621. [CrossRef]
29. Goodfellow, I.; Bengio, Y.; Courville, A. *Deep Learning*; MIT Press: Cambridge, MA, USA, 2016.
30. Castelldefels Images Video Monitoring Website. Available online: http://cooweb.cmima.csic.es/video-coo/images.jsp?site=CFA1-sam (accessed on 31 August 2021).

31. Meteocat. Metereological Forecast in Catalonia (Spain). Available online: https://en.meteocat.gencat.cat/?lang=en (accessed on 31 August 2021).
32. Statista. Leading City Destinations Worldwide in 2018, by Number of Overnight Visitors. 2020. Available online: https://www.statista.com/statistics/310355/overnight-visitors-to-top-city-destinations-worldwide/ (accessed on 25 October 2021).
33. Statista. Number of International Overnight Visitors in the Most Popular European City Destinations in 2016. 2017. Available online: https://es.statista.com/estadisticas/487720/turistas-internacionales-en-los-principales-destinos-europeos/ (accessed on 25 October 2021).
34. El impacto Socioeconómico de las Playas Metropolitanas, Area Metropolitana de Barcelona (AMB) and Institut d'Estudis Regionals i Metropolitans de Barcelona (IERMB), July 2021. Available online: https://www.amb.cat/es/web/territori/actualitat/publicacions/detall/-/publicacio/impacte-socioeconomic-de-les-platges-metropolitanes/11316283/11656 (accessed on 25 October 2021).
35. Ribas, F.; Simarro, G.; Arriaga, J.; Luque, P. Automatic Shoreline Detection from Video Images by Combining Information from Different Methods. *Remote Sens.* **2020**, *12*, 3717. [CrossRef]
36. Aarninkhof, S.G.J.; Turner, I.L.; Dronkers, T.D.T.; Caljouw, M.; Nipius, L. A Video-based Technique for mapping Intertidal Beach Bathymetry. *Coast. Eng.* **2003**, *49*, 275–289. [CrossRef]
37. Andriolo, U.; Mendes, D.; Taborda, R. Breaking Wave Height Estimation from Timex Images: Two Methods for Coastal Video Monitoring Systems. *Remote Sens.* **2020**, *12*, 204. [CrossRef]
38. Coastal Ocean Observatory. Available online: http://coo.icm.csic.es/ (accessed on 31 August 2021).
39. Institut of Marine Sciences. Available online: http://www.icm.csic.es/?q=en (accessed on 31 August 2021).
40. CountThings. Available online: https://countthings.com/ (accessed on 31 August 2021).
41. Hammerstrom, D. Working with Neural Networks. *IEEE Spectr.* **1993**, *30*, 46–53. [CrossRef]
42. Kingma, D.P.; Ba, J.L. Adam: A Method for Stochastic Optimization. In Proceedings of the 3rd International Conference for Learning Representations, San Diego, CA, USA, 7–9 May 2015; pp. 1–15.
43. Zhang, W.; Hu, W.; Wen, Y. Thermal Comfort Modeling for Smart Buildings: A Fine-Grained Deep Learning Approach. *IEEE Int. Things J.* **2019**, *6*, 2540–2549. [CrossRef]

applied sciences

MDPI

Article

On the Design of a Decision Support System for Robotic Equipment Adoption in Construction Processes

Carmen Marcher [1,2,*], Andrea Giusti [2] and Dominik T. Matt [1,2]

1 Faculty of Science and Technology, University of Bolzano, Piazza Università 5, 39100 Bolzano, Italy; dominik.matt@unibz.it
2 Fraunhofer Italia Research, via A.-Volta 13A, 39100 Bolzano, Italy; andrea.giusti@fraunhofer.it
* Correspondence: carmen.marcher@natec.unibz.it

Abstract: The construction sector is one of the major global economies and is characterised by low productivity and high inefficiencies, but could highly benefit from the introduction of robotic equipment in terms of productivity, safety, and quality. As the development and the availability of robotic solutions for the construction sector increases, the evaluation of their potential benefits compared to conventional processes that are currently adopted on construction sites becomes compelling. To this end, we exploit Bayesian decision theory and apply an axiomatic design guideline for the development of a decision-theoretic expert system that: (i) evaluates the utility of available alternatives based on evidence; (ii) accounts for uncertainty; and (iii) exploits both expert knowledge and preferences of the users. The development process is illustrated by means of exemplary use case scenarios that compare manual and robotic processes. A use case scenario that compares manual and robotic marking and spraying is chosen for describing the development process in detail. Findings show how decision making in equipment selection can be supported by means of dedicated systems for decision support, developed in collaboration with domain experts.

Keywords: equipment selection; construction robot; decision support system; axiomatic design; decision-theoretic expert system; construction industry; industry 4.0

Citation: Marcher, C.; Giusti, A.; Matt, D.T. On the Design of a Decision Support System for Robotic Equipment Adoption in Construction Processes. *Appl. Sci.* **2021**, *11*, 11415. https://doi.org/10.3390/app112311415

Academic Editors: João M. F. Rodrigues, Pedro J. S. Cardoso and Marta Chinnici

Received: 29 October 2021
Accepted: 29 November 2021
Published: 2 December 2021

1. Introduction

The broad adoption of automation and robotics is changing operations in many business sectors [1]. Even though the construction sector is one of the major economies, it suffers from inefficiencies and a low increase in productivity [2,3] and could, therefore, highly benefit from the introduction of automation and robotics. The adoption of robotic systems has the potential to increase safety, quality, productivity, and to reduce cost in construction processes [4–6]. During the last decades, the interest towards the development of robotic solutions for applications in the construction sector is constantly growing and their potential deployment is addressed in several works [5–8]. However, compared to other sectors, the construction industry can be considered as a traditional industry that is characterised by a lack of interest in innovation [9,10] and a clear opposition to changes [11]. Even though the awareness of potential benefits of automation and robotics in the construction industry is increasing, the adoption of such technologies can be judged as slow [5].

The purpose of this study is to facilitate the comparison of advanced technological solutions with conventional manual construction processes. For this purpose, we illustrate how a system for decision support in the field of robotic equipment adoption can be designed and developed. This can be achieved by performing a structured evaluation of the impact that advanced technologies may have on safety, quality, productivity, and cost, and by assessing the utility of their adoption compared to conventional manual processes in an unbiased way.

Current research shows that one way to support decision making in this field is the assessment of the actual performance of robots compared to traditional work practices on the construction site [12] and to provide adequate tools for supporting decision makers in the choice of replacing traditional processes with automated systems [13]. However, uncertainty and interdependencies, typical of construction operations, often hinder the definition of standardised approaches for decision-making [14]. Furthermore, when facing the decision of whether to replace conventional work methods with automated counterparts, the evaluation of alternatives should be able to reflect both preferences and knowledge of the users [15]. Equipment selection problems in construction can also be supported by different approaches and methods [16], as follows. The selection of cranes can be supported by multi-criteria decision-making methods that are able to evaluate project specific requirements, the characteristics of the equipment and economic factors [17,18]. Multicriteria decision-making methods that consider both qualitative and quantitative criteria are also employed for the selection of excavation machinery [19,20]. Construction machinery selection for infrastructure projects can be performed by using a decision support framework that evaluates risks and costs of the available alternatives [21]. In addition, the use of Building Information Modelling (BIM) is playing an important role in the field of automation and robotics in construction [22]. BIM is a methodology supporting the management of information within the construction sector. The result of the BIM methodology is a BIM model, a model that contains both geometric and semantic data of a building that can be employed along its whole lifecycle [23]. Current research shows how such BIM models can be integrated into robotic control systems to support the deployment of robotics in building construction and operation [24,25].

Nowadays, only few construction robots are actually used on real construction sites and comparisons to traditional or conventional processes are rarely available [26]. The performance of automated systems and traditional work methods for a given project can be assessed by showing how they perform on the level of single construction specific criteria or variables [12,15,27]. Analyses conducted in previous studies have shown that the key parameters considered when evaluating the utility of robotic systems compared to conventional processes are safety, quality, productivity, and cost [26]. Results of comparisons prove that construction robots have the capability to increase safety [12,26,28], productivity [12,26,28,29], and quality [12,26,29] on construction sites. Some studies prove that robots may have a positive impact on cost [5,8], while others find that extra costs can occur for their deployment on site [12,26]. These considerations show that the deployment of robotic equipment on construction sites can have both benefits and drawbacks. Although many potential benefits of construction robots are described in the above-mentioned works, other barriers can prevent their widespread adoption in practice, such as a low propensity towards changing existing work practices [5,11]. This conservative attitude can be changed by involving domain experts in the definition of potential application areas and the expected impacts of construction robots. Previous studies show how experts can be engaged in the strategic definition of high-priority applications and the evaluation of the potential impacts of construction robots in these fields [5].

The aim of this research is to illustrate the design of a decision support system (DSS) for robotic equipment adoption that compares conventional and robotic processes by evaluating their utility. To achieve this aim we apply an axiomatic-design guideline [27] that supports the design of decision-theoretic expert systems based on Bayesian decision theory to aid equipment selection in construction. Axiomatic design is a system design methodology that is successfully applied in engineering, business, software, and product development [30]. Decision-theoretic expert systems allow one to assess the utility of the available options based on evidence, preference statements and expert knowledge. The approach is illustrated by means of exemplary use case scenarios of a research project that aims at developing configurable collaborative modular robotic platforms targeting use cases in construction. The use case scenarios are defined in collaboration with domain experts of the construction group participating in the project. We choose a use case scenario that considers the employment of collaborative robots for semiautonomous or teleoperated marking and spraying for the detailed description of the approach.

As a result, this research shows how a DSS in the field of robotic equipment adoption in construction can be designed. In particular, Axiomatic design allows us to divide the complex design process of the decision theoretic expert system into small and manageable steps that can be easily replicated in additional use cases. The structured involvement of domain experts in every development step increases the reliability of the DSS, fills lack of data with expert knowledge, and can increase the acceptance of the system by the potential users.

The remainder of this article is organised as follows: in Section 2 we introduce the preliminaries on the materials and methods considered; in Section 3 we describe the development of the prototype and the obtained results; in Section 4 we discuss the results; and in Section 5 we draw the conclusions.

2. Preliminaries on Decision Theoretic Expert Systems

Systems that exhibit artificial intelligence performing intellectual demanding tasks restricted to a specific problem domain are defined as expert systems [31]. Expert systems that rely on probabilistic networks are called decision-theoretic expert systems. These systems perform reasoning under uncertainty while maximizing expected utilities of the outcomes, and give advice on the best rational decision considering evidence and preference statements [32].

Decision networks, often referred to as influence diagrams or Bayesian decision networks, provide a formalism for modelling and solving decision problems following the principle of Maximum Expected Utility (MEU). Decision networks can be described as an extension of Bayesian networks [32,33]. A Bayesian network consists of a qualitative and a quantitative part, a Directed Acyclic Graph (DAG) with an associated joint probability distribution. The construction of Bayesian networks involves two main steps [31]: (i) the identification of variables and causal relations between them for establishing the DAG, and (ii) the elicitation of the conditional probability distributions of the random variables. By extending Bayesian networks with actions and utilities we obtain decision networks. Decision networks consist of the following components [31–33]: (i) decision nodes represent the problem variables and refer to the decisions or choices that are available for the decision maker; (ii) chance nodes represent the random variables, also referred to as information variables that may be observed to provide information for solving the problem; (iii) utility nodes represent the utility function of an agent and assess the expected utility for available choices or actions; and (iv) arcs denote the influences and relations between variables.

3. Development of the Prototype

We follow the axiomatic design-based guideline that is presented in Table 1. The guideline is based on [27] and the development steps are defined as follows: (i) identification of the problem domain; (ii) implementation of the knowledge base including the definition of the qualitative and quantitative part of the decision model; (iii) implementation of the inference engine with computation of the decision that yields the MEU, and Value of Information (VOI) analysis; and (iv) definition of the functionalities that allow the user to interact with the system. In addition to the guideline provided, we evaluate if the system provides reasonable output.

3.1. Problem Domain

The problem domain concerns the execution of construction tasks that can be performed both with the collaborative robot to be developed within the research project or with a conventional manual construction method. We consider the following use cases (UCs):

- UC1: Collaborative semi-autonomous transport and delivery of material and tools.
- UC2: Supervised and collaborative drilling.

- UC3: Supervised and collaborative cutting.
- UC4: Semi-autonomous/teleoperated marking and spraying.
- UC5: Supervised/semi-autonomous documenting.

Within the project we will measure and verify the projected benefits and impact of developed technologies in the individual UCs through measurable Key Performance Indicators (KPIs) mutually agreed between the project partners. The results of the analysis of the UCs and related KPIs serve as a basis for defining the knowledge base of the DSS.

Table 1. Design guideline based on [27].

<table>
<tr><td colspan="3">3.1 Problem domain
Use of the collaborative robot or use of the conventional construction process?</td></tr>
<tr><td>3.2 Knowledgebase
3.2.1 Qualitative part</td><td>3.3 Inference Engine
3.3.1 MEU computation</td><td>3.4 User interaction</td></tr>
<tr><td>Definition of the variables to be considered in the evaluation and definition of the relations between them.</td><td>Computation and selection of the decision that yields the MEU.</td><td rowspan="3">Computation of results based on evidence and preferences.</td></tr>
<tr><td>3.2.2 Quantitative part</td><td>3.3.2 VOI analysis</td></tr>
<tr><td>Definition of the numbers that are necessary for performing the computations.</td><td>Computation of which information should be acquired by the user.</td></tr>
<tr><td colspan="3">3.5 Evaluation
Evaluation of reasonableness of the output of the system.</td></tr>
</table>

3.2. *Knowledgebase*

Within the knowledgebase, the qualitative and the quantitative parts of the decision network are defined. We elicit the knowledge related to the problem domain by collecting expert knowledge and by analysing available literature that addresses the comparison of construction robots and conventional construction processes.

The relevant literature in this field is mapped by performing a manual search of articles and conference papers on Elsevier's database Scopus, a peer-review database in the field of engineering sciences. The search is limited to English articles and conference publications in the publication period from 2011 to April 2021. The search keyword is "construction robot*". The screening of the abstracts and papers is performed to exclude articles with no to low relevance to our field of study. This reduces the data to 17 highly relevant items that are further analysed to extract the parameters that are relevant for our field of study. Out of these 17 remaining articles, only 11 do consider the comparison between conventional approaches and the use of robots for the execution of construction tasks and only nine contain relevant information in the field of robotic equipment adoption in construction processes.

This analysis allows one to define, in collaboration with domain experts, the variables to be considered for comparing robotic processes with conventional processes within our research. In Table 2 we provide the description of the variables to be considered in the assessment of the available options. The variables are presented along with the KPIs of the project, their relevance for the different domains, their consideration within the previously introduced UCs, and the supporting literature.

3.2.1. Qualitative Part

The elicitation of the qualitative part of the decision model is based on the list of variables presented in Table 2. Initially, we define the overall qualitative model, that considers all the UCs in which the robotic system shall be applied (Figure 1). The decision node (illustrated as a rectangle) represents the available choices, the chance nodes (illustrated as ovals) represent the random variables involved in the decision along with their relevance for the different UCs, the utility node (illustrated as a diamond) represents the utility function, and arcs represent the causal relations between the variables.

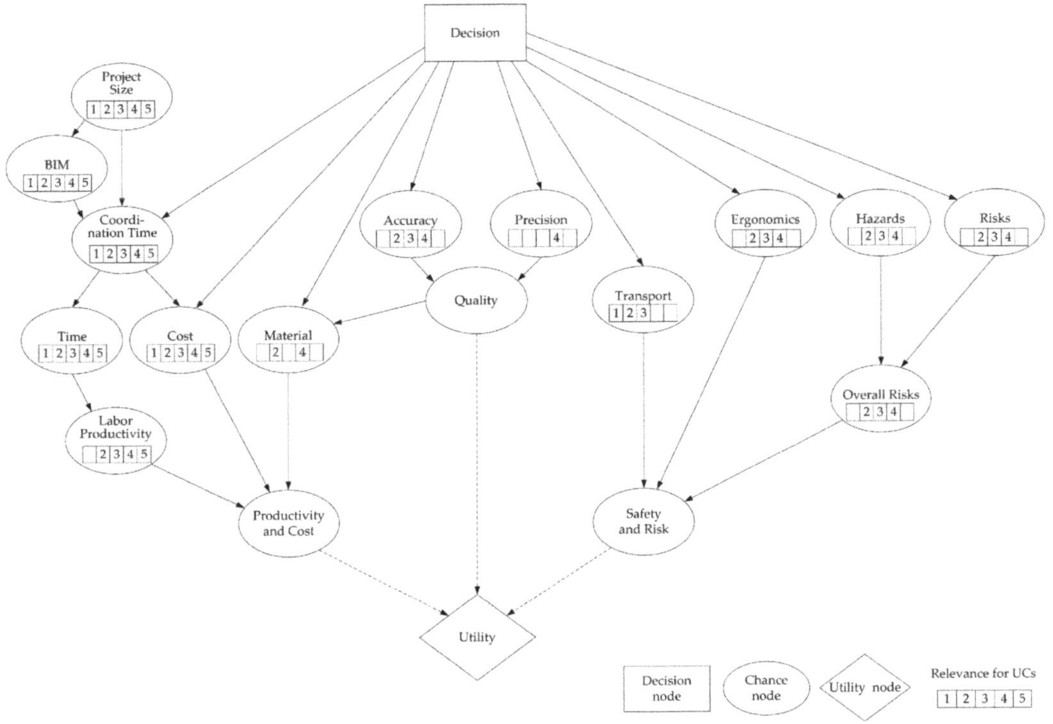

Figure 1. Decision model for UCs 1 to 5.

Table 2. List of variables to be considered in the evaluation.

KPI	Variable	Relevance	UC 1	UC 2	UC 3	UC 4	UC 5	Supporting Literature	Description
1	Time	Productivity and Cost	x	x	x	x	x	[12,29,34]	Process time needed to perform the task
2	Cost		x	x	x	x	x	[12,26,28,34–37]	Average cost needed to perform the task
3	Productivity			x	x	x	x	[28,29,35,36]	Labour productivity
4	Material			x		x		[36,37]	Consumption of material and resources needed to perform the task
-	Coordination time		x	x	x	x	x	[26]	Time needed for preparing the execution of the task
5	Accuracy	Quality		x	x	x		[12,26,28,29,35,36]	Number of errors
6	Precision					x			Quality of the performed work
7	Ergonomics	Safety and Risk		x	x	x		[12,26,37]	Reduction in unfavourable body postures during the execution of the task
8	Transports		x	x	x				Number of transport processes of heavy materials
9	Hazards			x	x	x		[6]	Time of exposure to hazards and use of protection equipment
10	Risks			x	x	x		[36,38]	Reduction in the time of ladder use and working at heights
11	Overall risks		x	x	x	x			Overall assessment of risks that can lead to accidents

Table 2. *Cont.*

KPI	Variable	Relevance	UC 1	2	3	4	5	Supporting Literature	Description
-	Project size	Project information	x	x	x	x	x	[28,37]	The project size can impact the decision of whether adopting a robot or not
-	BIM		x	x	x	x	x	[12,26]	The use of a BIM model is necessary for the deployment of the collaborative robot that is developed within the research project

For the sake of simplicity, and since the development process would be identical for all UCs, in the following we describe the development process of the prototype for UC4—marking and spraying. We define the decision model for UC4 in two steps. First, we eliminate the variables that are not directly involved in the decision. Then, we simplify the decision model for UC4 by combining variables. The resulting decision model for UC4 is presented in Figure 2. The final list of variables to be considered within the DSS, along with the states that they can assume, is presented in Table 3.

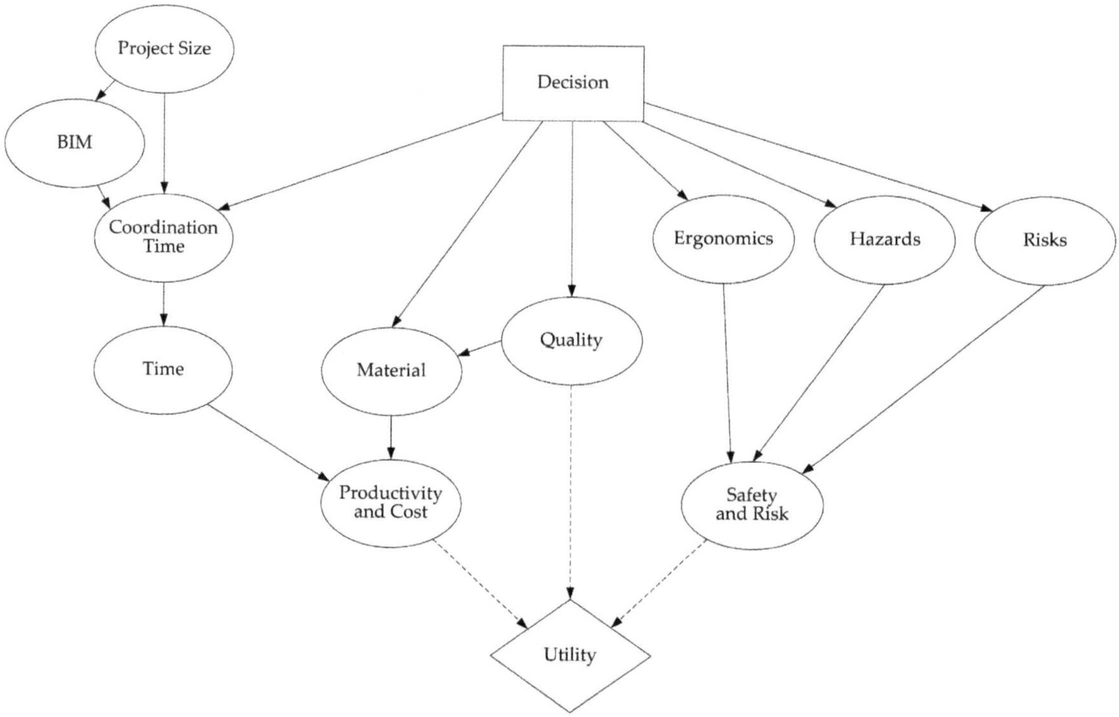

Figure 2. Simplified decision model for UC4—marking and spraying.

3.2.2. Quantitative Part

The elicitation of the quantitative part of the decision model involves the definition of the conditional probabilities, the utility function, and utilities in collaboration with the domain experts. We propose to elicit the conditional probability tables by mapping verbal statements of probability from "impossible" to "certain" to probabilities from 0 to 1 [31,39,40]. We defined the conditional probabilities for our use case scenario in collaboration with domain experts by performing educated guesses based on the expected performance of the collaborative robot in comparison to the conventional manual process.

Each variable has been studied singularly to associate a mutually agreed verbal statement of probability. In Table 4 we present one of the conditional probability tables of our use case scenario as an example. In particular, we see the conditional probabilities P(M | Q,D) of the variable material (M), given quality (Q), and decision (D).

Table 3. Variables to be considered for UC4—marking and spraying.

Variables	States of the Variables		
Decision	Collaborative robot	Conventional method	
Ergonomics	50% increase	unaltered	50% reduction
Hazards	80% reduction	unaltered	80% increase
Risks	50% reduction	unaltered	50% increase
Material	10% reduction	unaltered	10% increase
BIM	available	not available	
Coordination time	20% increase	unaltered	10% reduction
Project size	<10,000 m^3	>10,000 m^3	
Time	20% reduction	unaltered	20% increase
Safety and Risk	30% reduction in overall risk and increase in safety	unaltered	30% increase in overall risk and reduction in safety
Quality	20% reduction in errors and 30% reduction in variations	unaltered	20% increase in errors and 30% increase in variations
Productivity and Cost	20% increase in productivity and 10% reduction in cost	unaltered	20% reduction

Table 4. Conditional probability table of coordination time P(M | Q,D).

Quality (Q)	Decision (D)	Material (M)		
		Reduced	Unaltered	Increased
Increased	Conventional method	0.05	0.90	0.05
	Collaborative robot	0.90	0.05	0.05
Unaltered	Conventional method	0.10	0.80	0.10
	Collaborative robot	0.80	0.10	0.10
Reduced	Conventional method	0.00	0.50	0.50
	Collaborative robot	0.00	0.50	0.50

Utility assessment or preference elicitation concerns the definition of the utility function, necessary for the construction of the decision-theoretic expert system [32]. We define a utility function that allows one to capture the preferences of the users for our use case. Our utility function is chosen as:

$$U = p \times uP + q \times uQ + s \times uS \quad (1)$$

The parameters p, q, and s sum to one and represent the coefficients that allow the user to weight the utilities according to his preferences. uP, uQ, and uS represent the subjective utilities as defined in Table 5. Subjective utilities are assigned by ordering outcomes from worst to best [31]. We assigned utility 0 to the worst possible outcome, and utility 100 to the best possible outcome.

Table 5. Subjective utilities outcomes, where 0 is the worst outcome and 100 the best possible outcome.

Productivity	uP	Quality	uQ	Safety	uS
Reduced	0	Reduced	0	Reduced	0
Unaltered	100	Unaltered	100	Unaltered	100
Increased	100	Increased	100	Increased	100

3.3. Inference Engine

We use the previously defined decision model for performing the computations. For this purpose, the model is defined and implemented by using the python wrapper pyAgrum for the C++ aGrUM library for building and computing Bayesian networks [41] that allows one to perform computations based on the algorithms described in [42,43]. This implementation allows us to analyse the behaviour of the decision model and to verify if the model provides reasonable output.

3.3.1. MEU Computation

By applying the principle of MEU, we choose the decision that yields the highest expected utility [32]. The preferences are captured by the utility function $U(s)$ (1) that assigns a number to the different states (Table 5). The expected utility (EU) of a decision (d), given evidence (e) can be calculated by averaging the utility of the different outcomes with the probability (P) that the outcome can be achieved [32]:

$$EU(d\,|\mathrm{e}) = \sum_{s'} P(\mathrm{Result}(d) = s'|a, \mathrm{e}) \times U(s') \tag{2}$$

$$decision = argmax_d\ EU(d|\mathrm{e}) \tag{3}$$

In Table 6 we see the results of the MEU computation when we consider different weightings of preferences (p, q, or s). For equal weightings (p = q = s) we see that the best rational decision is the collaborative robot (R) with a MEU of 93.24.

Table 6. Variation of results due to user interaction (R = collaborative robot, C = conventional method).

Preferences [%]			Result		Results with Evidence								VPI	
					BIM				Project Size					
					Not Available		Available		Small		Large			
p	q	s	D	MEU	D	MEU	D	MEU	D	MEU	D	MEU	BIM	PS
33	33	33	R	93.24	C	95.56	R	95.28	C	95.71	R	95.18	4.72	4.74
100	0	0	C	92.99	C	92.70	C	93.38	C	93.20	C	92.83	1.77	1.77
80	10	10	C	93.79	C	93.57	C	94.10	C	93.95	C	93.66	1.92	1.92
60	20	20	C	94.58	C	94.42	C	94.82	C	94.71	C	94.49	2.07	2.07
50	25	25	C	94.98	C	94.85	R	94.15	C	95.09	R	94.03	2.09	2.12
40	30	30	C	95.38	C	95.27	R	94.83	C	95.46	R	94.72	2.22	2.25
0	100	0	R	94.34	R	94.99	R	95.15	R	93.57	R	95.08	5.65	5.65
10	80	10	R	94.01	C	95.16	R	95.19	C	95.21	R	95.11	5.38	5.38
20	60	20	R	93.68	C	95.33	R	95.23	C	95.42	R	95.14	5.10	5.11
25	50	25	R	93.51	C	95.41	R	95.24	C	95.53	R	95.16	4.96	4.98
30	40	30	R	93.35	C	95.50	R	95.26	C	95.64	R	95.17	4.81	4.84
0	0	100	R	99.91	R	99.91	R	99.91	R	99.91	R	99.91	0.00	0.00
10	10	80	R	97.91	R	97.19	R	98.52	R	97.23	R	98.49	1.06	1.06
20	20	60	R	95.90	R	94.48	R	97.13	R	94.56	R	97.07	2.14	2.14
25	25	50	R	94.90	R	93.13	R	96.43	R	93.22	R	96.36	2.68	2.68
30	30	40	R	93.90	C	95.90	R	95.74	C	96.04	R	95.65	4.22	4.24

3.3.2. VOI Analysis

VOI analysis allows one to define which information-gathering activities should be performed by the user. In our case, we calculate the Value of Perfect Information (VPI), also called value of clairvoyance, on the variables that are uncertain [33]. If exact evidence of variables can be obtained, VPI can be calculated as follows [32]:

$$VPI_e(E_j) = \left(\sum_k P(E_j = e_{jk}\,|\mathrm{e}) \times EU(\delta_{jk}|\mathrm{e}, E_j = e_{jk}) \right) - EU(\delta|\mathrm{e}) \tag{4}$$

where $EU(\delta \mid \mathbf{e})$ is the Expected Utility (*EU*) of the current best decision δ, $EU(\delta_{jk} \mid \mathbf{e}, E_j = e_{jk})$ is the *EU* of the new best decision δ with evidence E_j that is averaged over all the possible values e_{jk} that we may have for E_j. As *VPI* can be only computed for variables that are non-descendants of decision nodes [43], we can perform this computation for the variables BIM and Project size (PS). The EU of the new best action with evidence can be obtained by adding an influence arc from the uncertain variables BIM and PS to the decision node [33]. In Table 6, we see the result of the VOI computations when we consider different weightings of preferences (p, q, or s). For equal weightings (p = q = s), we see that *VPI*(PS) > *VPI* (BIM). This means that the best information to be acquired is the value of Project Size (PS).

3.4. User Interaction

The user can interact with the system by stating his preferences, by inserting evidence, and by answering to the questions of the inference engine to get advice on the best rational decision [27]. The user has the option to influence the advice through a targeted interaction with the system. The influence of the user interactions on the system, as well as the information that should be provided by the user, are shown in Table 6. We see how different preference statements (p, q, or s) lead to a different advice (D) of the system, how the advice changes if we add evidence on the chance nodes BIM and PS, and which would be the best information to be acquired by the user.

3.5. Evaluation

To evaluate whether the system provides reasonable output, the obtained results are analysed in collaboration with the domain experts that contributed to the development of the DSS following an axiomatic design guideline. For the development of the knowledgebase, we agreed on the qualitative and quantitative part of the decision model also involving the domain experts. The qualitative part concerns both the definition of the variables to be considered within the DSS and their relevance for the different use cases, as well as the structure of the complete decision model, and the simplified decision model for UC4—marking and spraying. The quantitative part concerns the definition of the numbers to be used for performing the computations within the inference engine. Here, we agreed with the domain experts on the utility function, preferences, and on the subjective utilities that are selected so that none of the available options can penalised or preferred by the system. Due to missing comparison values between robot and conventional method, we defined the conditional probability tables by performing educated guesses, by combining both experience and the expected performance of the system.

The behaviour due to user interaction is also analysed in collaboration with domain experts. For this reason, we explained how the inference engine is constructed and how the system is reasoning and reacting to user interventions. The changes in the system's advice due to the user's intervention are shown in Table 6:

- if we consider equal weightings of preferences without adding evidence the preferred solution is the collaborative robot. Results change if we set evidence on the chance nodes BIM and PS. The conventional system is suggested if the BIM model is not available or if we have a small project. The robotic system is suggested if a BIM model is available or if we have a large project. Looking at the *VPI* we see that the best information to be acquired is the project size.
- if we focus on productivity and cost, the preferred solution is mostly the conventional system. Additionally, different settings of preferences and evidence have an impact on the decision and on the best information to be acquired by the user.
- if we focus on quality, the preferred solution is mostly the collaborative robot. The conventional system is mostly preferred if the BIM model is not available or if we have a small project.
- if we focus on safety, the preferred solution is mostly the collaborative robot. Here, the conventional manual process is suggested for only one preference setting and when the BIM model is not available or if we have a small project.

The evaluation of the output with domain experts has been considered satisfactory.

4. Discussion

Previous studies have shown that impact of the introduction of robotic systems on the construction site requires a careful evaluation of their potential benefits and shortcomings on the construction process, which is very challenging due to the lack of available performance data of construction robots [12,26]. The lack of available data concerns both the conventional processes, where we have a lack of standardised work processes that cause uncertainty [14], and the processes that employ new technologies, which have not yet been sufficiently tested in real world scenarios [26].

Within this work, this challenge is addressed by developing a decision-theoretic-expert system that can perform rational reasoning and allows one to fill the lack of data with expert knowledge. The application of the guideline for the design of DSS [27], with a fixed development procedure, simplified the development process and made it possible to involve domain experts easily in the process. The close collaboration with the domain experts makes the underlying reasoning processes of the system more transparent and, therefore, increases the acceptance of the DSS. The increasing adoption of BIM in construction with the provision of digital models that contain both geometric and semantic data can be considered as a necessary condition for the efficient deployment of construction robots.

5. Conclusions

This work provides two contributions for decision support in the field of robotic equipment adoption in construction processes. On the one hand, it shows the applicability of an axiomatic design guideline for the collaborative design of DSS for robotic equipment adoption in construction processes, and it can serve as a basis for defining a software-tool that allows the systematic development of DSS. On the other hand, it confirms the ability of decision-theoretic expert systems to represent construction related decision problems in an adequate way. The adopted approach allows us to highlight both shortcomings and advantages of the robotic systems to be developed within the research project and conventional construction methods, by systematically evaluating their impacts on productivity and cost, quality, and safety and risk. The involvement of domain experts allows us to define a DSS that acts reasonably and to fill the gaps in the availability of data often found in decision problems that are related to construction execution processes.

The applicability of the approach is demonstrated by means of an exemplary UC of an ongoing EU-funded project. The conditional probabilities needed for the computations were defined by making educated guesses based on the target performance that the collaborative robot should reach in comparison to the conventional manual process. The values will be refined when tangible results of the testing activities will be available. Further, the DSS will be extended to all UCs of the project and allow the user to choose the different tasks that can be performed by the system.

Author Contributions: Conceptualization, C.M. and A.G.; methodology, C.M. and A.G.; investigation, C.M.; writing—original draft preparation, C.M. and A.G.; supervision, D.T.M. All authors have read and agreed to the published version of the manuscript.

Funding: This research has received funding from the European Union's Horizon 2020 research and innovation program under grant agreement No 101016007. (Project CONCERT—CONfigurable CollaborativE Robot Technologies).

Acknowledgments: The authors would like to thank the group guided by Lech Własak at Budimex, Poland, project partner in CONCERT, for collaborating in the development and evaluation of the prototype. Special thanks are also directed to Cinzia Slongo from Fraunhofer Italia for her support during the development and evaluation of the prototype.

Conflicts of Interest: The authors declare no conflict of interest. The funders had no role in the design of the study; in the collection, analyses, or interpretation of data; in the writing of the manuscript; or in the decision to publish the results.

References

1. Dirican, C. The Impacts of Robotics, Artificial Intelligence On Business and Economics. *Procedia—Soc. Behav. Sci.* **2015**, *195*, 564–573. [CrossRef]
2. Davila Delgado, J.M.; Oyedele, L.; Ajayi, A.; Akanbi, L.; Akinade, O.; Bilal, M.; Owolabi, H. Robotics and automated systems in construction: Understanding industry-specific challenges for adoption. *J. Build. Eng.* **2019**, *26*, 100868. [CrossRef]
3. Pasetti Monizza, G.; Bendetti, C.; Matt, D.T. Parametric and Generative Design techniques in mass-production environments as effective enablers of Industry 4.0 approaches in the Building Industry. *Autom. Constr.* **2018**, *92*, 270–285. [CrossRef]
4. Vähä, P.; Heikkilä, T.; Kilpeläinen, P.; Järviluoma, M.; Gambao, E. Extending automation of building construction—Survey on potential sensor technologies and robotic applications. *Autom. Constr.* **2013**, *36*, 168–178. [CrossRef]
5. Carra, G.; Argiolas, A.; Bellissima, A.; Niccolini, M.; Ragaglia, M. Robotics in the Construction Industry: State of the Art and Future Opportunities. In Proceedings of the 35th International Symposium on Automation and Robotics in Construction (ISARC), Berlin, Germany, 20–25 July 2018; pp. 866–873.
6. Aghimien, D.O.; Aigbavboa, C.O.; Oke, A.E.; Thwala, W.D. Mapping out research focus for robotics and automation research in construction-related studies: A bibliometric approach. *J. Eng. Des. Technol.* **2019**, *18*, 1063–1079. [CrossRef]
7. Son, H.; Kim, C.; Kim, H.; Han, S.H.; Kim, M.K. Trend analysis of research and development on automation and robotics technology in the construction industry. *KSCE J. Civ. Eng.* **2010**, *14*, 131–139. [CrossRef]
8. Bogue, R. What are the prospects for robots in the construction industry? *Ind. Robot. Int. J.* **2018**, *45*, 1–6. [CrossRef]
9. Deloitte Global Powers of Construction. 2018. Available online: https://www2.deloitte.com/gr/en/pages/energy-and-resources/articles/deloitte-global-powers-of-construction-2018.html (accessed on 10 July 2021).
10. Hampson, K.D.; Kraatz, J.A.; Sanchez, A.X. The Global Construction Industry and R&D. In *R&D Investment and Impact in the Global Construction Industry*; Routledge: New York, NY, USA, 2014; pp. 4–23.
11. Arayici, Y.; Coates, P. A system engineering perspective to knowledge transfer: A case study approach of BIM adoption. *Virtual Real.—Hum. Comput. Interact.* **2012**, *2006*, 179–206.
12. Brosque, C.; Skeie, G.; Fischer, M. Comparative Analysis of Manual and Robotic Concrete Drilling for Installation Hangers. *J. Constr. Eng. Manag.* **2021**, *147*. [CrossRef]
13. Marcher, C.; Giusti, A.; Schimanski, C.P.; Matt, D.T. Application of Decision Support Systems for Advanced Equipment Selection in Construction. In *Cooperative Design, Visualization, and Engineering*; Luo, Y., Ed.; Lecture Notes in Computer Science; Springer International Publishing: Cham, Switzerland, 2019; Volume 11792, pp. 229–235. ISBN 978-3-030-30948-0.
14. Dubois, A.; Gadde, L.-E. The construction industry as a loosely coupled system: Implications for productivity and innovation. *Constr. Manag. Econ.* **2002**, *20*, 621–631. [CrossRef]
15. Hastak, M. Advanced automation or conventional construction process? *Autom. Constr.* **1998**, *7*, 299–314. [CrossRef]
16. Marcher, C.; Giusti, A.; Matt, D.T. Decision Support in Building Construction: A Systematic Review of Methods and Application Areas. *Buildings* **2020**, *10*, 170. [CrossRef]
17. Marzouk, M.; Abubakr, A. Decision support for tower crane selection with building information models and genetic algorithms. *Autom. Constr.* **2016**, *61*, 1–15. [CrossRef]
18. Alshibani, A.; Elassir, H.; Al-Najjar, M.; Hamida, H. AHP based approach for crane selection of building construction in Saudi Arabia: A case study. In Proceedings of the Annual Conference–Canadian Society for Civil Engineering, Montreal, QC, Canada, 12–15 June 2019.
19. Temiz, I.; Calis, G. Selection of Construction Equipment by using Multi-criteria Decision Making Methods. In *Procedia Engineering*; Elsevier Ltd.: Amsterdam, The Netherlands, 2017; Volume 196, pp. 286–293.
20. Jankovic, I.; Djenadic, S.; Ignjatovic, D.; Jovancic, P.; Subaranovic, T.; Ristovic, I. Multi-criteria approach for selecting optimal dozer type in open-cast coal mining. *Energies* **2019**, *12*, 2245. [CrossRef]
21. Zeynalian, M.; Dehaghi, I.K. Choice of optimum combination of construction machinery using modified advanced programmatic risk analysis and management model. *Sci. Iran.* **2018**, *25*, 1015–1024. [CrossRef]
22. Kouch, A.M.; Illikainen, K.; Perälä, S. *Key Factors of an Initial BIM Implementation Framework for Small and Medium-sized Enterprises (SMEs)*; International Association for Automation and Robotics in Construction (IAARC): Taipei, Taiwan, 2018; pp. 904–912.
23. Azhar, S. Building Information Modeling (BIM): Trends, Benefits, Risks, and Challenges for the AEC Industry. *Leadersh. Manag. Eng.* **2011**, *11*, 241–252. [CrossRef]
24. Follini, C.; Magnago, V.; Freitag, K.; Terzer, M.; Marcher, C.; Riedl, M.; Giusti, A.; Matt, D.T. Bim-integrated collaborative robotics for application in building construction and maintenance. *Robotics* **2021**, *10*, 2. [CrossRef]
25. Giusti, A.; Magnago, V.; Siegele, D.; Terzer, M.; Follini, C.; Garbin, S.; Marcher, C.; Steiner, D.; Schweigkofler, A.; Riedl, M. BALTO: A BIM-Integrated Mobile Robot Manipulator for Precise and Autonomous Disinfection in Buildings against COVID-19. In Proceedings of the 2021 IEEE 17th International Conference on Automation Science and Engineering (CASE), Lyon, France, 23–27 August 2021; pp. 1730–1737.

26. Brosque, C.; Skeie, G.; Örn, J.; Jacobson, J.; Lau, T.; Fischer, M. Comparison of construction robots and traditional methods for drilling, drywall, and layout tasks. In Proceedings of the 2020 International Congress on Human-Computer Interaction, Optimization and Robotic Applications (HORA), Ankara, Turkey, 26–28 June 2020; pp. 1–14.
27. Marcher, C.; Rauch, E.; Giusti, A.; Matt, D.T. Decision Support Systems in Building Construction–An Axiomatic Design Approach. In Proceedings of the IOP Conference Series: Materials Science and Engineering, The 14th International Conference on Axiomatic Design (ICAD 2021), Lisbon, Portugal, 23–25 June 2021; Volume 1174, p. 012004.
28. Makovetskaya, E.; Deniskina, A.; Krylov, E.; Urumova, F. Organizational optimization of construction processes by virtue of robotization. In *E3S Web of Conferences*; Zheltenkov, A., Ed.; EDP Sciences: Les Ulis, France, 2019; Volume 91, p. 02036.
29. Yan, R.-J.; Kayacan, E.; Chen, I.-M.; Tiong, L.K.; Wu, J. QuicaBot: Quality Inspection and Assessment Robot. *IEEE Trans. Autom. Sci. Eng.* **2019**, *16*, 506–517. [CrossRef]
30. Suh, N.P. Axiomatic Design Theory for Systems. *Res. Eng. Des.* **1998**, *10*, 189–209. [CrossRef]
31. Kjærulff, U.B.; Madsen, A.L. *Bayesian Networks and Influence Diagrams: A Guide to Construction and Analysis*, 2nd ed.; Information Science and Statistics; Springer: New York, NY, SUA, 2013; ISBN 978-1-4614-5103-7.
32. Russell, S.J.; Norvig, P.; Davis, E. *Artificial Intelligence: A Modern Approach*, 3rd ed.; Prentice Hall Series in Artificial Intelligence; Prentice Hall: Upper Saddle River, NJ, USA, 2010; ISBN 978-0-13-604259-4.
33. Howard, R.A.; Matheson, J.E. Influence Diagrams. *Decis. Anal.* **2005**, *2*, 127–143. [CrossRef]
34. Liang, C.-J.; Kang, S.-C.; Lee, M.-H. RAS: A robotic assembly system for steel structure erection and assembly. *Int. J. Intell. Rob. Appl.* **2017**, *1*, 459–476. [CrossRef]
35. Taghavi, M.; Iturralde, K.; Bock, T. Cable-driven parallel robot for curtain wall modules automatic installation. In Proceedings of the 35th International Symposium on Automation and Robotics in Construction (ISARC), Berlin, Germany, 20–25 July 2018; pp. 396–403.
36. Hu, R.; Iturralde, K.; Linner, T.; Zhao, C.; Pan, W.; Pracucci, A.; Bock, T. A simple framework for the cost-benefit analysis of single-task construction robots based on a case study of a cable-driven facade installation robot. *Buildings* **2021**, *11*, 8. [CrossRef]
37. Dakhli, Z.; Lafhaj, Z. Robotic mechanical design for brick-laying automation. *Cogent Eng.* **2017**, *4*, 1361600. [CrossRef]
38. Lee, S.; Yu, S.; Choi, J.; Han, C. A methodology to quantitatively evaluate the safety of a glazing robot. *Appl. Ergon.* **2011**, *42*, 445–454. [CrossRef]
39. Renooij, S.; Witteman, C. Talking probabilities: Communicating probabilistic information with words and numbers. *Int. J. Approx. Reason.* **1999**, *22*, 169–194. [CrossRef]
40. van der Gaag, L.C.; Renooij, S.; Witteman, C.L.M.; Aleman, B.M.P.; Taal, B.G. Probabilities for a probabilistic network: A case study in oesophageal cancer. *Artif. Intell. Med.* **2002**, *25*, 123–148. [CrossRef]
41. Gonzales, C.; Torti, L.; Wuillemin, P.-H. aGrUM: A Graphical Universal Model Framework. In *Advances in Artificial Intelligence: From Theory to Practice*; Benferhat, S., Tabia, K., Ali, M., Eds.; Springer International Publishing: Cham, Switzerland, 2017; Volume 10351, pp. 171–177. ISBN 978-3-319-60044-4.
42. Liu, Q.; Ihler, A.T. Belief propagation for structured decision making. *arXiv* **2012**, arXiv:1210.4897v1.
43. Nilsson, D.; Lauritzen, S.L. Evaluating influence diagrams using LIMIDs. *arXiv* **2013**, arXiv:1301.3881v1.

applied sciences

MDPI

Article

Convolutional Neural Networks for Differential Diagnosis of Raynaud's Phenomenon Based on Hands Thermal Patterns

Chiara Filippini [1,*], Daniela Cardone [1], David Perpetuini [1], Antonio Maria Chiarelli [1], Giulio Gualdi [2], Paolo Amerio [2] and Arcangelo Merla [1]

1 Department of Neurosciences, Imaging and Clinical Sciences, University G. d'Annunzio of Chieti-Pescara, 66100 Chieti, Italy; d.cardone@unich.it (D.C.); david.perpetuini@unich.it (D.P.); antonio.chiarelli@unich.it (A.M.C.); arcangelo.merla@unich.it (A.M.)

2 Department of Medicine and Aging Science, Dermatologic Clinic, G. D'Annunzio University, 66100 Chieti, Italy; giuliogualdi@libero.it (G.G.); p.amerio@unich.it (P.A.)

* Correspondence: chiara.filippini@unich.it

Abstract: Raynaud's phenomenon (RP) is a microvessels' disorder resulting in transient ischemia. It can be either primary or secondary to connective tissue diseases, such as systemic sclerosis. The differentiation between primary and secondary to systemic sclerosis is of paramount importance to set the proper therapeutic strategy. Thus far, thermal infrared imaging has been employed to accomplish this task by monitoring the finger temperature response to a controlled cold challenge. A completely automated methodology based on deep convolutional neural network is here introduced with the purpose of being able to differentiate systemic sclerosis from primary RP patients by relying uniquely on thermal images of the hands acquired at rest. The classification performance of such a method was compared to that of a three-dimensional convolutional neural network model implemented to classify thermal images of the hands recorded during rewarming from a cold challenge. No significant differences were found between the two procedures, thus ensuring the possibility to avoid the cold challenge. Moreover, the convolutional neural network models were compared with standard feature-based approaches and showed higher performances, thus overcoming the limitations related to the feature extraction (e.g., biases introduced by the operator). Such automated procedures can constitute promising tools for large scale screening of primary RP and secondary to systemic sclerosis in clinical practice.

Keywords: deep learning; neural network; thermal imaging; Raynaud phenomenon; systemic sclerosis

Citation: Filippini, C.; Cardone, D.; Perpetuini, D.; Chiarelli, A.M.; Gualdi, G.; Amerio, P.; Merla, A. Convolutional Neural Networks for Differential Diagnosis of Raynaud's Phenomenon Based on Hands Thermal Patterns. *Appl. Sci.* **2021**, *11*, 3614. https://doi.org/10.3390/app11083614

Academic Editors: João M. F. Rodrigues, Pedro J. S. Cardoso, Marta Chinnici and Donato Cascio

Received: 29 March 2021
Accepted: 15 April 2021
Published: 16 April 2021

1. Introduction

Raynaud's phenomenon (RP) is a common vascular disorder consisting of recurrent, long-lasting, and episodic vasospasm of the fingers and toes often manifesting as discoloration and pain [1]. RP is typically induced by cold exposure and emotional stress [2]. It affects approximately 5–10% of the population (prevalently females) [3], with a worldwide distribution, although its prevalence is elevated in cold climates where the risk of exposure to low ambient temperatures is greater [4]. RP is classified as primary RP (PRP) if there is no known underlying illness and secondary when associated with a disorder detected upon assessment [5]. The distinction is important because prognosis, severity, and treatment can all be affected [6]. Secondary RP can be associated with many systemic rheumatic diseases. The most frequent association is with systemic sclerosis (SSc) [7].

SSc is a complex autoimmune connective tissue disease that is characterized by progressive generalized obliterative vasculopathy and widespread aberrant tissue fibrosis [8]. Although SSc is a heterogeneous disease, RP occurs in most patients, affecting ~96% of them [9]. RP is considered the most common and one of the earliest symptoms of this disease [10]. SSc-associated RP typically has a lag period that can last several years before additional SSc organ-specific disease manifestations emerge [10,11]. Whereas in primary or 'idiopathic' RP, tissue ischemia is transient or reversible, in secondary RP persistent tissue ischemia can occur, resulting in digital ulceration and/or gangrene [12].

Frequently in the early stages of disease there is not a clear-cut difference between PRP and secondary PR. Indeed, many RP patients have no sign of systemic disease, although they do present with subtle nailfold abnormalities. Conversely, the differential diagnosis of PRP versus secondary RP is of utmost importance to allow for the earliest successful treatment of this condition and the associated underlying disease. Since RP impacts the finger thermoregulatory system, the evaluation of the finger thermoregulatory impairment is crucial to distinguish between PRP and RP secondary to SSc [13]. To this aim, finger temperature can be monitored through thermal infrared (IR) imaging technique.

Thermal IR imaging is a contactless, non-invasive technique which provides a map of a body's superficial temperature by measuring its emitted infrared radiation [14–16]. It has been widely used in medicine to assess cutaneous temperature and its topographic distribution as well as to monitor the psychophysiological state of an individual [17,18]. Considering that the skin temperature depends on local blood perfusion and thermal tissue properties, thermal IR imaging provides important indirect information concerning circulation and thermoregulatory functionality of the cutaneous tissue [14]. This technique has been employed to differentiate primary from secondary RP, often by monitoring the finger response to a controlled cold challenge [19]. Indeed, SSc, PRP, and healthy controls (HC) show different thermal recovery to the same functional stimulation (i.e., cold challenge) [13]. Therefore, the differentiation among HC, SSc, and PRP was usually performed based on statistical analysis of simple descriptors of the cutaneous temperature recovery (e.g., lag time, time to reach a given recovery threshold) [19]. However, such an analysis procedure involves identifying the regions of interest on the fingers from which the statistical descriptors of the temperature recovery are computed as well as choosing the best descriptors to use for the classification. Hence, the intervention of an expert operator is required, thus introducing operator-dependent bias on the classification outcome.

Furthermore, since in PRP or SSc patients the blood vessels in the extremities are over-sensitive to changes in temperature, the administration of a cold challenge can be a very uncomfortable, and frequently painful, process. In addition, a major problem has been the lack of a standardized protocol that regularize such a challenges' administration procedure. It would therefore be desirable to develop an automated classification approach able to differentiate between PRP, SSc, and HC, thus limiting the human intervention and without the administration of a cold stimulus.

In this perspective, machine learning approaches are valuable to minimize or avoid the clinician's intervention and to speed up diagnosis. Indeed, such approaches are suitable for solving complex task and limiting human intervention. Presently, machine learning algorithms are gaining popularity across a wide range of innovative applications, such as smart houses [20] and autonomous vehicles [21] but also in physiological signal classification [22,23] or to support disease diagnosis [24,25]. Machine learning approaches can be feature-based or data-driven [26]. In the traditional feature-based machine learning approach, features are manually selected and extracted. The difficulty with this approach is that it is necessary to choose which features are important and, as the number of classes to classify increases, feature extraction can become cumbersome. On the other hand, data-driven methods provide a principled set of mathematical methods for extracting meaningful features from data [27]. Specifically, it learns from and makes predictions based on data. Those approaches are usually performed by employing advanced machine learning algorithm such as a deep neural network. Indeed, a deep neural network model exploits multiple layers of nonlinear information processing for feature extraction and transformation as well as for pattern analysis and classification [28]. Figure 1 shows the difference between the data driven and feature-based methods.

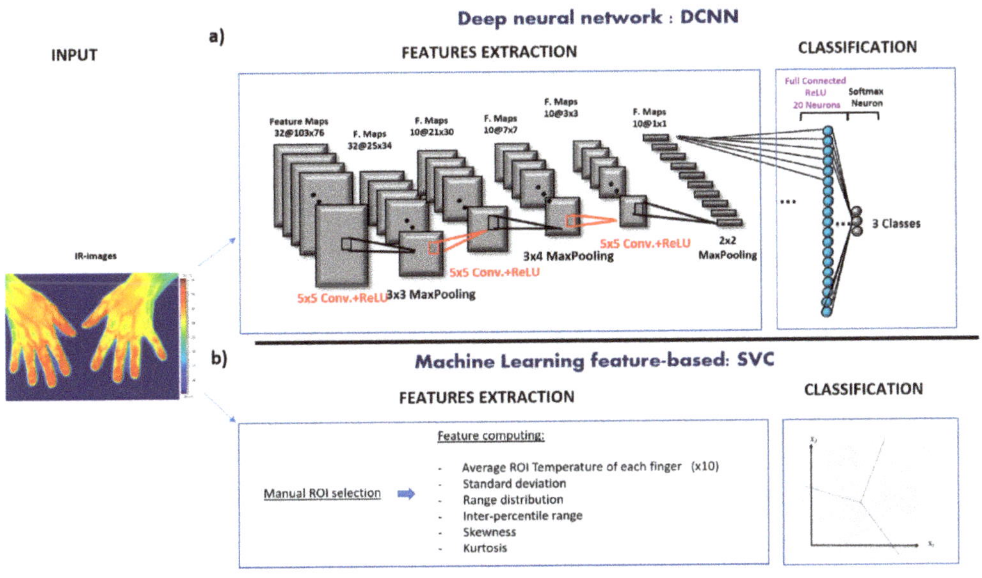

Figure 1. (**a**) Deep Convolutional Neural Network (DCNN) pipeline, which includes the structure of the convolution building block implemented. (**b**) Machine learning feature-based algorithm pipeline, which include the features extracted for the baseline analysis.

With respect to a deep neural network, deep convolutional neural networks (DCNNs) have become the leading architecture for most image classification tasks [29]. DCNNs make use of kernels (also known as filters), to detect features throughout an image. A kernel is a matrix of values, called weights [28]. They are typically composed of three types of layers: convolution, pooling, and fully connected layers. The first two perform feature extraction, whereas the third maps the extracted features into a final output, which is often a classifier. DCNNs find complex relationships by minimizing a cost function (a measure of error between the DCNN and real outputs) through the use of gradient descent approaches and a backpropagation algorithm [30]. DCNNs have been widely used in biomedical images analysis and they have reached excellent classification outcomes [31,32].

In this study, a novel automated DCNN classification methodology that aims to distinguish among PRP, SSc, and HC based on patient's hand thermal patterns measured at rest (i.e., without the administration of a cold challenge) is presented. The performances of the proposed classifier were compared to that of a DCNN model implemented to classify hands thermal images of participants undergoing the cold challenge procedure (CCP). Moreover, for both the analyses, the DCNN models' performances were compared to those of a feature-based machine learning approach. The comparison between different models allowed us to investigate the capability of the DCNN classifier based on the thermal image at rest to identify PRP, SSc, and HC with respect to approaches that require the administration of a cold challenge and the intervention of an expert operator.

2. Materials and Methods

2.1. Paticipants

The experimental session involved 36 participants: 13 healthy, 11 PRP, and 12 SSc participants (Table 1). Participation was strictly voluntary. Before the start of the experimental trials, the participants were adequately informed about the purpose and protocol of the study. All participants signed an informed consent form, which outlined the methods and the purposes of the experimentation in accordance with the Declaration of Helsinki [33]. The study was approved by the Institutional Review Board and Local Ethical Committee of the School of Medicine of the University of Chieti-Pescara (protocol code AC052514 12/07/2016).

Table 1. Physical characteristics of the participants.

Group	Sex (Female/Male)	Age (Avg ± std, Years)	Body Mass (Avg ± std, Kg)	Physically Active (%)
HC	7/6	48.5 ± 11.1	61.8 ± 5.9	42%
PRP	5/6	52.2 ± 10.3	64.5 ± 5.5	38%
SSc	5/7	50.8 ± 11.6	62.3 ± 6.2	35%

SSc patients were recruited from voluntary patients attending the Dermatologic Clinic of the University G. d'Annunzio, Chieti, Italy from December 2016 to February 2017. All the patients fulfilled the ACR-EULAR Collaborative Initiative Criteria for scleroderma [34]. PRP patients were recruited from the Capillaroscopy outpatients service of the Dermatologic Clinic of the University G. d'Annunzio, Chieti, Italy from December 2016 to February 2017. Healthy individuals were recruited from parents of dermatologic oncology patients attending the outpatients service of the Dermatologic Clinic of the University G. d'Annunzio, Chieti, Italy from December 2016 to February 2017. All healthy individuals did not disclose any history of vascular disease or present with any type of vascular disease or vasoactive drug.

PRP and SSc patients were a priori classified according to the criteria and the methods established in 2001 by the American College of Rheumatology [35,36]. All the patients received continuative vasodilator therapy for RP (Pentoxyfillin, calcium channel blockers) which was not discontinued for the purposes of the study. PRP and SSc patients' exclusion criteria were a history of bronchial asthma; renal or hepatic failure; hypotension; moderate or severe arterial hypertension; history of drug or alcohol abuse, smoking, gout, gastric ulcers, or cerebral or cardiac ischemic disease; and sympathectomy of the upper limb performed within 12 months of the beginning of the study. HC exclusion criteria were cigarette smoking; cardiovascular, or neurovascular disorders; hypertension; any overt dermatological or immunological disease; all types of therapeutic treatment; and history of drug or alcohol abuse. Moreover, participants were requested to refrain from vigorous exercise, caffeine, and alcohol for 4 hours prior to the assessment.

2.2. Procedure

Upon arrival, each participant was left in the experimental room for 15 min [37] to allow participants to achieve proper acclimatization to the room environmental conditions and the baseline skin temperature to stabilize. The recording room was set at a standardized temperature (23 ± 0.5 °C) and humidity (55%) by a thermostat according to the International Academy of Thermology guidelines [38]. Participants sat comfortably on a chair during acclimatization and measurement periods. The experimental paradigm consisted of the IR-images recording of the dorsal aspect of both hands, before and after the administration of a cold challenge. The first ones were necessary to obtain the baseline of the fingers' temperature and the remaining ones were useful to monitor the temperature recovery. The participant's hands were placed on a non-reflective, black surface, where the hands shape was drawn to maintain the same position before and after the cold stimulus. The cold challenge was administered by immersing the hands (protected from getting wet by thin, disposable latex gloves) for 2 min in a 3 L water bath maintained at 10 °C. After the cold challenge, the gloves were removed, and the hands were returned to their original position on the non-reflective surface. Each recording session lasted 23 min including both baseline (3 min), recovery phases (20 min), beside the time needed for the cold stress (2 min) that was not recorded. The images were acquired every 30 s.

2.3. Data Acquisition

A FLIR SC3000 digital thermal camera was used in the experiment. It is characterized by a Focal Plane Array of 320 × 240 Quantum Well Infrared Photodetectors, sensitive to the thermal radiation in the 8–9 μm band. The temperature sensitivity/noise equivalent temperature difference of the thermal camera is 0.02 K. The thermal camera was blackbody-calibrated to factory specifications by means of periodic calibration performed by the Quality Management Systems of FLIR, which is certified to comply with ISO 9001:2008. The process of blackbody calibration is described in [37]. Such a process is useful to remove noise-effects related to the sensor drift/shift dynamics and optical artefacts. In accordance to literature [39], cutaneous emissivity was considered as $\varepsilon \approx 0.98$. The thermal camera was placed 1.5 m distant from the hands' dorsum and it was placed perpendicular to the analyzed region [40].

The thermal images acquired were used to feed the DCNN models. In detail, the thermal image corresponding to the central part of the baseline period was chosen as representative of the resting condition for each participant. The whole CCP including the rewarming condition was instead represented by a sub-sample of thermal images constituted by extracting an image every three minutes.

2.4. Data Analysis

The data were analyzed considering the two conditions: thermal data acquired before the cold challenge (i.e., baseline data or hands' thermal images at rest) and data related to the whole CCP (i.e., baseline and recovery data). For each of the two conditions, a DCNN and a feature-based model such as the support vector classifier (SVC) were implemented to classify PRP, SSc, and Healthy participants. The performances of all the models were compared employing the McNemar–Bowker test. Figure 1 shows the DCNN and SVC processing pipeline implemented for the baseline analysis.

Concerning the feature-based approach for each finger of both hands, baseline and rewarming curves were obtained by averaging the temperature of the pixels within a specific region of interest. The regions of interest were identified as the nail-bed regions [19] (Figure 2). Displacement between images were corrected manually using anatomical landmarks based on the fingers profile [19].

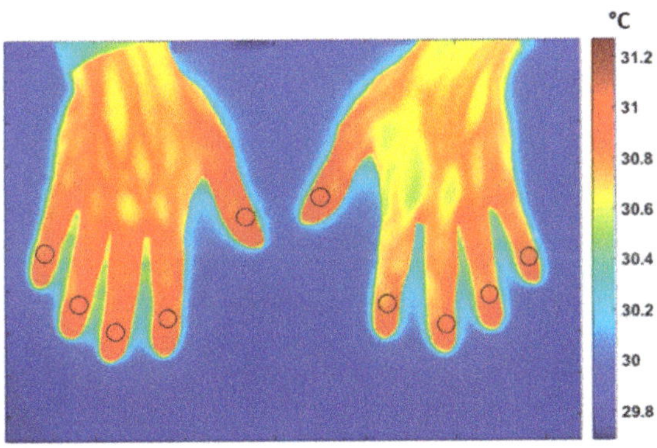

Figure 2. Raw thermal IR image of a representative healthy participant within the dataset. Black circles represent the regions of interest manually selected.

2.4.1. Baseline Analysis

DCNN Model

The IR images of the dorsum of both participants' hands during the baseline condition were used as the input of the DCNN model implemented to differentiate PRP from SSc and HC. The DCNN model was developed according to the following processing steps:

- **Input images preprocess.** Firstly, the IR images were down sampled from 320 × 240 to 80 × 107 pixels and normalized by subtracting the average value of the entire set of images. A Principal Components Analysis was then adopted to reduce the number of spatial features while keeping the information characterizing the object to be analyzed. The number of components kept was 20 with a cumulative explained variance ratio of 0.95.
- **Model architecture design.** The DCNN structure employed in this work was heuristically chosen in similarity with previously reported DCNN structures on biological images or signals classification [41]. The DCNN was composed of 3 convolutional layers, 3 pooling layers, and 1 fully connected prior to the output layer. The first convolutional layer consisted in 32 filters (size 5 × 5) applied to the input images to obtain 32 feature maps of the images. The other 2 convolutional layer were both composed of 10 filters (size 5 × 5). The activation function employed in all the 3 convolutional layers to add nonlinearity to the network, were the Rectified Linear Unit (ReLU) function. Then, as pooling layer (or down-sampling layer) a MaxPooling were chosen, where the largest element from the rectified feature map was retained. A filter size of 3 × 3, 3 × 4 and 2 × 2 for the 3 MaxPooling layers, respectively, was implemented to reduce the dimensionality of each feature to 1 after the 3 convolutional layers and before the fully connected layer. The fully connected layer consisted in 20 neurons employed to summarize information and compute the class score. Lastly, a softamx function was used in the output layer, which outputs a probability value from 0 to 1 for each of the 3 classification classes (PRP, SSc, and healthy). All the biases of the DCNN were initialized to a small constant, i.e., 0.1, whereas the weights were initialized in a pseudo-random manner employing a truncated normal distribution (standard deviation = 0.1). The DCNN architecture is shown in Figure 1a.
- **Model optimization.** The model optimization was primarily focused on how to reduce overfitting. Indeed, developing a DCNN model with a small sample size inevitably involves high risk of overfitting. A technique used to address overfitting is regularization [42]. In this study, a Ridge Regression (L2 regularization) was implemented by adding the sum of the squared values of the model coefficient (weights) as a penalty term to the loss function. The loss function employed in this study was the categorical cross-entropy, therefore after performing the regularization technique it resulted in the following Equation:

$$Loss_r = -\sum y_k \, \log \hat{y}_k \, + \frac{\lambda}{2} \sum {\omega_i}^2 \tag{1}$$

  where $\hat{y}_k$ is the kth scalar value of the model output, y_k is the corresponding target value, and the constant λ times the sum of the squared weight (ω) values is the regularization term. The λ value was set to 0.01. This was intended to reduce model complexity and make the model less prone to overfitting. In addition, instead of using a fixed learning rate hyperparameter for the presented model, which could lead the model to converge too quickly to a suboptimal solution, a tunable learning rate over the training process was implemented. In detail, a function was implemented to reduce learning rate by a factor of 0.1 once learning stop improving during at least 10 epochs.

- **Model evaluation.** To evaluate the model performance, a categorical cross-entropy was employed, which is suitable for multiclass classification task. The optimization procedure was iterated for 100 epochs with a batch size of 6 samples. To address the model generalization performance, a leave-one-out cross-validation procedure was performed [43]. The metrics used for evaluating the model was the accuracy, sensitivity, and specificity. Accuracy represents the percentage of correct predictions out of the total number of test samples. Sensitivity is the proportion of predicted true positives out of all patients with the disease, whereas specificity represent the percentage of the predicted true negatives out of all participants who do not have the disease [44]. Those metrics were performed by counting the number of correct DCNN predictors after an argmax evaluation of the DCNN output vector, averaged among the plateau iterations. This procedure was conducted following the leave one-out cross-validation. The overall sensitivity and specificity of the classifier were obtained by averaging the sensitivity and specificity of each class, respectively.

The described DCNN model was implemented in Python using Keras API with TensorFlow backend. For model evaluation, the scikit learn library was utilized. Figure 3 shows an example of the input images, respectively, for HC, SSc, and PRP participants, and the related feature maps resulting as output from the first two convolutional layers.

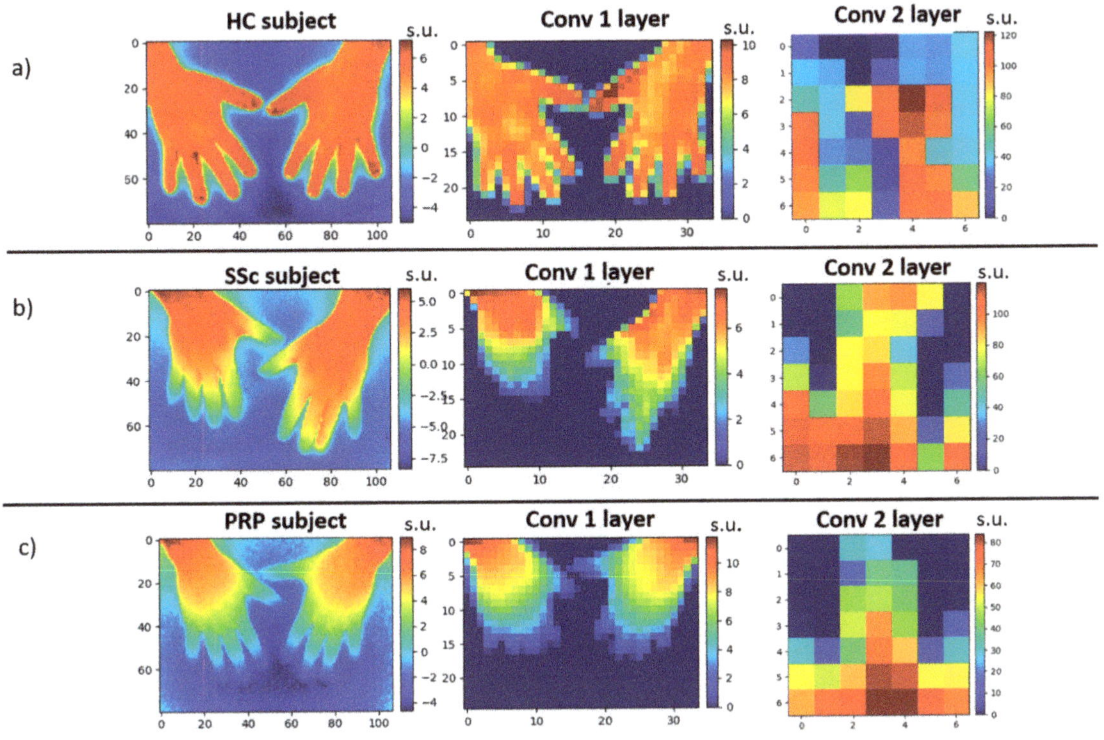

Figure 3. Deep Convolutional Neural Network (DCNN) feature maps. (**a**) Input image of a representative healthy participant within the dataset together with the features map resulting from the first two convolutional layers (s.u. = standardized units). (**b**) Input image of a representative systemic sclerosis (SSc) participant within the dataset and the features map resulting from the first two convolutional layers. (**c**) Input image of a representative primary Raynaud's phenomenon (PRP) participant within the dataset and the features map resulting from the first two convolutional layers. The images' axes units represent the images' pixels number.

Feature-Based Analysis

A feature extraction algorithm was then implemented to extract features of interest from the same IR-images of the baseline condition. In detail, a region of interest present on each fingers' nail bed was manually selected. For each region of interest, the temperature average value was extracted. Region of interests are shown in Figure 2. The extracted temperature data were then normalized by subtracting the average value of the entire dataset. In addition, for each participant, features related to the 10 fingers normalized temperature distribution were estimated. These features were: the standard deviation (STD), kurtosis, skewness, range (i.e., the difference between the largest and smallest values of the distribution), and the inter-quartile range (i.e., the difference between the third and first quartile). Figure 4 shows the group mean and standard deviation of the 10 fingers' temperature average value, and of all the other features z-score normalized.

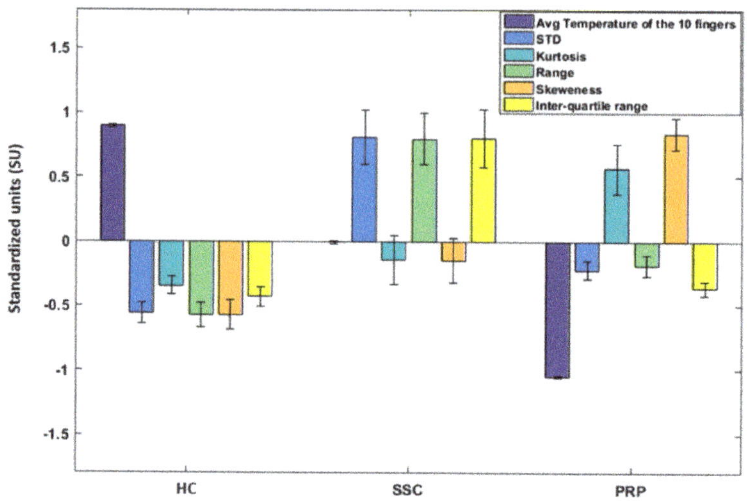

Figure 4. Group average values and the related standard deviation of each feature z-score normalized. The measurement units reported on the y-axis are the standardize units since the features shown are z-score normalized.

As a machine learning approach, the SVC was implemented. Specifically, the SVC was computed using radial basis function (RBF) as a kernel function (non-linear). Such a function has the following formula for two vectors u and v:

$$\text{RBF}(u, v) = \exp(-\gamma\ \|u-v\|^2) \tag{2}$$

where γ is a hyperparameter used as similarity measure between two data point. For the purpose of our study γ was set as follows:

$$\gamma = 1/(\text{number of features} = 15) \tag{3}$$

whereas the regularization hyperparameter C was set to 5. The SVC generalization capabilities were assessed through cross-validation. The same cross-validation procedure employed for the DCNN model were utilized. The metrics employed to be evaluated were the accuracy, sensitivity, and specificity. The z-score normalized features constituted the SVC inputs.

2.4.2. CCP Analysis

3-Dimensional DCNN

A 3-dimensional DCNN (3D-DCNN) model was implemented to classify the thermal images recorded during the whole CCP (i.e., baseline and recovery conditions) in the three classes (HC, SSc, and PRP). Due to the high computational load and the time required to analyze all the images of the recorded IR video (one every 30 s for 20 min, 40 images in total), a representative sample was considered for the successive analyses. This sample was composed of IR-images extracted every three minutes. The baseline image was also added to the set of images to better describe the rewarming phenomenon and in accordance with previous studies [13,19].

The 3D-DCNN differs from the 2D in the convolutional filter type. Indeed, it applies a three-dimensional filter to the dataset and the filter moves in three-directions (x, y, z) to calculate the low-level feature representations. Their output shape is a three-dimensional volume space. Architectures with volumetric (i.e., spatially 3D) convolutions have been successfully used in video analysis or 3D medical images [45,46]. In this case, time acts as the third dimension [47].

The 3D-DCNN model was implemented following the same processing step used to develop the DCNN model (Section 2.4.1). The batch of input images underwent the same preprocessing procedure as described in Section 2.4.1, whereas the model architecture design slightly changed. The number of 3D-filters in the three convolutional layers remained the same, whereas the filter sizes were, respectively, $5 \times 5 \times 4$, $3 \times 3 \times 2$, $3 \times 4 \times 3$, and the three MaxPooling, respectively, $3 \times 3 \times 1$, $3 \times 3 \times 1$, $3 \times 4 \times 2$. In this way the dimensionality of each feature was reduced to 1 before the fully connected layer. Finally, the model optimization and evaluation were conducted as previously described. The features map resulting as output from the first 2 convolutional layers together with the batch of input images, respectively for HC, SSc, and PRP participants are shown in Figure 5.

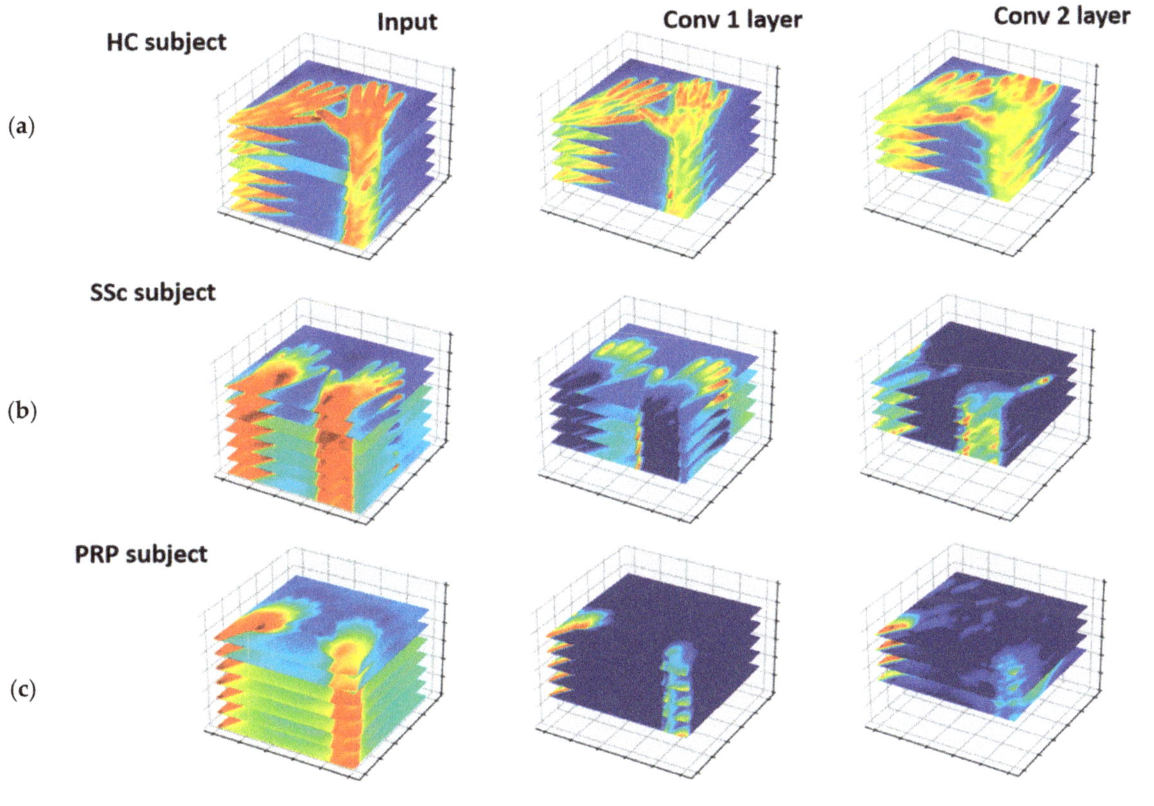

Figure 5. 3-Dimensional Deep Convolutional Neural Network (3D-DCNN) feature maps. (**a**) Batch of input images representative of the cold challenges procedure (CCP) data of a healthy participant randomly chosen, together with the features map resulting from the first two 3D-convolutional layers. (**b**) Batch of input images of a systemic sclerosis (SSc) participant randomly chosen, and the related features map resulting from the first two 3D-convolutional layers. (**c**) Batch of input images of a primary Raynaud's phenomenon (PRP) participant randomly chosen, and the related features map resulting from the first two 3D-convolutional layers.

Feature-Based Analysis

The feature-based analysis on the CCP data was performed by extracting features of interest from the fingers' temperature trend. Such a trend was obtained by averaging the temperatures of the selected regions of interest (see Figure 2) throughout the experiment. The temperature time courses of each finger of a representative HC, SSc, and PRP participant are shown in Figure 6.

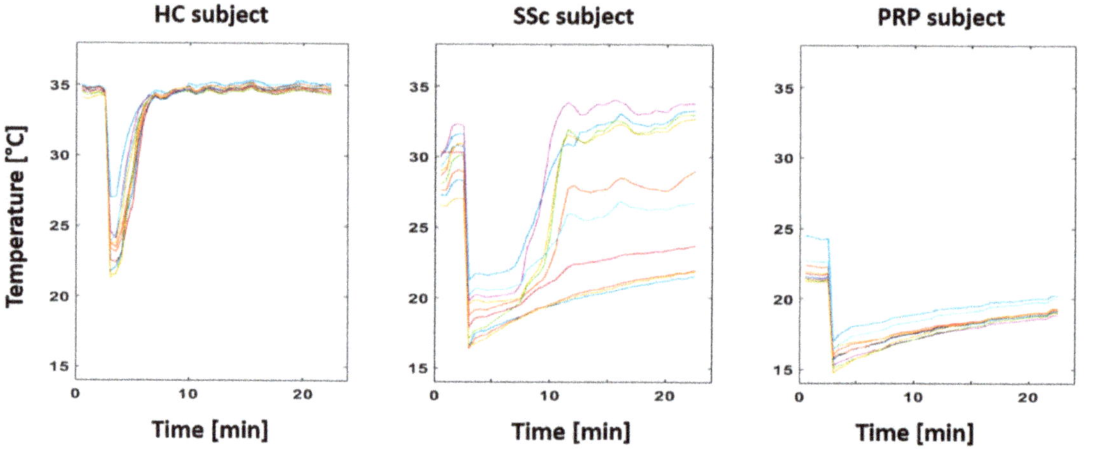

Figure 6. Time courses of the 10 fingers' thermal recovery after the cold challenge in healthy control (HC), systemic sclerosis (SSc), and primary Raynaud's phenomenon (PRP) representative participants.

The extracted temperature data were then normalized by subtracting the average value of the entire dataset. The features identified to characterize the dynamic response of the finger's temperature to the cold challenge were the following:

1. Delay time: the time required for the finger's temperature after the cold challenge to reach 50% of its final value (i.e., the recovery temperature after 20 min from the cold stress).
2. Rise time: the time required for the finger's temperature to rise from 10% to 90% of its final value.
3. Recovery time: time required for the finger's temperature to reach the 68% of the difference value between the baseline temperature and the temperature soon after the cold challenge.
4. Steady state error: the difference between the baseline temperature value and the final recovery temperature value.
5. Delta: difference between the finger's temperature on its final recovery point and the temperature soon after the stimulus.

These five features were collected from each of the participants' fingers, for a total of 50 features per participant. Indeed, these features are commonly used in literature to describe the fingers' recovery phase from a cold stimulus [13,19,48] in PR patients. The group mean of all the features averaged among fingers and the related standard deviation are reported in Figure 7. These features were z-score normalized and used as input to the SVC, which was preformed to classify HC, SSc, and PRP participants. The SVC was implemented using RBF as a kernel function with $\gamma = 1/50$, and the same hyperparameters were employed for the baseline classification.

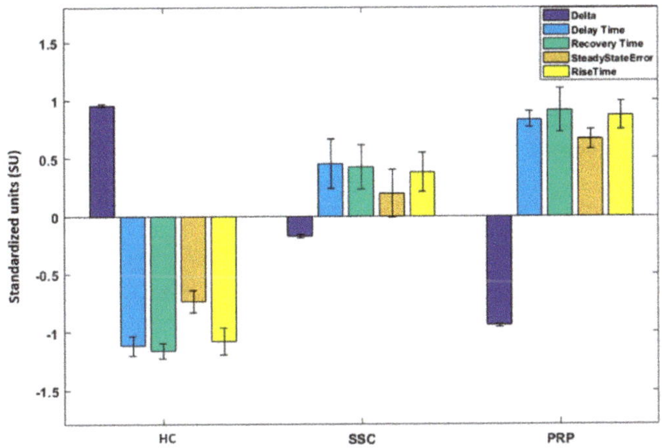

Figure 7. Group average values and the related standard deviation of each feature (z-score normalized) employed for the cold challenge procedure (CCP) analysis. The measurement units are reported as standardized units since the features shown are z-score normalized.

3. Results

3.1. Baseline Results

The DCNN average accuracy and the related standard error are reported in Figure 8 as a function of the training epoch for training and the testing set, respectively. No overfitting effect (decrease of the accuracy at increasing epoch) is visible in the testing set, proving the efficacy of the employed procedure. The DCNN accuracy in the test sample reached a plateau value of 0.84 ± 0.05 and a maximum value of 0.91 ± 0.04.

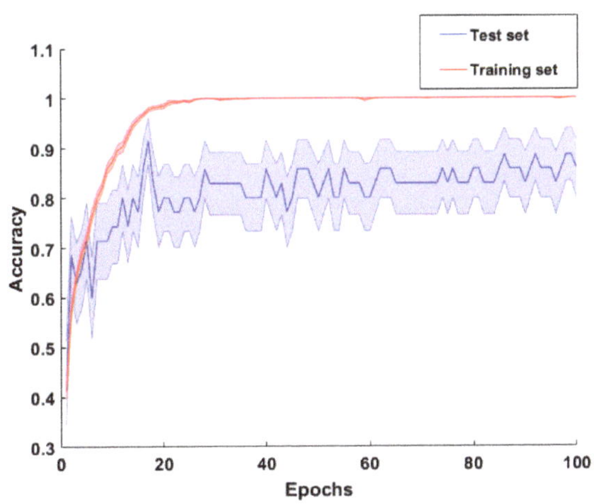

Figure 8. DCNN average (and related standard error) cross-validated accuracy as a function of the training epoch for the training and the testing set, respectively.

With respect to the feature-based analysis, performed by using the SVC, an accuracy of 0.77 was obtained. The sensitivity and specificity values for each of the 3 classes analyzed are shown in Table 2.

Table 2. Sensitivity and specificity of DCNN and SVC models on HC, SSc, and PRP participants' classification.

	DCNN		SVC	
Classes	**Sensitivity**	**Specificity**	**Sensitivity**	**Specificity**
HC	0.84	0.91	0.84	0.83
SSc	0.75	0.87	0.67	0.90
PRP	0.90	0.96	0.81	0.92

3.2. CCP Results

The accuracy trend of the 3D-DCNN model as a function of training epochs for training and test set is reported in Figure 9. The model did not show any overfitting effect. The plateau value of the average accuracy was reached at 0.88 ± 0.05.

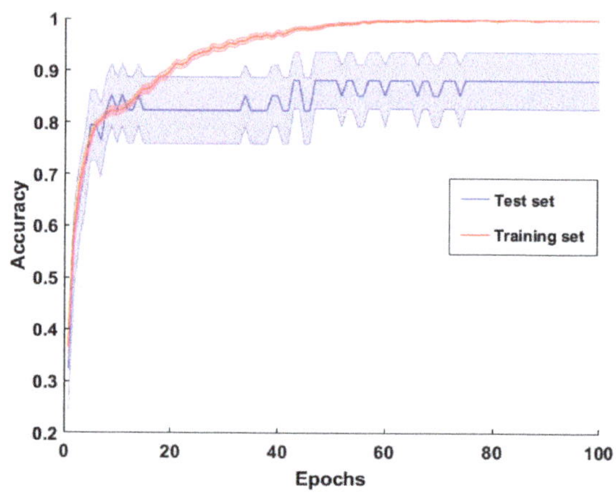

Figure 9. 3D-DCNN average (and related standard error) cross-validated accuracy as a function of the training epoch for the training and the testing set, respectively.

The SVC accuracy achieved a value of 0.83. The sensitivity and specificity for each of the three classes analyzed are reported in Table 3 for both the 3D-DCNN and SVC classifiers.

Table 3. Sensitivity and specificity of 3D-DCNN and SVC models on HC, SSc, and PRP participants' classification.

	3D-DCNN		SVC	
Classes	**Sensitivity**	**Specificity**	**Sensitivity**	**Specificity**
HC	0.92	0.91	1	0.95
SSc	0.83	0.92	0.58	0.91
PRP	0.90	1	0.81	0.84

3.3. Baseline vs. CCP Results

To test the statistical significance of the differences in the classifier performances, a McNemar–Bowker test was performed on the classifiers' prediction outcome [49]. Indeed, whereas McNemar's test requires that there are only two possible categories for each classification outcome to be tested, in the McNemar–Bowker test the outcome analyzed can be classified in more than two classes. No statistical difference was found between the classifiers' performance in the two experimental procedures (i.e., baseline vs. CCP). In detail, both comparisons between the DCNN and the 3D-DCNN performance and between the baseline SVC and the CCP SVC performance were found not significant (B = 2, df = 3, p = n.s. and B = 0.8, df = 3, p = n.s, respectively).

Within the procedures, the difference between the performances of the machine learning approaches employed were also tested and no statistically significant difference was reported. In detail, both comparisons between the DCNN and the SVC performance in the baseline condition (B = 0.6, df = 3, p = n.s.) and between the 3D-DCNN and the SVC in CCP (B = 2.7, df = 3, p = n.s.) were not significant. The overall performance metrics of both classifiers in both procedures are shown in Table 4.

Table 4. Overall accuracy, sensitivity, and specificity of DCNN and SVC models on baseline images classification, and of 3D-DCNN and SVC model in classification of the CCP images. Best performances are shown in bold.

	Baseline		**CCP**	
Metrics	**DCNN**	**SVC**	**3D-DCNN**	**SVC**
Accuracy	0.84	0.77	0.88	0.83
Sensitivity	0.83 + 0.08	0.77 + 0,09	0.88 + 0.04	0.80 + 0.20
Specificity	0.91 + 0.04	0.88 + 0.05	0.94 + 0.05	0.90 + 0.06

4. Discussion

RP is associated with characteristic abnormalities in the function and morphology of the vasculature, which can result in irreversible digital ischemia. Although RP is usually idiopathic (PRP), it can occur as part of an underlying disorder such as in SSc. The condition of vasoconstriction of PRP or SSc require different clinical treatments. It is therefore important to develop reliable methods able to differentiate between a PRP patient and an RP secondary to SSc patient with high specificity. To this end, thermal IR imaging has been widely used as a support to clinical diagnosis. However, the thermographic protocols often incorporate some form of a temperature challenge, usually cold, and need to be performed by an experienced operator. In this study, a completely automated methodology was developed to differentiate SSc from PRP patients by relying uniquely on IR images of the hands acquired at rest. The outcome of such an automated procedure was compared to those obtained with procedures involving the cold challenge and manual selection of the features. No statistically significant differences were found in the comparison, thus favoring the automated procedure performed at basal condition. The differences found between the classes of participants, and the procedures adopted are detailed in the following subsections.

4.1. Thermoregulatory Difference between Classes

Thermoregulation is a complex mechanism that is mostly regulated by the autonomic nervous system. Specifically, sympathetic cholinergic nerves mediate the cutaneous vasodilation in response to whole-body heating, whereas noradrenergic nerves are involved in the cutaneous vasoconstriction during whole-body cooling [50]. Concerning localized stimuli, skin warming induces cutaneous vascular responses due to temperature-sensitive afferent neurons and nitric oxide, whereas local cutaneous vasoconstriction in response to direct cooling of the skin is due to sensory and sympathetic noradrenergic nerves and non-neural mechanisms [50]. The thermal responses of an HC, PRP, and SSc participant's hands to a cold challenge are shown in Figure 6.

PRP is associated with abnormal responses to the environmental temperature and activity (i.e., vasospasm) related to vessels' diseases [51]. Typically, PRP produces symmetrical ischemia lasting 15–20 min, involving both hands with a more sensitive finger. Patients with PRP usually exhibit mild episodes that do not hamper daily activities and generally improve with aging [52,53], whereas SSc patients are affected by intense and frequent ischemic events, associated with ulcers in 25–39% of cases [54,55].

SSc's vascular reactivity and occlusive disease are related to a complex interaction between endothelial cells, smooth muscle cells, the extracellular matrix, and intravascular circulating factors [51]. The impairment of the endothelium induces overproduction of the vasoconstrictor endothelin-1 and underproduction of the vasodilator nitric oxide and prostacyclin. The diminished production of vasodilatory neuropeptides together with an over-regulation of the vascular smooth muscle adrenoreceptor (α2c-AR), leads to an abnormal vasoconstrictive response to stress or cold stimuli. Moreover, severe vasospasms induce repeated episodes of vasoconstriction, provoking occlusion of the microcirculation and injuries [56].

The findings of the present study further confirm the capability of thermal IR imaging to provide information regarding the physiology of the superficial blood circulation [57]. Moreover, the good classification performance obtained using images acquired at rest demonstrated that the vascular impairment associated to RP and SSc generated peculiar superficial temperature distribution of the hands even in the absence of an ongoing bout of the disease.

4.2. Comparisons with Previous Studies and Discussion on Machine Learning Results

Several research studies focused on the development of reliable methods able to differentiate between PRP patients and RP secondary to SSc patients with high specificity. Since changes in temperature pattern of the finger and toes are a clinical manifestation of RP, the evaluation of the finger thermoregulatory impairment is essential to detect the presence of the disease and to differentiate the two forms of this disorder. To this end, thermal IR imaging has been widely used as a support to clinical diagnosis. However, the thermographic protocols often incorporate some form of temperature challenge, usually cold, and need to be performed by an experienced operator. For instance, de Campos et al. employed thermal IR imaging together with cold stress procedure to classify RP patients [58]. The authors used a linear discriminant analysis as a classification method. The best result obtained on the classification of RP, in primary and secondary reached an accuracy of 0.80. Ismail et al., performed RP classification through the use of thermal IR imaging and cold challenge procedure [19]. The overall classification outcome of 0.87 of correctly classified participants was achieved by employing a multiple logistic regression algorithm followed by a receiver operating characteristic curve analysis. Moreover, studies such as Viana et al., demonstrated a statistically significant difference between the thermal IR time course of RP patients versus healthy ones during the recovery from a cold challenge [59]. Similar results were obtained in [2,60]. Recent studies explored the possibility to classify RP patients based on their hands' temperature in baseline condition [61,62]. Horikoshi et al. demonstrated that the baseline nail fold temperature was significantly lower in RP patients than in controls [48]. Martini et al., proved that at baseline, higher temperatures at the distal interphalangeal joint and lower temperatures at metacarpophalangeal joints were observed in PRP compared to the secondary RP [61]. A considerable step forward would be the implementation of a completely automated procedure to avoid the use of a cold challenge and limit human intervention. To these aims, an automated data-driven DCNNs classifier that allows differentiation among PRP, SSc, and HC based on their hands' baseline IR image is presented in this study. The classifier's performance was compared to those of a feature-based approach. In addition, the DCNN model for IR images at rest was compared to a 3D-DCNN model fed with the hands' IR images acquired during the whole CCP. Each implemented model was cross-checked for performances evaluation.

Regarding the basal condition, the DCNN implemented was able to classify PRP, SSc, and HC with a high degree of accuracy, i.e., an overall accuracy of 0.84, an overall sensitivity of 0.83, and specificity of 0.91. On the other hand, the feature-based approaches achieved an overall accuracy of 0.77, sensitivity of 0.77, and specificity of 0.88. With respect to the CCP, the 3D-DCNN model was able to classify PRP, SSc, and HC with an overall accuracy of 0.88, sensitivity of 0.88, and specificity of 0.94, whereas the SVC delivered a classification with an overall accuracy of 0.83, a sensitivity of 0.80, and specificity of

0.90. An investigatory analysis of the feature maps resulting from the convolutional layers (Figures 3 and 5) permits an insight on the feature extraction process performed by the DCNN and 3D-DCNN models. Both models seem to focus on the difference among fingers and dorsum temperature highlighting those fingers where the difference is smaller to differentiate between HC, SSc, and PRP participants.

Although the accuracy of the classification performed on the IR imaging recorded during CCP was better than the accuracy achieved by classifying baseline IR images only, the differences were not statistically significant. This result highlights the importance of the baseline temperature as a stable physiological variable that might provide insight into the etiology of RP. Most importantly, it provides a valid, pioneering solution for differentiating PRP, SSc, and HC with performance comparable to those reported in literature but in a completely automated way and without requiring the administration of a cold challenge.

With respect to the machine learning algorithms used in this study, even though no significant difference was found between DCNN and SVC accuracy, DCNN provided better performance and was much more agile. Indeed, manually identifying and extracting features can be time consuming as well as being influenced by human errors. Moreover, the use of the DCNN model for baseline IR images classification would avoid the need for a standardized protocol, as it would only require the acquisition of a single IR image during rest condition.

4.3. Study's Limitations and Future Directions

Finally, it is worth mentioning that despite the very promising results, the low sample size can be considered a limitation of the study. In fact, the classification outcome, might increase its performance with a larger study sample, relying it on a multivariate analysis approach. Although the sample size of the study could be considered rather small, the classification was performed implementing a leave-one-out cross-validation procedure, thus basically evaluating the out-of-sample performance. Hence, the results obtained are indeed generalizable. Increasing the sample numerosity may produce a further improvement of the classifier's performance by decreasing a possible in-sample overfitting issue. Moreover, a larger sample size will allow analyzing the effect of gender and age on the classification performance. Indeed, such factors affect the skin vasoconstriction /vasodilation capacity. Therefore, a future purpose is to validate the technique on a larger sample of participants. Furthermore, the recent introduction of low-cost IR cameras can increase the availability of thermal IR imaging. Considerations on the use of such low-cost thermal technology for RP patients classification are already addressed in the literature [62,63]. In addition, the introduction of such an automated procedure based on the recording of a single IR image of the hands at rest, will possibly overcome the implementation difficulties referred to by Maverakis et al. [64]. Besides, it is likely that thermography may become more widely used given the increased accessibility of the equipment. Both these aspects could pave the way toward the application of the proposed model in everyday clinical practice. However, in this perspective, it is worth noticing that during thermal imaging measurements, the environmental conditions must be controlled and standardized as much as possible to minimize physiological variability.

5. Conclusions

In this study, an innovative automatic procedure to differentiate PRP from SSc and HC based on machine learning algorithms and IR images of the hands was presented. Different classification procedures were compared, involving the use or not of a cold challenge and the manual or automated features selection. The results revealed that the use of a cold challenge did not statistically improve the classification accuracy and that an automated feature selection approach performed better than the manual ones. Indeed, this study demonstrated that, by employing a DCNN model and IR images of the hands in basal condition, it is possible to achieve high level of accuracy and specificity in the differentiation among PRP, SSc, and HC. This approach allowed for overcoming issues related to the administration of a cold stimulus to the patients, as well as to avoid biases in the classification introduced by the operator, thus representing a great improvement with respect to the standard procedure. However, due to the limited sample size, this study is intended to provide evidence of the feasibility of such an automated approach. Further studies are needed to corroborate the results on a large-scale population. Although in literature it is reported that thermographic examinations are useful for differentiating secondary RP, such as SSc, from PRP, to the best of our knowledge this is the first study employing such an automated approach for this task. Finally, the implemented model, together with the recent introduction of low-cost thermal systems, can provide a new, quick, automated, and accurate classification method suitable for everyday clinical practice environment.

Author Contributions: Conceptualization, A.M., P.A., and C.F.; methodology, A.M., P.A., C.F., D.C., D.P., and A.M.C.; software, C.F.; validation, D.C., D.P., A.M.C., and G.G.; formal analysis, C.F., D.C., D.P., and A.M.C.; investigation, C.F., D.C., D.P., and A.M.C.; resources, A.M., P.A., and G.G.; writing—original draft preparation, C.F., D.C., and D.P.; writing—review and editing, A.M., P.A., G.G., and A.M.C.; supervision, A.M., P.A., and G.G.; project administration, A.M. All authors have read and agreed to the published version of the manuscript.

Funding: This research received no external funding.

Institutional Review Board Statement: The study was conducted according to the guidelines of the Declaration of Helsinki and approved by the Institutional Review Board and Local Ethical Committee of the School of Medicine of the University of Chieti-Pescara (protocol code AC052514 12/07/2016).

Informed Consent Statement: Informed consent was obtained from all participants involved in the study. Written informed consent has been obtained from the patients to publish this paper.

Data Availability Statement: The data presented in this study are available on request from the corresponding author. The data are not publicly available due to privacy issues.

Conflicts of Interest: The authors declare no conflict of interest. The funders had no role in the design of the study; in the collection, analyses, or interpretation of data; in the writing of the manuscript; or in the decision to publish the results.

References

1. Prete, M.; Fatone, M.C.; Favoino, E.; Perosa, F. Raynaud's Phenomenon: From Molecular Pathogenesis to Therapy. *Autoimmun. Rev.* **2014**, *13*, 655–667. [CrossRef]
2. Mariotti, A.; Grossi, G.; Amerio, P.; Orlando, G.; Mattei, P.A.; Tulli, A.; Romani, G.L.; Merla, A. Finger Thermoregulatory Model Assessing Functional Impairment in Raynaud's Phenomenon. *Ann. Biomed. Eng.* **2009**, *37*, 2631–2639. [CrossRef]
3. Ruaro, B.; Smith, V.; Sulli, A.; Pizzorni, C.; Tardito, S.; Patané, M.; Paolino, S.; Cutolo, M. Innovations in the Assessment of Primary and Secondary Raynaud's Phenomenon. *Front. Pharmacol.* **2019**, *10*. [CrossRef]
4. Maricq, H.R.; Carpentier, P.H.; Weinrich, M.C.; Keil, J.E.; Palesch, Y.; Biro, C.; Vionnet-Fuasset, M.; Jiguet, M.; Valter, I. Geographic Variation in the Prevalence of Raynaud's Phenomenon: A 5 Region Comparison. *J. Rheumatol.* **1997**, *24*, 879–889.
5. Hughes, M.; Allanore, Y.; Chung, L.; Pauling, J.D.; Denton, C.P.; Matucci-Cerinic, M. Raynaud Phenomenon and Digital Ulcers in Systemic Sclerosis. *Nat. Rev. Rheumatol.* **2020**, *16*, 208–221. [CrossRef] [PubMed]
6. Herrick, A.L. The Pathogenesis, Diagnosis and Treatment of Raynaud Phenomenon. *Nat. Rev. Rheumatol.* **2012**, *8*, 469–479. [CrossRef]
7. Kahaleh, M.B. Raynaud Phenomenon and the Vascular Disease in Scleroderma. *Curr. Opin. Rheumatol.* **2004**, *16*, 718–722. [CrossRef]
8. Abraham, D.J.; Varga, J. Scleroderma: From Cell and Molecular Mechanisms to Disease Models. *Trends Immunol.* **2005**, *26*, 587–595. [CrossRef]
9. Pauling, J.D.; Domsic, R.T.; Saketkoo, L.A.; Almeida, C.; Withey, J.; Jay, H.; Frech, T.M.; Ingegnoli, F.; Dures, E.; Robson, J.; et al. Multinational Qualitative Research Study Exploring the Patient Experience of Raynaud's Phenomenon in Systemic Sclerosis. *Arthritis Care Res.* **2018**, *70*, 1373–1384. [CrossRef] [PubMed]
10. Walker, U.A.; Tyndall, A.; Czirják, L.; Denton, C.; Farge-Bancel, D.; Kowal-Bielecka, O.; Müller-Ladner, U.; Bocelli-Tyndall, C.; Matucci-Cerinic, M. Clinical Risk Assessment of Organ Manifestations in Systemic Sclerosis: A Report from the EULAR Scleroderma Trials and Research Group Database. *Ann. Rheum. Dis.* **2007**, *66*, 754–763. [CrossRef]
11. Matucci-Cerinic, M.; Kahaleh, B.; Wigley, F.M. Evidence That Systemic Sclerosis Is a Vascular Disease. *Arthritis Rheum.* **2013**, *65*, 1953–1962. [CrossRef] [PubMed]

12. McMahan, Z.H.; Wigley, F.M. Raynaud's Phenomenon and Digital Ischemia: A Practical Approach to Risk Stratification, Diagnosis and Management. *Int. J. Clin. Rheumtol.* **2010**, *5*, 355–370. [CrossRef]
13. Merla, A.; Donato, L.D.; Luzio, S.D.; Farina, G.; Pisarri, S.; Proietti, M.; Salsano, F.; Romani, G.L. Infrared Functional Imaging Applied to Raynaud's Phenomenon. *IEEE Eng. Med. Biol. Mag.* **2002**, *21*, 73–79. [CrossRef]
14. Sousa, E.; Vardasca, R.; Teixeira, S.; Seixas, A.; Mendes, J.; Costa-Ferreira, A. A Review on the Application of Medical Infrared Thermal Imaging in Hands. *Infrared Phys. Technol.* **2017**, *85*, 315–323. [CrossRef]
15. Quesada, J.I.P. *Application of Infrared Thermography in Sports Science*; Springer: Berlin, Germany, 2017; ISBN 3-319-47409-X.
16. Perpetuini, D.; Filippini, C.; Cardone, D.; Merla, A. An Overview of Thermal Infrared Imaging-Based Screenings during Pandemic Emergencies. *Int. J. Environ. Res. Public Health* **2021**, *18*, 3286. [CrossRef] [PubMed]
17. Filippini, C.; Perpetuini, D.; Cardone, D.; Chiarelli, A.M.; Merla, A. Thermal Infrared Imaging-Based Affective Computing and Its Application to Facilitate Human Robot Interaction: A Review. *Appl. Sci.* **2020**, *10*, 2924. [CrossRef]
18. Filippini, C.; Spadolini, E.; Cardone, D.; Bianchi, D.; Preziuso, M.; Sciarretta, C.; del Cimmuto, V.; Lisciani, D.; Merla, A. Facilitating the Child–Robot Interaction by Endowing the Robot with the Capability of Understanding the Child Engagement: The Case of Mio Amico Robot. *Int. J. Soc. Robot.* **2020**, 1–13. [CrossRef]
19. Ismail, E.; Orlando, G.; Corradini, M.L.; Amerio, P.; Romani, G.L.; Merla, A. Differential Diagnosis of Raynaud's Phenomenon Based on Modeling of Finger Thermoregulation. *Phys. Meas.* **2014**, *35*, 703. [CrossRef]
20. Chand, G.; Ali, M.; Barmada, B.; Liesaputra, V.; Ramirez-Prado, G. Tracking a Person's Behaviour in a Smart House. In *Proceedings of the International Conference on Service-Oriented Computing*; Springer: Cham, Switzerland, 2018; pp. 241–252.
21. Cardone, D.; Perpetuini, D.; Filippini, C.; Spadolini, E.; Mancini, L.; Chiarelli, A.M.; Merla, A. Driver Stress State Evaluation by Means of Thermal Imaging: A Supervised Machine Learning Approach Based on ECG Signal. *Appl. Sci.* **2020**, *10*, 5673. [CrossRef]
22. Paszkiel, S. Using neural networks for classification of the changes in the EEG signal based on facial expressions. In *Analysis and Classification of EEG Signals for Brain–Computer Interfaces*; Springer: Cham, Switzerland, 2020; pp. 41–69.
23. Paszkiel, S. *Analysis and Classification of EEG Signals for Brain-Computer Interfaces*; Springer: Berlin, Germany, 2020; ISBN 3-030-30580-5.
24. Filippini, C.; Perpetuini, D.; Cardone, D.; Chiarelli, A.M.; Merla, A. *Thermal Infrared Imaging and Artificial Intelligence Techniques Can Support Mild Alzheimer Disease Diagnosis*; CEUR Workshop Proceedings: Aachen, Germany, 2020; Volume 2804, pp. 31–39.
25. Perpetuini, D.; Chiarelli, A.M.; Filippini, C.; Cardone, D.; Croce, P.; Rotunno, L.; Anzoletti, N.; Zito, M.; Zappasodi, F.; Merla, A. Working Memory Decline in Alzheimer's Disease Is Detected by Complexity Analysis of Multimodal EEG-FNIRS. *Entropy* **2020**, *22*, 1380. [CrossRef]
26. Bikmukhametov, T.; Jäschke, J. Combining Machine Learning and Process Engineering Physics towards Enhanced Accuracy and Explainability of Data-Driven Models. *Comput. Chem. Eng.* **2020**, *138*, 106834. [CrossRef]
27. Shang, C.; You, F. Data Analytics and Machine Learning for Smart Process Manufacturing: Recent Advances and Perspectives in the Big Data Era. *Engineering* **2019**, *5*, 1010–1016. [CrossRef]
28. Rawat, W.; Wang, Z. Deep Convolutional Neural Networks for Image Classification: A Comprehensive Review. *Neural Comput.* **2017**, *29*, 2352–2449. [CrossRef] [PubMed]
29. Khan, A.; Sohail, A.; Zahoora, U.; Qureshi, A.S. A Survey of the Recent Architectures of Deep Convolutional Neural Networks. *Artif. Intell. Rev.* **2020**, *53*, 5455–5516. [CrossRef]
30. Hecht-Nielsen, R.I. 3-Theory of the Backpropagation Neural Network. In *Neural Networks for Perception*; Academic Press: Cambridge, MA, USA, 1992; pp. 65–93.
31. Ronneberger, O.; Fischer, P.; Brox, T. U-Net: Convolutional Networks for Biomedical Image Segmentation. In *Proceedings of the International Conference on Medical Image Computing and Computer-Assisted Intervention*; Springer: Berlin, Germany, 2015; pp. 234–241.
32. Li, Q.; Cai, W.; Wang, X.; Zhou, Y.; Feng, D.D.; Chen, M. Medical Image Classification with Convolutional Neural Network. In Proceedings of the 13th International Conference on Control Automation Robotics & Vision (ICARCV), Singapore, 10–12 December 2014; IEEE: Washington, DC, USA, 2014; pp. 844–848.
33. World Medical Association Declaration of Helsinki. Recommendations Guiding Physicians in Biomedical Research Involving Human Subjects. *JAMA* **1997**, *277*, 925–926. [CrossRef]
34. Van Den Hoogen, F.; Khanna, D.; Fransen, J.; Johnson, S.R.; Baron, M.; Tyndall, A.; Matucci-Cerinic, M.; Naden, R.P.; Medsger, T.A., Jr.; Carreira, P.E. 2013 Classification Criteria for Systemic Sclerosis: An American College of Rheumatology/European League against Rheumatism Collaborative Initiative. *Arthritis Rheum.* **2013**, *65*, 2737–2747. [CrossRef]
35. Le Roy, E.C.; Medsger, T.A. Criteria for the Classification of Early Systemic Sclerosis. *J. Rheum.* **2001**, *28*, 1573–1576.
36. Goundry, B.; Bell, L.; Langtree, M.; Moorthy, A. Diagnosis and Management of Raynaud's Phenomenon. *BMJ* **2012**, *344*. [CrossRef]
37. Cardone, D.; Merla, A. New Frontiers for Applications of Thermal Infrared Imaging Devices: Computational Psychopshysiology in the Neurosciences. *Sensors* **2017**, *17*, 1042. [CrossRef]
38. Thermology Guidelines, Standards and Protocols in Clinical Thermography Imaging. Available online: https://www.researchgate.net/publication/273755657_Thermology_guidelines_standards_and_protocols_in_clinical_thermography_imaging (accessed on 22 September 2020).
39. Bernard, V.; Staffa, E.; Mornstein, V.; Bourek, A. Infrared Camera Assessment of Skin Surface Temperature–Effect of Emissivity. *Phys. Med.* **2013**, *29*, 583–591. [CrossRef]

40. Moreira, D.G.; Costello, J.T.; Brito, C.J.; Adamczyk, J.G.; Ammer, K.; Bach, A.J.; Costa, C.M.; Eglin, C.; Fernandes, A.A.; Fernández-Cuevas, I. Thermographic Imaging in Sports and Exercise Medicine: A Delphi Study and Consensus Statement on the Measurement of Human Skin Temperature. *J. Therm. Biol.* **2017**, *69*, 155–162. [CrossRef]
41. Chiarelli, A.M.; Bianco, F.; Perpetuini, D.; Bucciarelli, V.; Filippini, C.; Cardone, D.; Zappasodi, F.; Gallina, S.; Merla, A. Data-Driven Assessment of Cardiovascular Ageing through Multisite Photoplethysmography and Electrocardiography. *Med. Eng. Phys.* **2019**, *73*, 39–50. [CrossRef]
42. Murugan, P.; Durairaj, S. Regularization and Optimization Strategies in Deep Convolutional Neural Network. *arXiv* **2017**, arXiv:1712.04711.
43. Kohavi, R. A Study of Cross-Validation and Bootstrap for Accuracy Estimation and Model Selection. In Proceedings of the International Joint Conference on AI Palais de Congres, Montreal, QC, Canada, 20–25 August 1995; Volume 14, pp. 1137–1145.
44. Shreffler, J.; Huecker, M.R. Diagnostic Testing Accuracy: Sensitivity, Specificity, Predictive Values and Likelihood Ratios. In *StatPearls*; StatPearls Publishing: Treasure Island, FL, USA, 2020.
45. Ji, S.; Xu, W.; Yang, M.; Yu, K. 3D Convolutional Neural Networks for Human Action Recognition. *IEEE Trans. Pattern Anal. Mach. Intell.* **2012**, *35*, 221–231. [CrossRef]
46. Jin, T.; Cui, H.; Zeng, S.; Wang, X. Learning Deep Spatial Lung Features by 3D Convolutional Neural Network for Early Cancer Detection. In *Proceedings of the 2017 International Conference on Digital Image Computing: Techniques and Applications (DICTA)*; IEEE: Washington, DC, USA, 2017; pp. 1–6.
47. Maturana, D.; Scherer, S. Voxnet: A 3d Convolutional Neural Network for Real-Time Object Recognition. In Proceedings of the 2015 IEEE/RSJ International Conference on Intelligent Robots and Systems (IROS); IEEE: Washington, DC, USA, 2015; pp. 922–928.
48. Horikoshi, M.; Inokuma, S.; Kijima, Y.; Kobuna, M.; Miura, Y.; Okada, R.; Kobayashi, S. Thermal Disparity between Fingers after Cold-Water Immersion of Hands: A Useful Indicator of Disturbed Peripheral Circulation in Raynaud Phenomenon Patients. *Intern Med.* **2016**, *55*, 461–466. [CrossRef]
49. Lachenbruch, P.A.; Lynch, C.J. Assessing Screening Tests: Extensions of McNemar's Test. *Stat. Med.* **1998**, *17*, 2207–2217. [CrossRef]
50. Kellogg, D.L., Jr. In Vivo Mechanisms of Cutaneous Vasodilation and Vasoconstriction in Humans during Thermoregulatory Challenges. *J. Appl. Phys.* **2006**, *100*, 1709–1718. [CrossRef] [PubMed]
51. Wigley, F.M. Vascular Disease in Scleroderma. *Clin. Rev. Allergy Immun.* **2009**, *36*, 150–175. [CrossRef]
52. Suter, L.G.; Murabito, J.M.; Felson, D.T.; Fraenkel, L. The Incidence and Natural History of Raynaud's Phenomenon in the Community. *Arthritis Rheum.* **2005**, *52*, 1259–1263. [CrossRef]
53. Carpentier, P.H.; Satger, B.; Poensin, D.; Maricq, H.R. Incidence and Natural History of Raynaud Phenomenon: A Long-Term Follow-up (14 Years) of a Random Sample from the General Population. *J. Vasc. Surg.* **2006**, *44*, 1023–1028. [CrossRef] [PubMed]
54. Tiso, F.; Favaro, M.; Ciprian, L.; Cardarelli, S.; Rizzo, M.; Tonello, M.; Ruffatti, A.; Cozzi, F. Digital Ulcers in a Cohort of 333 Scleroderma Patients. *Reumatismo* **2007**, *59*, 215–220. [CrossRef] [PubMed]
55. Nihtyanova, S.I.; Brough, G.M.; Black, C.M.; Denton, C.P. Clinical Burden of Digital Vasculopathy in Limited and Diffuse Cutaneous Systemic Sclerosis. *Ann. Rheum. Dis.* **2008**, *67*, 120–123. [CrossRef]
56. Guiducci, S.; Giacomelli, R.; Cerinic, M.M. Vascular Complications of Scleroderma. *Autoimmun. Rev.* **2007**, *6*, 520–523. [CrossRef]
57. Love, T.J. Thermography as an Indicator of Blood Perfusion. *Ann. N. Y. Acad. Sci.* **1980**, *335*, 429–437. [CrossRef]
58. De Campos, M.F.; Ripka, W.L.; Campos, D.; Heimbecher, C.T.; Esmanhoto, E.; Ulbricht, L. Raynaud's Phenomenon Differentiating After Cold Stress Using Thermal Parameters from Fingers. In *Proceedings of the XXVI Brazilian Congress on Biomedical Engineering*; Springer: Berlin/Heidelberg, Germany, 2019; pp. 869–874.
59. Viana, J.R.; Campos, D.; Ulbricht, L.; Sato, G.Y.; Ripka, W.L. Thermography for the Detection of Secondary Raynaud's Phenomenon by Means of the Distal-Dorsal Distance. In *Proceedings of the 2020 42nd Annual International Conference of the IEEE Engineering in Medicine & Biology Society (EMBC)*; IEEE: Washington, DC, USA, 2020; pp. 1528–1531.
60. Lim, M.J.; Kwon, S.R.; Jung, K.-H.; Joo, K.; Park, S.-G.; Park, W. Digital Thermography of the Fingers and Toes in Raynaud's Phenomenon. *J. Korean Med. Sci.* **2014**, *29*, 502. [CrossRef]
61. Martini, G.; Cappella, M.; Culpo, R.; Vittadello, F.; Sprocati, M.; Zulian, F. Infrared Thermography in Children: A Reliable Tool for Differential Diagnosis of Peripheral Microvascular Dysfunction and Raynaud's Phenomenon? *Pediatr. Rheum.* **2019**, *17*, 1–9. [CrossRef]
62. Herrick, A.L.; Murray, A. The Role of Capillaroscopy and Thermography in the Assessment and Management of Raynaud's Phenomenon. *Autoimmun. Rev.* **2018**, *17*, 465–472. [CrossRef]
63. Herrick, A.L.; Dinsdale, G.; Murray, A. New Perspectives in the Imaging of Raynaud's Phenomenon. *Eur. J. Rheum.* **2020**, *7*, S212–S221. [CrossRef] [PubMed]
64. Maverakis, E.; Patel, F.; Kronenberg, D.G.; Chung, L.; Fiorentino, D.; Allanore, Y.; Guiducci, S.; Hesselstrand, R.; Hummers, L.K.; Duong, C.; et al. International Consensus Criteria for the Diagnosis of Raynaud's Phenomenon. *J. Autoimmun.* **2014**, *48–49*, 60–65. [CrossRef] [PubMed]

applied sciences

Article

Investigating Issues and Needs of Dyslexic Students at University: Proof of Concept of an Artificial Intelligence and Virtual Reality-Based Supporting Platform and Preliminary Results

Andrea Zingoni [1,*], Juri Taborri [1], Valentina Panetti [1], Simone Bonechi [2], Pilar Aparicio-Martínez [3], Sara Pinzi [4] and Giuseppe Calabrò [1]

1 Department of Economics, Engineering, Society and Business Organization (DEIM), University of Tuscia, 01100 Viterbo, Italy; juri.taborri@unitus.it (J.T.); valentina.panetti@unitus.it (V.P.); giuseppe.calabro@unitus.it (G.C.)
2 Department of Computer Science, University of Pisa, 56100 Pisa, Italy; simone.bonechi@unitus.it
3 Department of Nursing, Physiotherapy and Pharmacology, University of Córdoba, 14001 Córdoba, Spain; n32apmap@uco.es
4 Department of Chemistry, Physics and Applied Thermodynamic, University of Córdoba, 14001 Córdoba, Spain; sara.pinzi@uco.es
* Correspondence: andrea.zingoni@unitus.it

Citation: Zingoni, A.; Taborri, J.; Panetti, V.; Bonechi, S.; Aparicio-Martínez, P.; Pinzi, S.; Calabrò, G. Investigating Issues and Needs of Dyslexic Students at University: Proof of Concept of an Artificial Intelligence and Virtual Reality-Based Supporting Platform and Preliminary Results. *Appl. Sci.* **2021**, *11*, 4624. https://doi.org/10.3390/app11104624

Academic Editors: João M. F. Rodrigues, Pedro J. S. Cardoso and Marta Chinnici

Received: 20 April 2021
Accepted: 14 May 2021
Published: 19 May 2021

Featured Application: The outcomes of this work can represent a turning point toward a more and more inclusive university environment for dyslexic students, at the same time showing the full potential of artificial intelligence and virtual reality in dealing with issues related to education.

Abstract: Specific learning disorders affect a significant portion of the population. A total of 80% of its instances are dyslexia, which causes significant difficulties in learning skills related to reading, memorizing and the exposition of concepts. Whereas great efforts have been made to diagnose dyslexia and to mitigate its effects at primary and secondary school, little has been done at the university level. This has resulted in a sensibly high rate of abandonment or even of failures to enroll. The VRAIlexia project was created to face this problem by creating and popularizing an innovative method of teaching that is inclusive for dyslexic students. The core of the project is BESPECIAL, a software platform based on artificial intelligence and virtual reality that is capable of understanding the main issues experienced by dyslexic students and to provide them with ad hoc digital support methodologies in order to ease the difficulties they face in their academic studies. The aim of this paper is to present the conceptual design of BESPECIAL, highlighting the role of each module that composes it and the potential of the whole platform to fulfil the aims of VRAIlexia. Preliminary results obtained from a sample of about 700 dyslexic students are also reported, which clearly show the main issues and needs that dyslexic students experience and these will be used as guidelines for the final implementation of BESPECIAL.

Keywords: specific learning disorders; dyslexia; artificial intelligence; virtual reality; adaptive learning; inclusive teaching

1. Introduction

According to the classification of the World Health Organization (WHO) [1], specific learning disorders (SLDs) are neurodevelopmental disorders characterized by significant and persistent difficulties in learning skills, which may include reading, writing and performing calculations. This leads to an incomplete automation of such processes, which is likely to affect scholarly and academic life significantly and even generate forms of psychological distress, especially when the problem is not detected early enough [2,3]. Nevertheless, following the work of [1], SLDs can be grouped into four categories, depending on the impaired learning skills: dyslexia, which affects skills related to reading; dysgraphia, which affects skills related to writing; dyscalculia, which affects skills related to arithmetic; and a fourth group including all those disorders that affect other skills. Among them, dyslexia is absolutely the most common [4]. The difficulties associated with it involve not only word reading accuracy and reading fluency but also, as a consequence, comprehension, memorization, concepts exposition [1] and the ability to take notes, compose text and organize the study activity [5]. It is straightforward to understand that the learning process of a dyslexic student is very likely to be compromised.

Over the years, a lot has been done to diagnose dyslexia. Diagnosis has usually relied on some specific tests that aim to quantify reading difficulties, jointly with clinical tools that measure cognitive abilities. Dyslexia is diagnosed for reading performances that, in terms of speed and accuracy, are below the fifth percentile or below two standard deviations with respect to the mean, in the presence of normal cognitive abilities. The focus of such tests has been especially on primary school students. In the past few years, however, tools that are also targeted to secondary school and university students have been created, including the LSC-SUA test [6] and the Adult Dyslexia Battery (BDA) [7]. Classical and widely used diagnostic tests consist of reading aloud meaningful and meaningless words. More novel approaches, however, are based on silent reading, fused passages (namely, reporting when spaces between words are missing) and dys-words (namely, reading words that contain numerous spelling errors but are still recognizable as if they were spelled correctly) [6].

The advent and the wide spread of information technology (IT) has also positively impacted the problem of dyslexia diagnosis and novel and interesting approaches exploiting digital technologies have been proposed. These can be broadly grouped into two categories [8]: The first one comprises those approaches based on neurological data analysis, which aim at spotting the differences in brain anatomy, organization and functioning that correlate with the presence of the typical symptoms caused by dyslexia [9,10] by employing modern screening techniques and novel algorithms to enhance their output. Significant evidence of anomalies in dyslexic people's cerebral morphology and operative processes has been found through the analysis of 3D scans of the brain obtained with magnetic resonance imaging (MRI) [11–15] and functional magnetic resonance imaging (fMRI) [16–19], respectively. In addition, different behaviors between dyslexic and non-dyslexic subjects have been observed in the frequency [20], entropy [21] and spatial activation patterns [22,23] of electroencephalogram (EEG), a widely used technique to assess human concentration [24]. Interesting results have also been obtained with the tracking of the movements of the eye during the act of reading (easily and cheaply performable at the state-of-the art level thanks to the progress in the design and production of eye-trackers [25] and tracking techniques [26–28]), which have demonstrated that the saccades of the readers differ in number and amplitude, depending on if they have dyslexia or not [29–32]. The second category consists of those approaches that revise and improve classical testing methodologies, using the potential offered by IT. These, in turn, can be divided into three groups, on the basis of the aspects they aim at improving: administration of the tests, choice of the most predictive ones and analysis of the results. The first group is focused on presenting the most consolidated tests for dyslexia diagnosis by means of platforms or digital tools that allow the facilitation and speeding up of the collection of the necessary data [33–36]. These platforms/tools can also provide specialists with real-time information and help them monitor the specific objectives step by step, in order to constantly provide the chance to compare results and revise previous evaluations, increasing the probability of a correct final diagnosis [33]. In addition, their capability to gather and store data can also be exploited for research purposes [35]. This kind of approach is particularly useful for the assessment of dyslexia during childhood, since it allows the administration of tests as a set of serious games that are appealing to the children and hide the sensation of being under evaluation [33–35]. In light of its peculiar features, virtual reality (VR) has proven to be a powerful tool to achieve this goal and, in fact, its use in this area is increasing [36,37]. VR also provides a control environment that submerges the user in a controlled but relaxed structure, making it possible to carry out screening tests and decreasing emotional distress [38]. The second group aims at selecting the best tests among the available ones [39,40]. To do this, fuzzy logic, artificial intelligence (AI) and genetic algorithms are generally employed to reach a more significant joint interpretation of the scores of the typical dyslexia screening tasks [39] and to exclude the less predictive ones [40]. The third group is instead focused on jointly analyzing several dyslexia assessment tests, in order to build an automatic predictor of the presence/absence of the disorder. Again, AI perfectly suits this purpose, both for the large amount of data provided by the tests and considering the capability of machine learning (ML) algorithms to find significant relationships between the tested features and obtained results. Numerous promising works have used AI profitably; for example, in [41], the results of the Gibson test of brain skills were used to train an artificial neural network (ANN) and an accuracy predictor of almost 90% was implemented. A similar approach was followed in the works of [42,43], but instead relying, respectively, on human–computer interaction measures and on a self-evaluation questionnaire about difficulties in speaking, reading, spelling and writing. In the works of [44,45], instead, state-of-the-art ML algorithms were trained on a wide battery of tests, exceeding 90% accuracy.

It is not surprising that a much has been done to improve the diagnosis of dyslexia, since it is universally acknowledged that recognizing it early, especially before the beginning of the school, is crucial to help affected people fill their learning gap [46–48]. However, we are still far from having a rigorous, systematic and widespread methodology to spot it at pre-school age [49]; therefore, the development of supporting strategies and tools for dyslexic people whose condition is found late is of paramount importance. With this in mind, several methodologies have been developed to improve reading skills in terms of both accuracy and speed. For example, in the work of [50], a training method based on phonetic instructions was presented and defined as the only statistically effective method. In the work of [51], a combination of the cognitive training of executive functions with a phonological-based treatment has proven a significant method for the rehabilitation of dyslexic subjects. Other approaches, instead, focused on using a rhythmic background [52] or even music [53] for a sublexical training. It is interesting to point out that the vast majority of these studies agree that providing support to dyslexic people also helps them in building a personal and creative study method that follows their own learning styles, thus being even more effective. Although the efforts in this direction are multiplying year after year, not enough has been done yet. One of the largest gaps is the integration of IT, which seems particularly suitable for the purpose but has not yet been fully exploited.

2. Related Works

As mentioned, IT solutions have not yet reached their great potential in supporting students with dyslexia, even if the efforts in this direction are gradually increasing. The works of [54,55] raise the problem of proposing an ontology to facilitate the development of e-learning tools aligned with the needs of dyslexic students, but no practical supporting solutions are introduced. In the works of [56,57], instead, the readability of websites for dyslexic people is analyzed and improved, but this can be considered only as a first step toward their full inclusion in the learning system. A second step is performed in [58], where web applications are also taken into account, but only the preliminary phase of a dyslexia-friendly collaborative learning system is presented. A further advance has been made with the introduction of specific and sophisticated tools aimed at increasing the impaired skills. For example, in the work of [59], using a computer platform equipped with speech synthesis and eye tracking was a great boost to the comprehension of a text by dyslexic subjects. Several works have explored the use of AI, demonstrating that it is likely to be one of the most effective instruments to face the problem. The ability of AI to predict future situations by learning from available data makes it a powerful tool to analyze a great amount of information. Further, its improvement over the years has led to very effective techniques capable of dealing with a lot of different data at the same time. Nowadays, multivariate, quantitative, qualitative and ranked data can be used together to train AI algorithms that can automatically find hidden relationships among them, leading to a deeper understanding of the examined phenomenon. It appears clear how AI can be used to support people affected by dyslexia. In the work of [60], a supervised ANN is used to model the reading ability of dyslexic primary school students, in order to give specific support to each one. Network training is performed on the data about the main deficits caused by dyslexia and the results demonstrate that only personalized models can correctly profile dyslexic students and be helpful for them. In the work of [61], a hidden Markov model predicts the difficulties in learning the Malay language in primary school students affected by dyslexia by tracking their mistakes in solving phonology, spelling, reading and writing exercises. The results are not reported but, again, the need for individual support interventions is pointed out. A similar approach is adopted in the work of [62] but, in this case, students' behavior is also considered in training the model. A computer-assisted learning system is developed on the basis of the output predictions and a 60% improving of dyslexics' skills with respect to classical supporting tools was achieved. The aim of the works of [63,64] is to implement an assistive learning platform for primary and secondary school students. The former is focused on helping the process of reading. First, optical character recognition (OCR) is employed to capture the text; then, an ANN identifies two classes of words: easy and difficult ones; finally, the second ones are highlighted, spelled, pronounced or accompanied by images and synonyms in order to make their comprehension easier. ML techniques are used to decide which kind of support is better, depending on the words encountered. The latter, instead, is focused on both reading and writing abilities. A support vector machine (SVM) predictor is constantly trained on the basis of the dyslexic students' scores in serious games and exercises, whose types and difficulty levels are adaptively changed, depending on the obtained performance. Improvements of students' skills and engagement in learning—along with parents' and therapists' capabilities of monitoring progress—are reported. As previously mentioned, a powerful IT tool in the education field is VR [65], thanks to the opportunity it provides to present totally customizable and appealing contents (study material, activities to be performed, tests, etc.), which can be received by the users in an immersive environment—typically created by means of wearable helmets, provided of near-eye displays (NED) and headphones—, increasing their engagement and, thus, their attention. In addition, it allows monitoring easily progresses and obstacles encountered during the learning process. Virtual environments have been included in education since the beginning of the century, especially to reproduce real-life situations [66]. More recently, they have been employed also for training purposes [67]. Previous works have investigated their application in several learning tasks, like improving linguistic abilities and assess the attention of the reader, by also relying on eye-trackers included in VR helmets [68]. In addition, the results of the study of [69] have indicated that VR can improve memory and audio–visual abilities. Its application has also been prompted by the advances in the dedicated hardware technology, which have led to the development of cheap but effective devices. Several works have thus explored the use of VR to increase dyslexic students' memory performance [70] and reading abilities [69], obtaining better performance than using classical or less immersive training methods. It is worth nothing that the vast majority of the works related to supporting dyslexic learners are targeted to pre-school, primary and secondary school age. Methodologies and tools to help them at university level are almost totally missing or provide just a partial aid [71]. However, as discussed in depth in the next chapter, dyslexic university students often experience the typical problems given by this disorder and see their academic career slowed down or even spoiled.

The project VRAIlexia (virtual reality and artificial intelligence for dyslexia) has been launched to face this problem. Starting from the necessity to untap dyslexic students' potential and enhance their strengths, it aims to develop learning tools and services for to ensure to them equal access and opportunity of success during their career and their lifelong learning experience. Among the several activities provided for by VRAIlexia project, the main one consists of the design and implementation of the software platform

BESPECIAL, the role of which is to provide dyslexic students at university with digital supporting tools that are specific for each of them and, thus, much more effective, in order to decisively reduce the problems they usually encounter and facilitate their academic career. To do this, BESPECIAL will be trained right on the individual issues and needs of the students, exploiting the powerful means of AI—to learn which are the best tools for each and deliver them automatically—and VR— to administer evaluation test and, at the same time, monitor progress and weak points and provide a constant feedback to the AI.

In this paper, the conceptual design of BESPECIAL will be presented, focusing on the background situations of dyslexic students, which have suggested the necessary intervention of VRAIlexia project and the design choices of the software platform (Section 3); the role of each module of the platform (Section 4); the preliminary results that will be used as a guideline for the future implementations (Section 5). Section 6 concludes and discusses the next steps.

3. Background, Motivation and Purpose of the VRAIlexia Project

The decision to develop the VRAIlexia project and implement the BESPECIAL platform was born from a case study (referred as "Tuscia case study") conducted at the University of Tuscia (an Italian academic institution, whose headquarters are located in Viterbo, Italy) concerning the number, distribution and academic outcomes of students with SLDs. The results, considered jointly with the global statistics about the topic, pointed out some crucial issues about the academic opportunities of such students.

According to the works of [4,72], between 6.25% and 12.50% of the world population experiences SLDs, which means around 875 million people worldwide. However, the datum is likely to be underestimated, since some evidence [73] makes the percentage increase to up to 21.25%, that is, around 1.6 billion people. Such numbers give an idea of the scale of this phenomenon. Among them, about 80% are dyslexic, though the remaining 20% suffer from other SLDs [4]. It is also important to note that two or more disorders may occur together. However, dyslexia alone is the most common learning disorder, with a percentage of between 35% and 50% of SLD cases [74].

An investigation by Italian Ministry of Education and Ministry of University and Research that was carried out in 2019 [75] reported a similar distribution of the SLD typologies within the students of national primary and secondary schools; of all the students with SLDs, 34% are affected only by dyslexia (Figure 1).

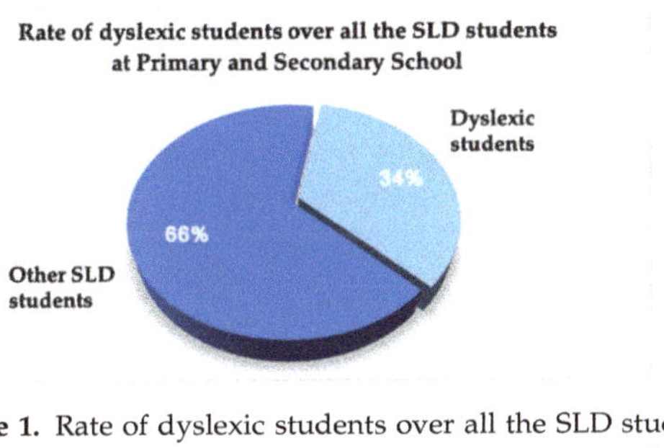

Figure 1. Rate of dyslexic students over all the SLD students attending Italian primary and secondary school, according to the 2019 report of Ministry of Education and Ministry of University and Research [75].

The report also showed that students with SLDs constitute 3.2% of the total students at primary and secondary school, whereas at university, they constitute only 1.2% (Figure 2a). This drastic decrease clearly highlights how university can become an insurmountable wall for them and how some actions in this regard are necessary to mitigate the main problems they experience during academic life. According to the Tuscia case study, the percentage of university students with SLDs is slightly higher (Figure 2b), but consistent with [75].

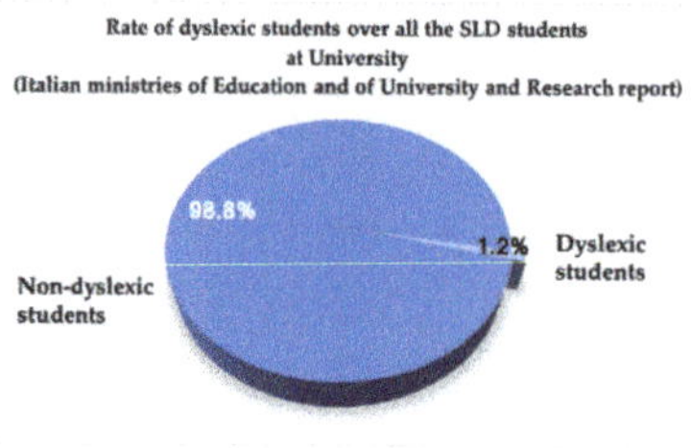

(a)

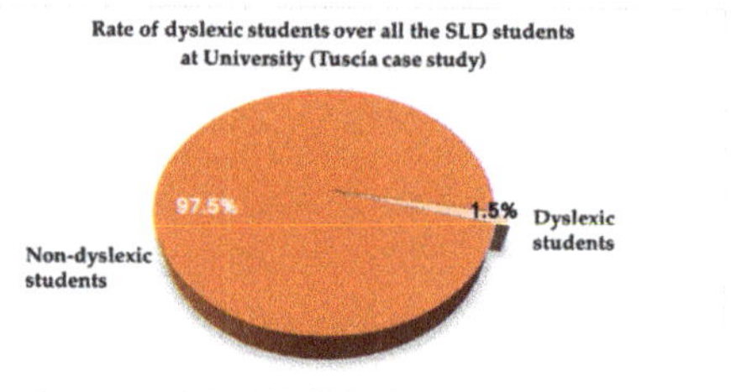

(b)

Figure 2. Rate of dyslexic students attending university according to the 2019 report of Ministry of Education and Ministry of University and Research [75] (**a**) and to Tuscia case study (**b**).

The rate of dyslexic students over all the students with SLDs is also in line with both primary/secondary school datum and the statistic at global level, showing that this disorder affects 39% of academic population (Figure 3) and is the most frequent one.

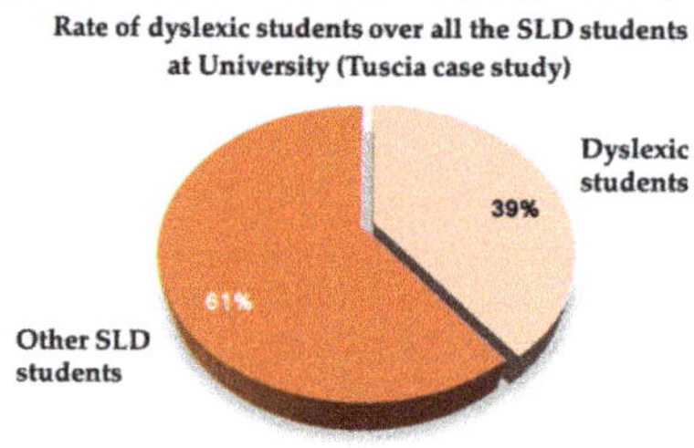

Figure 3. Rate of dyslexic students over all the SLD students attending university, according to the Tuscia case study.

Another interesting fact is that dyslexic students attending a degree course in the sciences constitute 44.3% of the total number of students with SLDs. Instead, they constitute only 25.8% in degree courses in humanities (Figure 4). This matches with the typical problems that dyslexics experience, which have an adverse impact on tasks that are more commonly required in the second branch, like reading, memorizing, exposing concepts aloud, etc. The conspicuous presence of such tasks is likely to discourage the enrolment in humanities courses or leads to their abandonment. Further evidence comes from the graph in Figure 5, which shows the average number of university credits in the European Credit Transfer and Accumulation System (ECTS) achieved per year by dyslexic and non-dyslexic students, both globally and specifically for sciences and humanities. The global data show that dyslexic people accumulate, on average, six ECTS less, which correspond, approximately, to one exam per year lost compared to non-dyslexics. This illustrates a situation of hardship for dyslexic students that must be faced thoroughly, in order to guarantee equal opportunities to them. The specific data for sciences and humanities confirm the above regarding the greater difficulties encountered by dyslexic students in the second branch. The difference in the achieved ECTS rises up to 6.7 for humanities, compared to 4.6 for sciences.

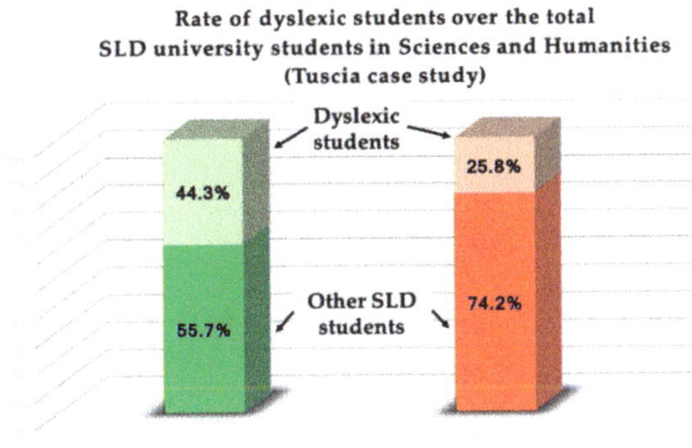

Figure 4. Rate of dyslexic students over the total number of SLD university students in sciences and humanities, according to the Tuscia case study.

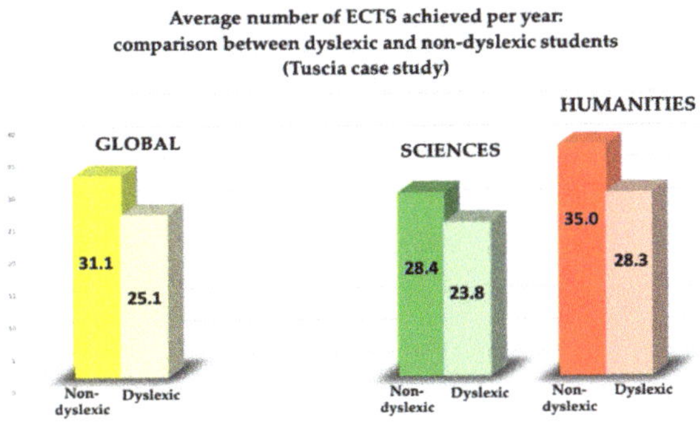

Figure 5. Average number of ECTS achieved per year: comparison between dyslexic and non-dyslexic students globally and specifically for sciences and humanities, according to the Tuscia case study.

Summarizing the statistics shown above, four main facts stand out: (i) people with SLDs are a significant percentage of the global population; (ii) the vast majority of them experience dyslexia; (iii) the presence of this disorder affects the academic career in terms of both graduation time dilation and abandonments; (iv) this problem strikes mainly degree courses in humanities.

The VRAIlexia project starts from these assumptions and, within the first three years of the project, intends to achieve five tangible outcomes to overcome all the main difficulties encountered by dyslexic students at university and eliminate, or at least reduce, the gap with respect to non-dyslexic students. More specifically, in addition to the software platform BESPECIAL, which represents both the core of the project and the focus of this paper, the other four outcomes are the ones listed below.

1. A battery of tests in VR to assess the skills of dyslexic students, which will allow teachers to better and more easily understand the issues of them.
2. An online shared repository containing all the digital modules, resources, tools and any other kind of material that can be useful to implement innovative teaching and learning methods.
3. A training path consisting in two events, the first one addressed to the improvement of dyslexia awareness from teachers and the second one addressed to dyslexic students to enhance their self-entrepreneurial mindsets. Both events will be organized by experts of various disciplines, according to the Universal Design Learning methodology [76].
4. A memorandum of understanding for the creation of common inclusion strategies among European higher education institutions.

It is easy to understand how these outcomes have the potentiality to represent a valid opportunity to eliminate or at least mitigate significantly the above-reported issues, related to the enrolment of dyslexic students in university. In addition, they will allow to define a standardized approach, which is likely to increase students' motivation to complete their career and to move the first steps in the work market. The high level of intricacy of the SLDs spectrum requires not dealing with all the disorders at the same time, but focusing on each singularly. Since dyslexia is absolutely the prevailing SLD among the world population, VRAIlexia will focus on it. In addition, the intrinsic diversity in teaching/learning different disciplines [77] suggests thinking of specific methodologies for each of them. Humanities being the disciplines more challenging for dyslexic students, only them will be considered within VRAIlexia. An extension to the other SLD types and to scientific disciplines has been already projected and is likely to be performed at a later time.

The previously listed assumptions, jointly with the information obtained from the literature analysis, have also been taken into account in the design of BESPECIAL that, as mentioned, is in charge of providing dyslexic students with ad hoc strategies and tools to support them during university, so as to try to mitigate the career slowdown and abandonments phenomenon. In order to produce specific material for each student (a fundamental factor to achieve the goal [60,61]), the platform will be implemented starting from the individual issues they experience and according to their precise needs. Given the focus on humanities, the produced material and methodologies will cover the typical problems related to the main tasks of these disciplines. In addition, the BESPECIAL output is not meant to be delivered only to students in the form of digital tools, but also to teachers and university institutions in the form of strategies and best practice, which will simplify the academic path for dyslexics, making it really inclusive for everyone.

4. Conceptual Design of BESPECIAL

To deal with the complexity of the task BESPECIAL must perform, it is advisable to develop it in two stages.

4.1. First Development Stage

In the first stage, the basics of BESPECIAL will be implemented, so as to build the backbone of the platform. The kernel consists of an AI-based module that will be employed to assess the dyslexia status of the users, starting from both their clinical reports and a questionnaire about the problems they feel while studying, plus the solutions they deem to be helpful. From the results of the assessment, the best supporting tools and strategies will be predicted. Then, the former will be used to digitalize the study material and provided directly to the users, whereas the latter will be passed to the teachers and, in general, to the university institutions to enable them to help students in the best way by implementing standardized guidelines. The training of the AI module will be performed from a large database of clinical reports of dyslexic students and from their answers to the above-mentioned questionnaire. This allows simultaneous consideration of both the evaluation made by the experts and the self-evaluation of dyslexic subjects. At this level, the provided material will not yet be specific for each one of the students singularly, but still they will be divided in some broad categories created on the basis of the relations between their issues and their needs, which will have been spotted by the AI. The group of tools and strategies that best fits those students belonging to a specific category will be assigned to them. This preliminary classification also allows the simplifying the work of the AI assessment module of the second stage, which will have to switch from being category-specific to be student-specific.

The block diagram of the first development stage of BESPECIAL is depicted in Figure 6, whereas its modules and steps are described in detail in the following subsections.

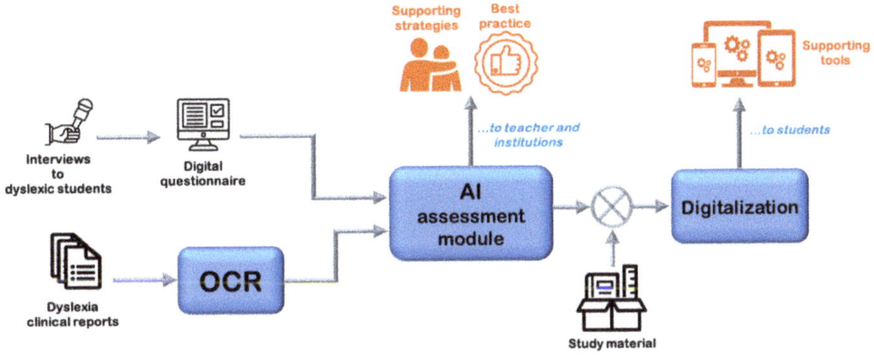

Figure 6. Block diagram of BESPECIAL at its first stage.

4.1.1. Interviews to Dyslexic Students

The very first step for the development of BESPECIAL consists of carrying out semi-structured interviews with dyslexic students with the aim of investigating aspects concerning metacognition and learning methods, which will be used to build a self-assessment questionnaire that will allow students to describe their study issues and the supporting strategies and tools that each one finds most useful. A sample of 20 dyslexic university students will thus be interviewed on a voluntary basis. The distribution between male and female subjects will be random. Students will be asked to answer to a group of questions concerning their study method, the main difficulties they experienced in their university learning path and the tools and strategies they have applied and found helpful. Finally, the answers will be analyzed, in order to create a list of typical issues and needs of dyslexic university students, which will be then used to design the questionnaire. The obtained data will not be aggregated at this step, so as to maintain a wider analysis capability. Possible aggregations can be done later.

4.1.2. Digital Questionnaire

On the basis of the information gathered from the interviews, a questionnaire will be created and then digitalized and hosted online, so as to significantly speed up the collection of the data about the self-evaluation of issues and needs of dyslexic students. The questionnaire will be organized as follows. After a few demographical questions (age, gender, etc.), information about the high school and university career and about the dyslexia status and history will be asked. The answers will be useful mostly for the objectives of VRAIlexia other than BESPECIAL implementation. The software platform, however, could also benefit from this information by finding unexpected relations between it and dyslexic students' problematics and needs, which could be interesting additional outcomes of the project. Then, three groups of questions will be asked: (i) which have been the main issues experienced during the last years of the learning path; (ii) which have been the most useful supporting tools; (iii) which have been the most useful supporting strategies. Each group will be organized in multiple choice questions, each of which concerning one of the issues/tools/strategies that emerged from the interviews to dyslexic students. The choice consists of a score from 1 (very little) to 5 (very much), depending on the severity of the issue or on the utility of the supporting methodology, plus the option "not experienced" (for the issues) or "not useful" (for the tools and strategies). In addition, the groups of questions about supporting methodologies will also present the options "never tried" and "don't know". This will allow the AI to distinguish the reason why a certain methodology is not regarded as helpful. In Figure 7, two pictures showing two of the above-mentioned inquiries from the questionnaire are reported by way of an example.

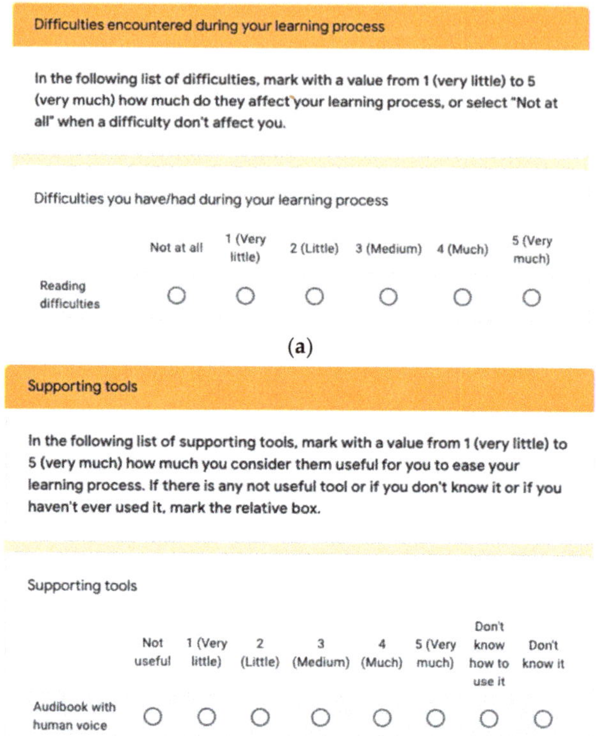

(**b**)

Figure 7. Example of two inquiries of the questionnaire, about the issues experienced (**a**) and the supporting methodologies considered as useful (**b**) by dyslexic students.

The answers to these three groups of questions will be the main input of BESPECIAL. A large amount of them will, indeed, be used initially to train the AI assessment module and build the predictor of the most suitable supporting methodologies, given the issues experienced by dyslexic students. Later, instead, the answers of each student will be read to formulate the prediction.

4.1.3. Application of Optical Character Recognition to the Dyslexia Clinical Report of the Students

The clinical reports of dyslexic students are the second input to BESPECIAL. From them, the information given by the experts about the status of the disorder for each student will be collected. In order to automate the extraction of this information, an algorithm based on optical character recognition (OCR) will be devised. Medical reports are generally not released in a digital format, requiring to be photographed to be passed to a computer, or, however, their format does not permit easy modifications (like pdf). The platform must, thus, be able to take files in different formats as input and interpret them. The system will then use the OCR algorithm provided by Google Vision API to recognize and extract the text of the medical reports from such files. After that, the three typical steps of text pre-processing will be applied to the text extracted with OCR. These are: removing stop words, tokenization and stemming. The first one aims at removing commonly used words that are not useful to characterize the content of the document, such as articles and conjunctions. The second one, instead, splits the text into single words (tokens), so as to ease the content analysis. Finally, the third one aims at reducing each word to its root, allowing inflected forms of the same word to be grouped and then treated as a single element. Once the clinical reports will be processed by the OCR module, useful information can be extracted from them. For example, it may be possible to understand the type of dyslexia of each subject or their specific needs.

4.1.4. AI Assessment Module

The input coming from the questionnaire and the clinical reports will be passed to the AI assessment module, first to train it and then, after the training will be completed, to provide those pieces of information about BESPECIAL users that allow predicting the most suitable supporting material for each of them. To develop the module, ML techniques will be used. At this stage, indeed, it is not advisable to rely on deep learning (DL), since the input (namely, the issues experienced by dyslexic students) and the output (the supporting tools and strategies) variables are not a huge number. Using DL is, thus, likely to result in data overfitting. At the second stage, instead, when a much higher number of variables will have to be processed, the possibility to switch to DL techniques will be taken into account. The good practice of testing several ML algorithms will be followed. In particular, supervised algorithms will be considered, since the questionnaire and the clinical reports provide labeled output variables, for which the label is given by the supporting tools and strategies. The typical and state-of-the-art set of ML algorithms will be used. It comprises naïve Bayes, logistic regression, k-nearest neighbors, random forest, gradient boosting tree, SVM and ANN [78]. After training them and build a predictor for each, cross-validation will be carried out, so as to choose the best performing one. Accuracy and precision metrics (like AUC, F-measure, etc. [79]) will be taken into account as performance criteria. To take account of the possible presence of highly variable data, the algorithms testing will be run on different database setups, namely, rearranging data in different ways. For example, a different ranking scales for the answers to the questionnaire could be adopted, like scores from 1 to 3 (obtained by opportunely aggregating the original scores from 1 to 5), or the questions could be considered as "yes/no" ones (by collapsing the original 1 to 5 scores into "yes" and keeping the "no" answer). Similarly, different ways to aggregate and rank the information gained from the clinical reports will be considered. Furthermore, protocols to manage possible missing data will be developed. This step is necessary for two reasons. The main one is that clinical reports are compiled by human beings and guidelines on how to do it correctly are still lacking. This results in documents that are often and considerably incomplete. The other reason is that the questionnaire features the option "never tried" and "don't know" among the answers about the most useful supporting methodologies. Obviously, these answers cannot be considered the same as "not useful" and, thus, they must be treated as missing data. All the typical strategies to address this issue will be applied, like simple deletion, imputation and prediction and the techniques that will give the best result will be implemented. Once the AI module will have made its prediction, the results will be split into two parts. The most suitable tools for the user will be passed to the digitalization module, whereas the most suitable strategies will be one of the final output of BESPECIAL platform, which will be provided to the institutions and the teachers to orient their interventions in favor of the students.

4.1.5. Digitalization of the Supporting Tools

As previously mentioned, the tools that best fit the students' needs, according to the AI module prediction, will be passed to another module that is in charge of their digitalization. This operation will allow full exploitation of the benefits offered by the modern IT devices (PCs, tablets, smartphones, e-books, etc.), by creating supporting material that is not only easy-to-use and appealing for the users, but also customizable and portable. The digitalization will involve the study material of the university courses, which will also be passed as input to the module and which the tools will be applied to. To define the techniques to be employed to achieve the goal, it is of course necessary to know which supporting tools must be implemented. Nevertheless, some steps will almost certainly be required in any case, given the nature of the typical study material of humanities branch.

The first one is the transformation of the texts into ASCII format, which is a necessary initial operation, in order to have the possibility to treat them digitally. Again, OCR will be used for this purpose. Then, the ASCII strings will be passed to a language detection algorithm. This will allow improvement of the accuracy of the OCR output, by inferring the specific feature of the detected language. Once a text is readable by a computer, the possibility for the user to modify some display settings will be implemented, following the suggestions in [80]. In particular, they will be able to specify style, size and spacing of the font and to communicate which syllables and words cause reading or understanding difficulties, so as to have them highlighted in different colors. The selected syllables/words will be identified within the text by means of regular expressions, which are sequences of characters that specifies patterns of search, thus finding all the repetitions of the targeted string. The audio reproduction of the documents will also be implemented by relying on speech synthesis algorithms based on DL models. The detection of the language that has been obtained previously will also be useful to this end. To meet the user's needs better, the possibility to set preferences on how the audio should be played will be given. These will include the choice of the speaking voice and the audio speed. The digitalized tools applied to the study material constitute the second final output of BESPECIAL. It will be provided to the students in order to support them during their academic career.

4.2. Second Development Stage

In the second stage, the features developed in the first one will be enhanced, leading to the final version of BESPECIAL platform, which is shown in Figure 8.

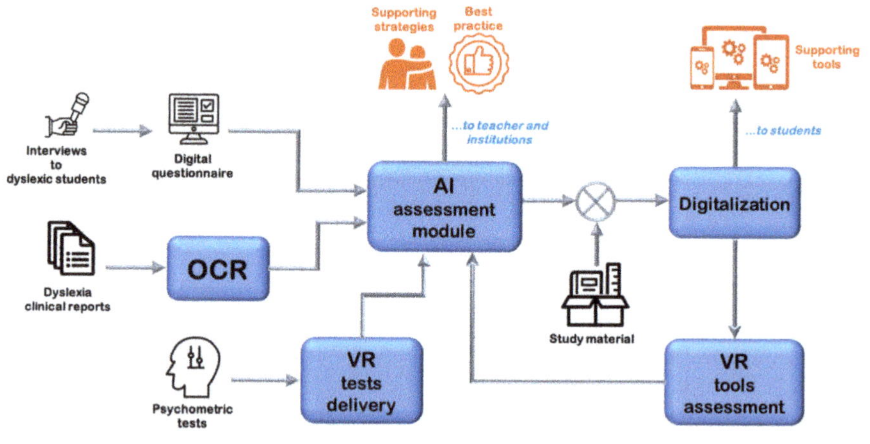

Figure 8. Block diagram of the final implementation of BESPECIAL.

In particular, two aspects will be improved, namely, the input data collection and the level of specificity of the provided supporting material for each student. Concerning the former, a new input source will be introduced, namely, the results of a battery of psychometric tests, which will be delivered using VR to ease their accessibility and improve students' engagement. Concerning the latter, a further assessment module will be added, which will evaluate the response of the users to the digital supporting tools and will give a feedback about their usefulness to the AI module, in order to guide its prediction towards the creation of an increasingly customized material for each one. This step will allow us to switch from category-specific to student-specific supporting tools, which should give a decisive contribution to facilitate the career of dyslexic students. Again, VR will be exploited, given the possibility it offers to easily and quickly monitor users' progress, skills and weak points.

An in-depth description of the new features introduced in this second and last implementation stage is presented in detail in the next subsections.

4.2.1. Psychometric Tests and Their Delivery via VR

As previously mentioned, a new input from dyslexic students will be collected in the final version of BESPECIALBESPECIAL, that is, the information coming from a battery of psychometric tests. The introduction of psychometric tests has been deemed necessary in order to have also an objective initial profile of the users, allowing comparison of different people in a unitary manner. In addition, the tests will enable the progress of dyslexic students to be tracked by monitoring the obtained scores over time. It is worth noting that not only learning performance, but also psychological aspects related to the learning process, such as anxiety, self-esteem, self-efficacy, strategies and motivations for studying will be assessed. Psychometric tests will be divided into five batteries. The first one is the previously mentioned BDA. It is targeted at 16 to 30 years old dyslexics and consists of 11 unpublished tests that evaluate three kinds of skills, namely, reading, writing and texts comprehension. In particular, BDA investigates the flexibility of the process of reading aloud and in silent mode, providing specific tasks to assess the degree of automation when writing and introduces a culture free, multiple choice answer-free reading comprehension trial. Its results allow the creation of detailed operating profiles of the assessed skills. The second battery is the State-Anxiety Inventory (STAI-Y), devised in the middle of the 1960s and revised in the early 1980s [81]. It is split into two scales (Y1 and Y2), which evaluate state anxiety and trait anxiety, respectively. The assessment of the former is performed through questions related to how the subject feels at the time of administering the questionnaire, whereas the assessment of the latter relies on questions that investigate how the subject feels habitually. For the purpose of BESPECIAL, only Y1 will be performed. The third battery measures the General Self-Efficacy Scale [82] and consists of 10 items aimed at explaining various cognitive and motivational aspects related to learning, including the impact of positive experiences and successes, perseverance in commitment, optimism and the development of interests in specific cultural and professional fields. The fourth battery is the Questionnaire on Learning Processes [83], in the D version (QPA-D), which is the one targeted at university students. It uses three scales that investigate the following aspects.

1. Intrinsic motivation for learning (MI), in which those who are engaged in learning regularly perform their school duties and progress harmoniously in all disciplines obtain high scores.
2. Metacognition and self-regulated learning (MA), in which those who are aware of their cognitive processes and, hence, manage their learning process effectively obtain high scores.
3. Learning strategies (SA), in which those that are capable to adopt good and effective strategies to process contents and information obtain high scores.

The fifth battery is the Rosenberg Self-Esteem Scale (R-SES) [84], which is aimed at measuring self-esteem and consists of 10 items related to overall feelings of self-worth or self-acceptance. The items are answered on a four-point scale ranging from strongly agree to strongly disagree.

An intrinsic problem of the introduced psychometric tests lies in their format, which tends to be too formal and repetitive. Moreover, such tests require a lot of time to be completed. This causes people to get bored easily and, thus, to lose concentration and perform them without commitment. This issue is even more severe for dyslexics, whose problems affect exactly concentration and understanding. To overcome this, psychometric tests will be delivered by relying on VR. Its characteristics, indeed, allow recreating an interactive environment with high fidelity, thus presenting any material in a playful and appealing way, which is likely to improve engagement and attention [38]. In addition, tests performing time will be reduced and the obtained results will be collected and stored more easily and in a convenient way for their use in the AI module. Two main approaches can be followed to deliver contents via VR, depending on the type of devices used, namely, immersive virtual reality (IVR) and desktop virtual reality (DVR). IVR is capable of offering a complete immersion of the user in the virtual environments by using specific hardware known as a VR headset, which consists of a helmet provided with two head-mounted displays and headphones. The displays allow for 3D vision and isolate the user's sight from the real world, as well as headphones for the user's hearing. Coherence between the user's and virtual images movements are ensured by inertial sensors equipped on the helmet. Hand controllers are also provided often, in order to improve the interaction with the virtual world. The result is the projection of multiple images that configure a room-size venue for experiencing VR and creating a flow through the scenario, thus generating the full immersion experience. Unfortunately, IVR suffer a major limitation due to the high costs and the low accessibility of the needed devices. Conversely, DVR does not need complex and expensive hardware to be played, since it runs on common devices, like personal computers, tablets and smartphones, which are more sustainable for higher education institutions [85]. As a counterpart, it can provide only a static and less interactive experience, making the sense of complete immersion be lost. Some cheap helmets, into which a smartphone can be inserted, however, can partially render it affordable. The majority of the previous studies are based on IVR [86]. Nevertheless, some novel works are focused on DVR environments [85]. Inspired by the second ones, BESPECIAL VR psychometric tests will be created by relying on DVR. They will thus, be suitable for ubiquitous devices, like smartphones and tablets, allowing for a wider and quicker spread. JSON and PhoneGap framework will be used to develop the VR module, including PHP and HTML, combined with MySQL, CSS, JavaScript and JQuery, as complementary base languages. The created VR environment will present the tests in a calming background, made of soft colors and arcade game elements. Cheap cardboard helmets for common smartphones to be inserted into will be used to reduce the field of view of the user, in order to enhance the degree of immersion perceived. The results of psychometric tests will then be passed to the AI assessment module and used jointly with the results of the questionnaire and the information extracted from the clinical reports, first to train the AI and, then, to predict which is the best supporting material for each user.

4.2.2. VR Tools Assessment Module

The second module based on VR will have in charge to assess the response of each dyslexic student to the provided digital supporting tools. The assessment result will then be passed to the AI module, which will combine it with the BESPECIAL input data, so as to create completely ad hoc tools to be exploited during their academic career. Analogously to the previous subsection concerning psychometric tests, in this case, the evaluation process can also be long and boring for the users. Using VR can, thus, make it less heavy, ensuring a much higher level of engagement and, therefore, a more accurate assessment. In addition, VR itself can become an interesting supporting tool, enhancing the interest of the students, promoting autonomy and increasing self-esteem. A further application is that it can create a virtual environment where the main difficulties and the emotional distress of dyslexic students is simulated, enabling teachers and students without learning disorders to experience it, in order to improve understanding of the dyslexia and teacher-to-student empathy. This can provide useful strategies to enrich the BESPECIAL outcome of the best practice. In this phase, the IVR approach will be preferred to DVR one, since tools assessment will last much longer than psychometric tests and, thus, an increased level of engagement is required. Furthermore, fewer students will participate in the training phase; therefore, the higher cost of the needed devices will be compensated by the low number. VR will also be particularly useful in this phase because of its capability to collect and store data quickly. A comparison between the use of DVR and IVR in the two different phases is shown in Table 1, jointly with the main advantages and drawbacks.

Table 1. Use of the two VR approaches (DVR and IVR) within BESPECIAL.

VR Approach	Devices Needed	Targeted Users	Aim	Pros	Cons
DVR	Smartphone + Cardboard headset	Dyslexic Students	Psychometric tests delivering	Accessible hardware Low cost devices	Lack of immersivity
IVR	VR specific headset	Dyslexic and non dyslexic students	Supporting tools assessment	Completely immersive experience	Complex hardware required
		Teachers	Creation of an emphatic experience of dyslexia	Improved capability of data collection	Expensive devices

4.2.3. New Version of the AI Assessment Module

In this second stage of BESPECIAL, not only the input data from clinical reports, questionnaire and psychometric tests, but also the feedback from the VR tools assessment module will be used to predict the best supporting tools for each dyslexic student. The AI assessment module will thus have to be modified accordingly. Even if it is not possible to design the new version of the module before knowing which these tools are and, thus, how the VR module will evaluate them, the amount of data on which the AI will be trained is expected to be great. Big data and DL techniques will therefore be likely to be employed, to manage it and to build the predictor.

5. Results

At present, the interviews of dyslexic students have been carried out and the questionnaire has been created and released online. The following tables report the three groups of questions that will be passed as input to BESPECIALBESPECIAL, that is, those concerning the issues of dyslexic students (Table 2) and those concerning the tools (Table 3) and the strategies (Table 4) considered helpful. Note that they are derived directly from the information gathered through the interviews and that they will be presented to the users in the form shown in Figure 7. To date, more than 800 answers to the questionnaire in Italian have been collected. An analogous collection will be made in other languages, namely, Spanish, English, French, Portuguese, Greek and Dutch, since the partners of the VRAIlexia project are from countries where these languages are spoken. An extension to further languages will be evaluated after the end of the project. Initially, the data will be considered separately for each language. It has been shown, indeed, that the issues related to dyslexia can change considerably from a language to another [87]. Nevertheless, possible analogies will be investigated.

In addition to the answers to the questionnaire, a large number of clinical reports has been collected and OCR has been applied to each, in order to extract useful data for the AI module. By analyzing the reports, however, it was realized that, unfortunately, only a few contain the necessary information about the issues and needs of people affected by dyslexia. The vast majority are limited to a coarse division into wide categories like "impairment in writing" or "impairment in reading", etc., or even to a simple statement about the presence or the absence of the disorder. Thus, it was decided to only use report information to: (i) verify automatically if a subject is affected by dyslexia and, therefore, if their answers to the questionnaire can be inserted in the AI training; (ii) make an initial sketch of the categories in which AI will group dyslexic people together to provide them ad hoc supporting material. These categories will then be refined by using the data from the questionnaire and, later, also from the psychometric tests.

Table 2. List of the questions about dyslexic students' issues, asked within the BESPECIAL questionnaire.

Have You Ever Experienced One or More of These Issues during Your Academic Career?		
Reading difficulties	Text comprehension difficulties	Uncommon words understanding
Concentration difficulty while studying	Concentration difficulty during in-class lessons	Concentration difficulty during online lessons
Verbal short-term memory impairment	Verbal long-term memory impairment (memory loss during exams)	Study scheduling
Note-taking difficulties	Lack of time to prepare exams	

Table 3. List of the questions about the most useful supporting tools for dyslexic students, asked within the BESPECIAL questionnaire.

Are One or More of These Tools Useful to Support You with Dyslexia Related Issues?		
Audiobook with human voice	Audiobook with artificial voice	Words in different colors
Clear layout of the study material	Highlighted keywords	Digital concept maps
Digital schemes	Summaries	Digital Tutor
Use of images for words memorization and understanding	Use of images for concepts memorization	Audio recording of the lessons
Video lessons	Integrating study material using internet	

Table 4. List of the questions about the most useful supporting strategies for dyslexic students, asked within the BESPECIAL questionnaire.

Are One or More of These Strategies Useful to Support You with Dyslexia Related Issues?		
Someone that reads the study material	Repeating studied material	Study groups
Tutor	Participating or creating students' associations to exchange information	On-line lessons availability
Pauses during lessons	Lessons slides availability	Recording lessons
Early availability of courses programme	Dividing exams in multiple shorter modules	Only written exams
Only oral exams		

A preliminary analysis was made on the collected data. A total of 807 students with SLDs sent their clinical reports. Using the above-described approach with OCR, 114 of them were discarded, since dyslexia did not appear among their disorders. The remaining 693 students completed the questionnaire, providing information about their difficulties related to dyslexia (see Table 2) and the supporting tools and strategies they find useful (see Tables 3 and 4). Figure A1 (hosted in Appendix A for better clarity) shows the results about the importance given by the students to each issue they have encountered during their university career. The first detachable fact is that low scores (less importance) are less frequent than high scores (more importance). This demonstrates how dyslexic people, on average, still experience a lot of difficulties in higher education and, thus, how supporting them is of paramount importance. Further evidence comes from the number of answers affirming that a specific issue is not present, which is always less than all the other answers for all the assessed issues. To better understand which are the most affecting issues, the average score from 1 (very little experienced issues) to 5 (very much experienced issues) was calculated for each issue and reported in Table 5 jointly with the percentage of students that had not experience that issue.

Table 5. Average score of the experiencing of dyslexia-related issues, given by the students in the questionnaire, from 1 (very little experienced) to 5 (very much experienced) and percentage of students that has not experience them.

Issue	Average Score	Not Experienced by (%)
Reading difficulties	3.18	8.9%
Text comprehension difficulties	3.18	6.5%
Uncommon words understanding	3.30	7.2%
Concentration difficulty while studying	3.76	2.6%
Concentration difficulty during in-class lessons	3.07	7.5%
Concentration difficulty during online lessons	3.76	5.8%
Verbal short-term memory impairment	3.46	3.2%
Verbal long-term memory impairment	3.35	4.0%
Study scheduling	3.37	11.1%
Note-taking difficulties	3.32	7.2%
Lack of time to prepare exams	3.57	3.5%

It turns out that concentration is the most strongly striking difficulty, but only when the students are alone, that is, while studying and during online lessons. The problem is indeed less felt during lessons in the classroom. Verbal memory impairments also strike quite strongly, especially in the short-term, as well as scheduling problems, which prompt students to ask for more time to prepare the exams. Reading and text comprehension difficulties are present, but weigh less significantly. This is probably because the majority of the students have already found compensative learning strategies by university age.

In Figure A2 (see Appendix A), the results of the students' self-assessment about the most useful supporting tools within the list of Table 3 are reported. The average score for each tool and the percentage of students that have not found it suitable for their need are shown in Table 6. Highlighted keywords and a clear layout of the study material are the supporting tools that best fit dyslexic students' needs, followed by the use of images, summaries, concept maps and schemes. This points out that classical study material, generally consisting of books with long and visually monotonous texts, is likely to be a considerable obstacle in presence of dyslexia, even at university age. A more straight-to-the-point material, in which the basic concepts and their relations are summed up and emphasized in a simple and clear way, alternative to large verbal explanations, should thus be provided. Audio recording lessons have also been indicated as a useful supporting tool, confirming that the auditory channel can be preferred to the visual one, as shown in [52,53]. Despite all of this, in this case, a monotonous presentation is not helpful, as the low score of audiobooks (which are generally less lively than the speech of a teacher during a lesson) demonstrate. In particular, the use of audiobooks read by an artificial voice obtained the lowest score and more than half of the students that participated in the questionnaire have found it useless. This pairs with the 25.4% of unfavorable opinions by the digital tutor in giving a clear indication of how interacting with machines instead of human beings should be avoided when teaching people affected by dyslexia. It is worth noting that using different colors for the words of a text, which is strongly recommended in childhood [55,56], is still a helpful supporting tool at university age for almost 9 out of 10 people, but is less important compared to other tools.

Table 6. Average score given by dyslexic students in the questionnaire to the usefulness of each supporting tool, from 1 (very little useful) to 5 (very much useful) and percentage of students that found it useless.

Supporting Tool	Average Score	Not Useful for (%)
Audiobook with human voice	3.25	26.8%
Audiobook with artificial voice	2.28	51.7%
Words in different colors	3.61	10.2%
Clear layout of the study material	4.02	4.6%
Highlighted keywords	4.24	2.5%
Digital concept maps	3.80	7.9%
Digital schemes	3.80	8.0%
Summaries	3.94	6.1%
Digital Tutor	3.34	25.4%
Use of images for words memorization and understanding	3.90	7.1%
Use of images for concepts memorization	4.00	4.8%
Audio recording of the lessons	3.82	6.2%
Video lessons	3.67	9.2%
Integrating study material using internet	3.65	7.8%

The same analysis made for the supporting tools was also carried out for the supporting strategies listed in Table 4. Results are reported in Figure A3 (see Appendix A) and in Table 7.

Table 7. Average score given by dyslexic students in the questionnaire to the usefulness of each supporting strategies, from 1 (very little useful) to 5 (very much useful) and percentage of students that found it useless.

Supporting Tool	Average Score	Not Useful for (%)
Someone that reads the study material	4.05	3.3%
Repeating studied material	4.16	2.7%
Study groups	3.39	11.0%
Tutor	3.56	10.4%
Participating or creating students' associations to exchange information	3.86	5.8%
On-line lessons availability	4.22	1.4%
Pauses during lessons	4.49	1.0%
Lessons slides availability	4.14	3.3%
Recording lessons	4.04	3.8%
Early availability of courses programme	4.15	2.5%
Dividing exams in multiple shorter modules	3.27	19.0%
Only written exams	3.17	15.3%
Only oral exams	3.69	13.3%

The very first observation that deserves to be made is that, on average, the score achieved by supporting strategies is higher than the one achieved by supporting tools. This fact highlights the necessity to combine both supporting methodologies in order to be capable of helping dyslexic students during their academic career. Conversely, the vast majority of approaches proposed until now [56–65] have only taken into account digital tools, neglecting to also create, in parallel, a list of the best practices to be followed to provide support to dyslexic students. The supporting strategies that obtained the highest scores are taking pauses during lessons, hosting lessons online, repeating the studied material and making the program and the slides of the course available, which are considered useful by about 96 to 99% of the participants in the questionnaire. Having someone read the study material and recording the lessons also achieved high scores. This, again, underlines the importance of the auditory channel in the learning process of dyslexic students. A lower score was obtained by tutors and study groups, even if about 9 of every

10 students recognize a certain utility in such supporting strategies. Finally, oral exams are preferred to written ones (3.69 points against 3.17) and dividing them into shorter modules is not considered as fundamental (19% of the students thinks it is useless, a high percentage compared to the other strategies).

6. Conclusions and Next Steps

In this paper, the conceptual design of a supporting software platform (BESPECIAL) to help dyslexic students during their academic career has been presented jointly with preliminary results. The functioning of BESPECIAL and its role within the wider project (VRAIlexia), aimed at mitigating the difficulties encountered by dyslexic students at university, have been addressed in detail, in order to show a proof of concept of the overall methodology that will lead to its final implementation.

The platform takes as input the clinical reports of dyslexia, the answers to a self-assessment questionnaire and the results of a battery of psychometric tests to extract useful information about the issues and needs of dyslexic students at university. By relying on AI, it will be capable of predicting automatically which of the supporting methodologies are most suitable for each student, in terms of both best practices to be followed by teachers and institutions and digital tools to make the study material more accessible, given the difficulties they encountered during their academic career. A two-stage implementation of the platform is planned. In the first stage, AI will be trained with a significant number of the above-mentioned input data, so as to create a preliminary version of the predictor that can be considered category-specific, since it will be based on statistical data. In the second stage, each student will test the digital supporting tools and their reactions, improvements and residual difficulties will be evaluated. The results of the evaluation will feed back the AI, so as to improve the predictor by transforming it from category- to student-specific, as strongly recommended by experts to most usefully support people with dyslexia. The evaluation modules will be implemented in VR, given its capability to deliver material in a more engaging way and, at the same time, to easily gather the required information. This should help avoid problems such as reading impairments and lack of concentration, which often strike people with dyslexia. VR will also be used to simulate the difficulties encountered by them, in order to make teachers more conscious about the phenomenon and allow them intervene in order to mitigate it.

At present, data from about 700 dyslexic students have been collected. A preliminary analysis has been carried out and first results about the most severe difficulties and the most helpful supporting tools and strategies have been obtained. In particular, concentration when alone is the issue that most affects dyslexic students, followed by memory impairments, especially in short-term and scheduling problems. Moreover, in general, all the presented issues significantly affect (average score higher than 3 out of 5 for all of them) the majority of dyslexic students (at least more than 88.9% of the students experience each issues), highlighting that providing them with proper support is fundamental. Concerning the supporting methodologies, highlighted keywords, a clear layout of the study material and use of images, summaries, concept maps and schemes have been proven to be the most useful tools, with an average score equal or higher than 3.8 out of 5. Conversely, audiobooks read by artificial voices and digital tutors are considered useless by about 1 in every 2 and 1 in every 4 students, respectively, underlining that interacting with machines instead of human beings should be avoided. Taking pauses during lessons, hosting lessons online, repeating the studied material and making the program and the slides of the course available, instead are considered the most suitable supporting strategies (average score higher than 4 out of 5) and should, thus, be taken into account as the best practices to follow. Furthermore, the auditory channel should be exploited more, by automatically or manually recording lessons, using audiobooks with a human voice, having someone read aloud the study material and preferring oral exams to written ones. Finally, it is worth noting that supporting strategies obtained, on average, a higher score than supporting tools. This should prompt us to overcome the numerous approaches that have relied exclusively on digital tools and have not proposed any best practices to be shared with institutions and teachers. Such results will be exploited as useful guidelines to improve the conceptual and technical choices that will be made for the final implementation of BESPECIAL.

The next implementation step will be the training of the AI module responsible for the automatic prediction of the most useful tools and strategies, which will be performed once further data are available. The tools will be digitalized in the meantime. Then, the development of the second and last stage will begin by adding the VR functionalities to deliver psychometric tests and assess the digital tools.

Future work will concern the extension of the platform to several languages and to scientific disciplines and the support to other SLDs, aiming at reaching a full inclusion of all students in academic life. In addition, the data collected through the questionnaire about demographical characteristics and dyslexia history of the students will be exploited both to create social statistics to be widespread, in order to raise awareness on dyslexia phenomenon, and to perform further analysis that may improve the AI predictor.

Author Contributions: Conceptualization, A.Z., J.T. and G.C.; methodology, A.Z., J.T., V.P., S.B., P.A.-M., S.P. and G.C.; design of the software, A.Z., J.T, S.B., P.A.-M. and S.P.; formal analysis, A.Z. and J.T.; investigation, A.Z., J.T., V.P. and G.C.; writing—original draft preparation, A.Z.; writing—review and editing, A.Z., J.T., V.P., S.B., P.A.-M., S.P. and G.C.; supervision A.Z., J.T. and G.C.; funding acquisition, G.C. All authors have read and agreed to the published version of the manuscript.

Funding: This research was co-funded by European Union Committee within the Erasmus+ Programme 2014–2020—Key Action 2: Strategic Partnership Project (Agreement n. 2020-1-IT02-K203-080006—P.I.: Prof. Giuseppe Calabrò).

Institutional Review Board Statement: Ethical review and approval were waived for this study, since no experimentation has been carried out on human beings. The involvement of humans has been limited to the completion of a questionnaire, for which informed consent has been regularly obtained.

Informed Consent Statement: Informed consent was obtained from all subjects involved in the study.

Acknowledgments: The VRAIlexia project has received funding from the European Committee and the authors would like to acknowledge them for the financial support to this study. In addition, the authors want to acknowledge Estro s.r.l. for its support in the design of BESPECIAL and, in particular, of the tools digitalization module.

Conflicts of Interest: The authors declare no conflict of interest.

Appendix A

This appendix hosts some of the figures mentioned in Section 5, which are quite large. They have been thus reported here for the sake of clarity.

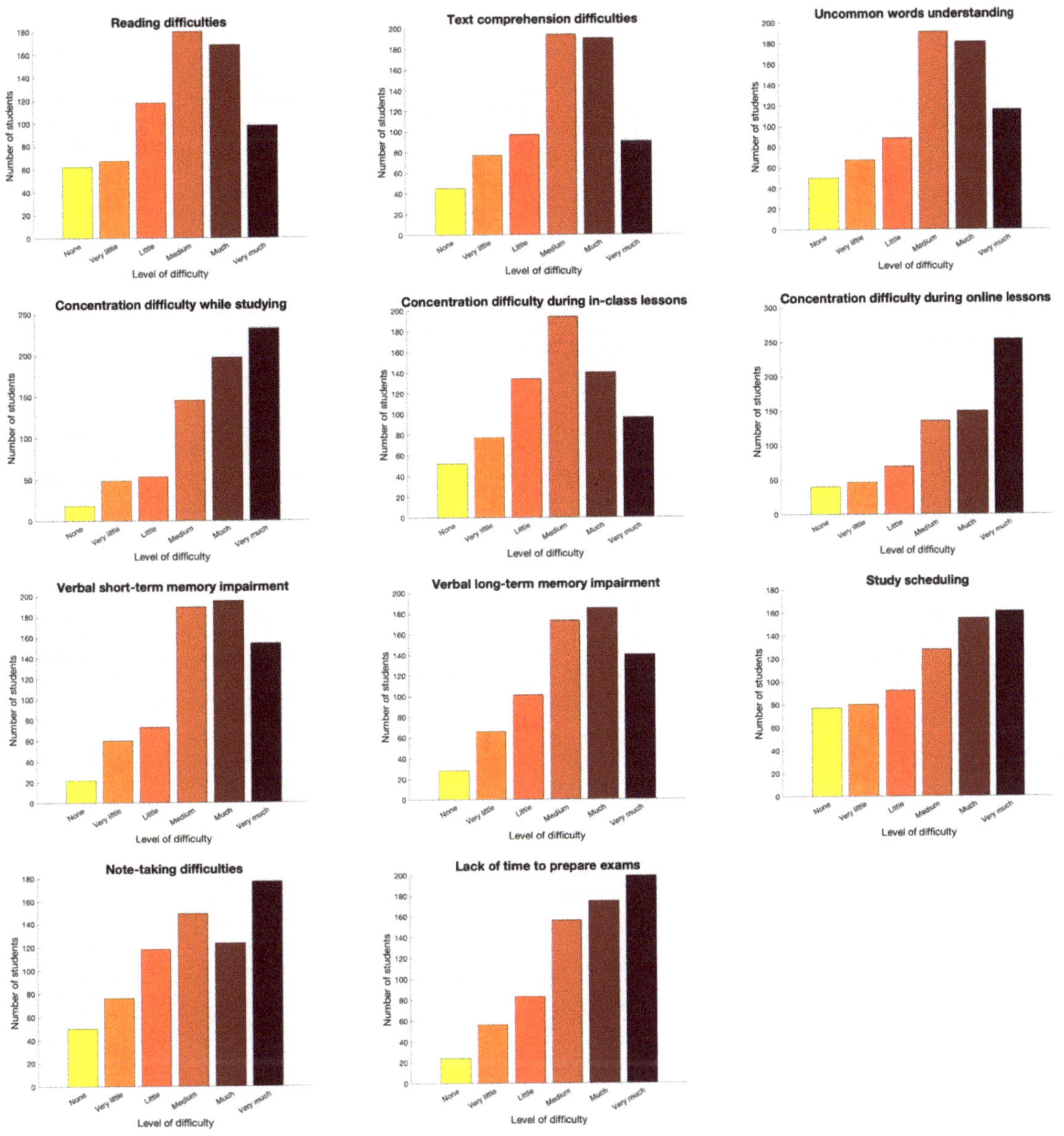

Figure A1. Histograms of the scores given by dyslexic students to the importance of each issue present in the questionnaire.

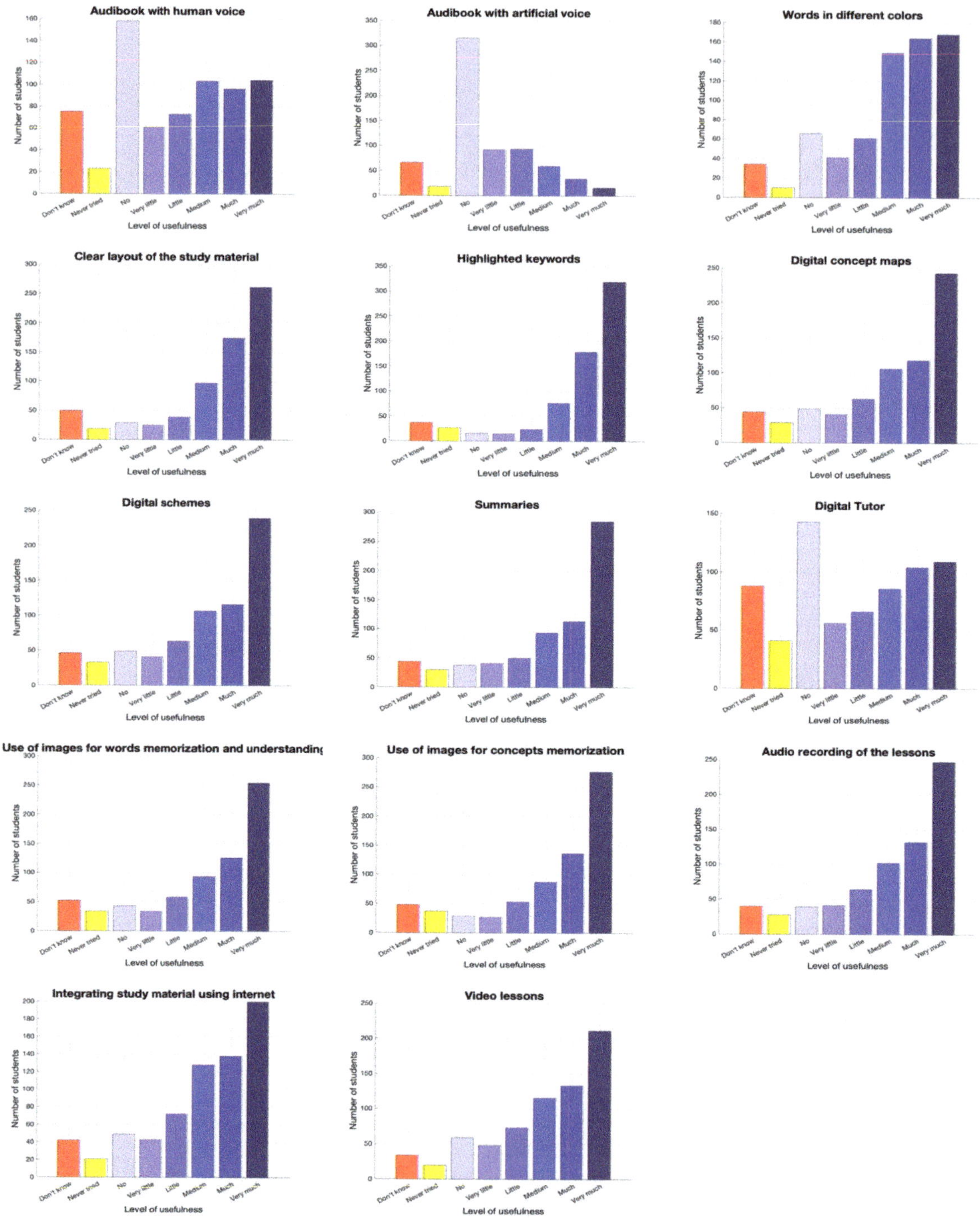

Figure A2. Histograms of the scores given by dyslexic students to the usefulness of each supporting tool present in the questionnaire.

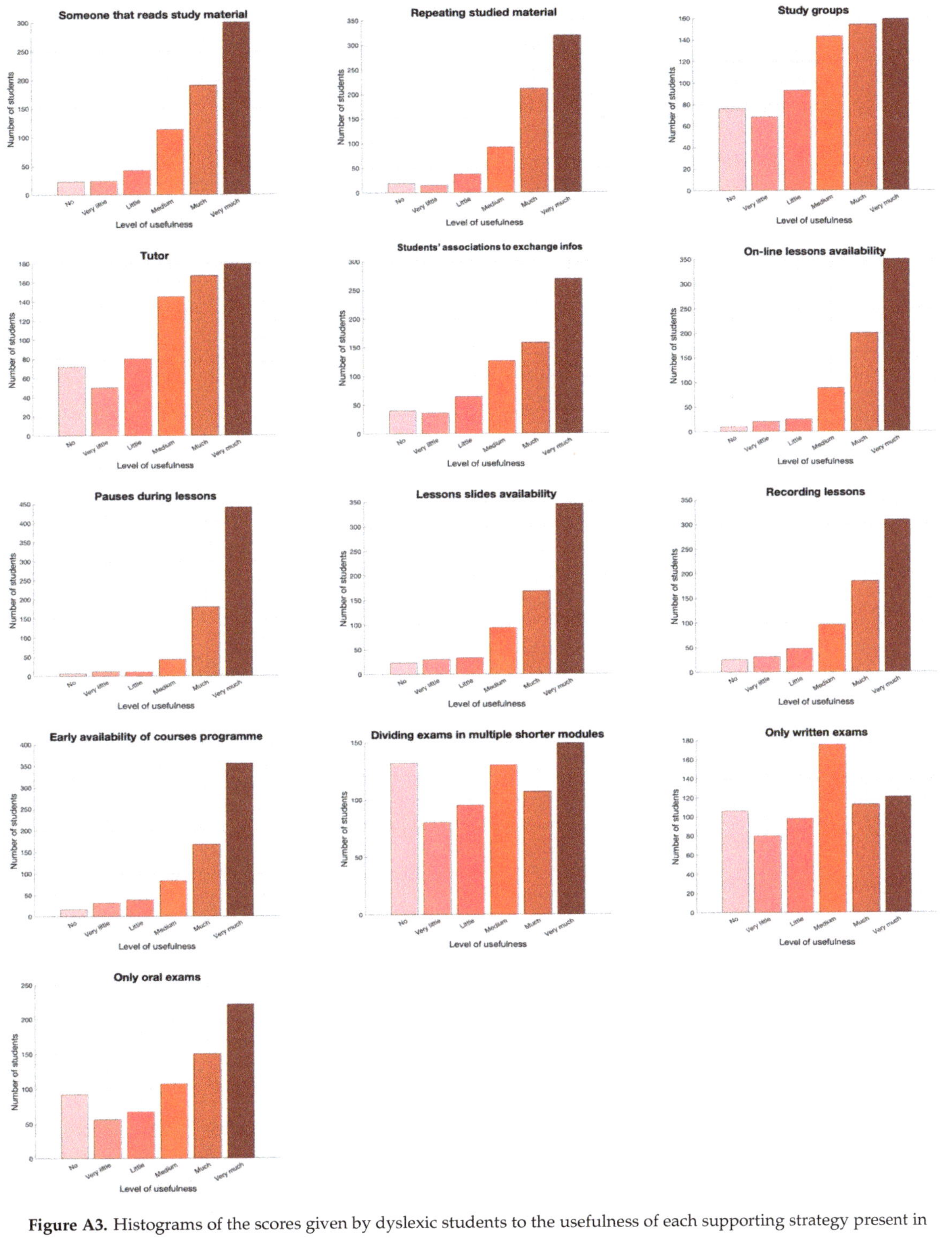

Figure A3. Histograms of the scores given by dyslexic students to the usefulness of each supporting strategy present in the questionnaire.

References

1. World Health Organization. 6A03-Developmental Learning Disorder. In *International Classification of Diseases and Related Health Problems (ICD-11)*, 11th ed.; World Health Organization: Geneva, Switzerland, 2005; Available online: https://icd.who.int/browse11/l-m/en#/http://id.who.int/icd/entity/2099676649 (accessed on 14 February 2021).
2. Carawan, L.W.; Nalavany, B.A.; Jenkins, C. Emotional experience with dyslexia and self-esteem: The protective role of perceived family support in late adulthood. *Aging Ment. Health* **2016**, *20*, 284–294. [CrossRef]
3. Ghisi, M.; Bottesi, G.; Re, A.M.; Cerea, S.; Mammarella, I.C. Socioemotional Features and Resilience in Italian University Students with and without Dyslexia. *Front. Psychol.* **2016**, *7*, 478. [CrossRef]
4. Kohli, A.; Sharma, S.; Padhy, S.K. Specific Learning Disabilities: Issues that Remain Unanswered. *Indian J. Psychol. Med.* **2018**, *40*, 399–405. [CrossRef]
5. Mortimore, T.; Crozier, R. Dyslexia and difficulties with study skills in higher education. *Stud. High. Educ.* **2006**, *31*, 235–251. [CrossRef]
6. Cornoldi, C. La dislessia evolutiva. In *I Disturbi Dell'Apprendimento*; Il Mulino: Bologna, Italy, 2019; pp. 109–111, 126–131.
7. Cassandro, C.; Manassero, A.; Scarpa, A.; Landi, V.; Aschero, G.; Lovallo, S.; Velardo, P.; de Luca, P.; Albera, A.; Albera, P.; et al. Auditory-Verbal Processing Disorder and Dyslexia in Adulthood. *Transl. Med. UniSa* **2019**, *20*, 28–31.
8. Perera, H.; Shiratuddin, M.F.; Wong, K.W. Review of the Role of Modern Computational Technologies in the Detection of Dyslexia. *Inf. Sci. Appl. (ICISA)* **2016**, *376*, 1465–1475. [CrossRef]
9. Mohamad, S.; Mansor, W.; Lee, K.Y. Review of neurological techniques of diagnosing dyslexia in children. In Proceedings of the 2013 IEEE 3rd International Conference on System Engineering and Technology, Shah Alam, Malaysia, 19–20 August 2013; pp. 389–393. [CrossRef]
10. Paszkiel, S.; Szpulak, P. Methods of Acquisition, Archiving and Biomedical Data Analysis of Brain Functioning. In Proceedings of the 3rd International Scientific Conference on Brain-Computer Interfaces, Opole, Poland, 13–14 March 2018; pp. 158–171. [CrossRef]
11. El-Baz, A.; Casanova, M.; Gimel'farb, G.; Mott, M.; Switala, A. An MRI-based diagnostic framework for early diagnosis of dyslexia. *Int. J. Comput. Assist. Radiol. Surg.* **2008**, 3. [CrossRef]
12. Brown, W.E.; Eliez, S.; Menon, V.; Rumsey, J.M.; White, C.D.; Reiss, A.L. Preliminary evidence of widespread morphological variations of the brain in dyslexia. *Neurology* **2001**, *56*, 781–783. [CrossRef] [PubMed]
13. Steinbrink, C.; Vogt, K.; Kastrup, A.; Müller, H.P.; Juengling, F.D.; Kassubek, J.; Riecker, A. The contribution of white and gray matter differences to developmental dyslexia: Insights from DTI and VBM at 3.0 T. *Neuropsychologia* **2008**, *46*, 3170–3178. [CrossRef]
14. Kronbichler, M.; Wimmer, H.; Staffen, W.; Hutzler, F.; Mair, A.; Ladurner, G. Developmental dyslexia: Gray matter abnormalities in the occipitotemporal cortex. *Hum. Brain Mapp.* **2008**, *29*, 613–625. [CrossRef] [PubMed]
15. Elnakib, A.; Casanova, M.; Gimelrfarb, G.; Switala, A.E.; El-Baz, A. Dyslexia Diagnostics by 3-D Shape Analysis of the Corpus Callosum. *IEEE Trans. Inf. Technol. Biomed.* **2012**, *4*, 700–708. [CrossRef]
16. Temple, E.; Poldrack, R.A.; Protopapas, A.; Nagarajan, S.; Salz, T.; Tallal, P.; Merzenich, M.M.; Gabrieli, J.D.E. Disruption of the neural response to rapid acoustic stimuli in dyslexia: Evidence from functional MRI. *Proc. Natl. Acad. Sci. USA* **2000**, *97*, 13907–13912. [CrossRef]
17. Pugh, K.R.; Mencl, W.E.; Jenner, A.R.; Katz, L.; Frost, S.J.; Lee, J.R.; Shaywitz, S.E.; Shaywitz, B.A. Functional neuroimaging studies of reading and reading disability (developmental dyslexia). *Ment. Retard. Dev. Disabil. Res. Rev.* **2000**, *6*, 207–213. [CrossRef]
18. Soo-Yeon, J.; Najarian, K. A modified maximum correlation modeling method for fMRI brain mapping; application for detecting dyslexia. In Proceedings of the IEEE International Conference on Bioinformatics and Biomeidcine Workshops, Philadelphia, PA, USA, 3–5 November 2008; pp. 64–69. [CrossRef]
19. Berninger, V.W.; Richards, T.L.; Abbott, R.D. Differential diagnosis of dysgraphia, dyslexia, and OWL LD: Behavioral and neuroimaging evidence. *Read. Writ.* **2015**, *28*, 1119–1153. [CrossRef] [PubMed]
20. Fadzal, C.W.; Mansor, W.; Lee, K.Y.; Mohamad, S.; Amirin, S. Frequency analysis of EEG signal generated from dyslexic children. In Proceedings of the 2012 International Symposium on Computer Applications and Industrial Electronics (ISCAIE), Kota Kinabalu, Malaysia, 3–4 December 2012; pp. 202–204. [CrossRef]
21. Andreadis, I.; Giannakakis, G.; Papageorgiou, C.; Nikita, K. Detecting Complexity Abnormalities in Dyslexia Measuring Approximate Entropy of Electroencephalographic Signals. In Proceedings of the 2009 Annual International Conference of the IEEE Engineering in Medicine and Biology Society, Minneapolis, MN, USA, 3–6 September 2009; pp. 6292–6295. [CrossRef]
22. Arns, M.; Peters, S.; Bretler, R.; Verhoeven, L. Different brain activation patterns in dyslexic children: Evidence from EEG power and coherence patterns for the double-deficit theory of dyslexia. *J. Integr. Neurosci.* **2007**, *6*, 175–190. [CrossRef] [PubMed]
23. Perera, H.; Shiratuddin, M.F.; Wong, K.W.; Fullarton, K. EEG signal analysis of passage reading and rapid automatized naming between adults with dyslexia and normal controls. In Proceedings of the 2017 IEEE 8th International Conference on Software Engineering and Service Science (ICSESS), Beijing, China, 24–26 November 2017; pp. 104–108. [CrossRef]
24. Paszkiel, S.; Hunek, W.; Shylenko, A. Project and Simulation of a Portable Device for Measuring Bioelectrical Signals from the Brain for States Consciousness Verification with Visualization on LEDs. In *Challenges in Automation, Robotics and Measurement Techniques*; Springer: Cham, Switzerland, 2016; Volume 440. [CrossRef]

25. Sharafi, Z.; Soh, Z.; Guéhéneuc, Y. A systematic literature review on the usage of eye-tracking in software engineering. *Inf. Softw. Technol.* **2015**, *67*, 79–107. [CrossRef]
26. Zingoni, A.; Diani, M.; Corsini, G. A Flexible Algorithm for Detecting Challenging Moving Objects in Real-Time within IR Video Sequences. *Remote Sens.* **2017**, *9*, 1128. [CrossRef]
27. Zingoni, A.; Diani, M.; Corsini, G. Real-time moving objects detection and tracking from airborne infrared camera. In Proceedings of the SPIE Security + Defence, Warsaw, Poland, 5 October 2017; p. 10434. [CrossRef]
28. Zingoni, A.; Diani, M.; Corsini, G. Tutorial: Dealing with rotation matrices and translation vectors in image-based applications: A tutorial. *IEEE Aerosp. Electron. Syst. Mag.* **2019**, *34*, 38–53. [CrossRef]
29. De Luca, M.; Borrelli, M.; Judica, A.; Spinelli, D.; Zoccolotti, P. Reading Words and Pseudowords: An Eye Movement Study of Developmental Dyslexia. *Brain Lang.* **2002**, *80*, 617–626. [CrossRef]
30. Bellocchi, S.; Muneaux, M.; Bastien-Toniazzo, M.; Ducrot, S. I can read it in your eyes: What eye movements tell us about visuo-attentional processes in developmental dyslexia. *Res. Dev. Disabil.* **2013**, *34*, 452–460. [CrossRef]
31. Macaš, M.; Lhotská, L.; Novák, D. Hidden Markov models for analysis of eye movements of dyslexic children. In Proceedings of the 18th International Conference on Digital Signal Processing (DSP), Fira, Greece, 1–3 July 2013; pp. 1–5. [CrossRef]
32. Rello, L.; Ballesteros, M. Detecting readers with dyslexia using machine learning with eye tracking measures. In Proceedings of the 12th Web for All Conference (W4A'15), Florence Italy, 18–20 May 2015; pp. 1–8. [CrossRef]
33. Bartolomé, N.A.; Zorrilla, A.M.; Zapirain, B.G. Dyslexia diagnosis in reading stage though the use of games at school. In Proceedings of the 2012 17th International Conference on Computer Games (CGAMES), Louisville, KY, USA, 30 July–1 August 2012; pp. 12–17. [CrossRef]
34. Gaggi, O.; Palazzi, C.E.; Ciman, M.; Galiazzo, G.; Franceschini, S.; Ruffino, M.; Gori, S.; Facoetti, A. Serious Games for Early Identification of Developmental Dyslexia. *Comput. Entertain.* **2017**, *15*. [CrossRef]
35. Lyytinen, H.; Erskine, J.; Hämäläinen, J.; Torppa, M.; Ronimus, M. Dyslexia—Early Identification and Prevention: Highlights from the Jyväskylä Longitudinal Study of Dyslexia. *Curr. Dev. Disord. Rep.* **2015**, *2*, 330–338. [CrossRef] [PubMed]
36. Kalyvioti, K.; Mikropoulos, T.A. Virtual Environments and Dyslexia: A Literature Review. *Procedia Comput. Sci.* **2014**, *27*, 138–147. [CrossRef]
37. Rizzo, A.; Parsons, T.D.; Kenny, P.; Buckwalter, J.G. Using Virtual Reality for Clinical Assessment and Intervention. In *Handbook of Technology in Psychology, Psychiatry, and Neurology: Theory, Research, and Practice*; L'Abate, P., Ed.; Nova Science Publishers: Hauppage, NY, USA, 2012.
38. Jivraj, B.A.; Schaeffer, E.; Bone, J.N.; Stunden, C.; Habib, E.; Jacob, J.; Mulpuri, K. The Use of Virtual Reality in Reducing Anxiety during Cast Removal: A Randomized Controlled Trial. *J. Child. Orthop.* **2020**, *14*, 574–580. [CrossRef] [PubMed]
39. Palacios, A.M.; Sánchez, L.; Couso, I. Diagnosis of dyslexia with low quality data with genetic fuzzy systems. *Int. J. Approx. Reason.* **2010**, *8*, 993–1009. [CrossRef]
40. Le Jan, G.; Le Bouquin-Jeannès, R.; Costet, N.; Trolès, N.; Scalart, P.; Pichancourt, D.; Faucon, G.; Gombert, J.E. Multivariate predictive model for dyslexia diagnosis. *Ann. Dyslexia* **2011**, *61*, 1–20. [CrossRef] [PubMed]
41. Al-Barhamtoshy, H.M.; Motaweh, D.M. Diagnosis of Dyslexia using computation analysis. In Proceedings of the 2017 International Conference on Informatics, Health & Technology (ICIHT), Riyadh, Saudi Arabia, 21–23 February 2017; pp. 1–7. [CrossRef]
42. Rello, L.; Romero, E.; Rauschenberger, M.; Ali, A.; Williams, K.; Bigham, J.P.; Cushen White, N. Screening Dyslexia for English Using HCI Measures and Machine Learning. In Proceedings of the 2018 International Conference on Digital Health (DH'18), Lyon, France, 23–26 April 2018; pp. 80–84. [CrossRef]
43. Maitrei, K.; Prasad, T.V. Identifying Dyslexic Students by Using Artificial Neural Networks. In Proceedings of the 2010 World Congress on Engineering (WCE 2010), London, UK, 30 June–2 July 2010; Volume 1.
44. Varol, H.A.; Mani, S.; Compton, D.L.; Fuchs, L.S.; Fuchs, D. Early prediction of reading disability using machine learning. In Proceedings of the AMIA Annual Symposium, San Francisco, CA, USA, 14 November 2009; pp. 667–671.
45. Palacios, A.; Sánchez, L.; Couso, I.; Destercke, S. An extension of the FURIA classification algorithm to low quality data through fuzzy rankings and its application to the early diagnosis of dyslexia. *Neurocomputing* **2016**, *176*, 60–71. [CrossRef]
46. Glazzard, J. The impact of dyslexia on pupils' self-esteem. *Supporting Learn.* **2010**, *25*, 63–69. [CrossRef]
47. Battistutta, L.; Commissaire, E.; Steffgen, G. Impact of the Time of Diagnosis on the Perceived Competence of Adolescents with Dyslexia. *Learn. Disabil. Q.* **2018**, *41*. [CrossRef]
48. Sanfilippo, J.; Ness, M.; Petscher, Y.; Rappaport, L.; Zuckerman, B.; Gaab, N. Reintroducing Dyslexia: Early Identification and Implications for Pediatric Practice. *Pediatrics* **2020**, *146*, e20193046. [CrossRef]
49. Barbiero, C.; Lonciari, I.; Montico, M.; Monasta, L.; Penge, R.; Vio, C.; Tressoldi, P.E.; Ferluga, V.; Bigoni, A.; Tullio, A.; et al. The submerged dyslexia iceberg: How many school children are not diagnosed? Results from an Italian study. *PLoS ONE* **2012**, *7*, e48082. [CrossRef] [PubMed]
50. Galuschka, K.; Use, E.; Krick, K.; Schulte-Körne, G. Effectiveness of Treatment Approaches for Children and Adolescents with Reading Disabilities: A Meta-Analysis of Randomized Controlled Trials. *PLoS ONE* **2014**, *9*, e89900. [CrossRef]
51. Pasqualotto, A.; Venuti, P. A Multifactorial Model of Dyslexia: Evidence from Executive Functions and Phonological-based Treatments. *Learn. Disabil. Res. Pract.* **2020**, *35*, 150–164. [CrossRef]
52. Bonacina, S.; Cancer, A.; Lanzi, P.L.; Lorusso, M.L.; Antonietti, A. Improving reading skills in students with dyslexia: The efficacy of a sublexical training with rhythmic background. *Front. Psychol.* **2015**, *6*, 1510. [CrossRef]

53. Cogo-Moreira, H.; Andriolo, R.B.; Yazigi, L.; Ploubidis, G.B.; Brandão de Ávila, C.R.; Mari, J.J. Music education for improving reading skills in children and adolescents with dyslexia. *Cochrane Database Syst. Rev.* **2012**. [CrossRef]
54. Wang, R.; Chen, L.; Solheim, I.; Schulz, T.; Ayesh, A. Conceptual Motivation Modeling for Students with Dyslexia for Enhanced Assistive Learning. In Proceedings of the 2017 ACM Workshop on Intelligent Interfaces for Ubiquitous and Smart Learning (SmartLearn'17), Limassol, Cyprus, 13 March 2017; pp. 11–18. [CrossRef]
55. Yaquob Alsobhi, A.; Khan, N.; Rahanu, H. Personalised Learning Materials Based on Dyslexia Types: Ontological Approach. *Procedia Comput. Sci.* **2015**, *60*, 113–121. [CrossRef]
56. Pang, L.; Jen, C.C. Inclusive dyslexia-friendly collaborative online learning environment: Malaysia case study. *Educ. Inf. Technol.* **2018**, *23*, 1023–1042. [CrossRef]
57. Saidah Sarpudin, S.N.; Zambri, S. Web readability for students with Dyslexia: Malaysian case study. In Proceedings of the 2014 3rd International Conference on User Science and Engineering (i-USEr), Shah Alam, Malaysia, 2–5 September 2014; pp. 192–197. [CrossRef]
58. Alhabashneh, M.; Abu-Salih, B.; Knight, S. Impact of Web 2.0 Technology on Students with Learning Difficulties: A State-of-the-Art and Future Challenges. In Proceedings of the 2018 32nd International Conference on Advanced Information Networking and Applications Workshops (WAINA), Krakow, Poland, 16–18 May 2018; pp. 693–697. [CrossRef]
59. Schiavo, G.; Mana, N.; Mich, O.; Zancanaro, M.; Job, R. Attention-driven read-aloud technology increases reading comprehension in children with reading disabilities. *J. Comput. Assist. Learn.* **2021**. [CrossRef]
60. Perry, C.; Zorzi, M.; Ziegler, J.C. Understanding Dyslexia through Personalized Large-Scale Computational Models. *Psychol. Sci.* **2019**, *30*, 386–395. [CrossRef]
61. Siti Suhaila, A.H.; Admodisastro, A.N.; Abd Ghani, A.A. Computer-based learning model to improve learning of the malay language amongst dyslexic primary school students. In Proceedings of the Asia Pacific HCI and UX Design Symposium, Melbourne, Vic, Australia, 7 December 2015. [CrossRef]
62. Ndombo, D.M.; Ojo, S.; Osunmakinde, I.O. An intelligent integrative assistive system for dyslexic learners. *J. Assist. Technol.* **2013**, *7*. [CrossRef]
63. Rajapakse, S.; Polwattage, D.; Guruge, U.; Jayathilaka, I.; Edirisinghe, T.; Thelijjagoda, S. ALEXZA: A Mobile Application For Dyslexics Utilizing Artificial Intelligence And Machine Learning Concepts. In Proceedings of the 2018 3rd International Conference on Information Technology Research (ICITR), Moratuwa, Sri Lanka, 5–7 December 2018; pp. 1–6. [CrossRef]
64. Thelijjagoda, S.; Chandrasiri, M.; Hewathudalla, D.; Ranasinghe, P.; Wickramanayake, I. The Hope: An Interactive Mobile Solution to Overcome the Writing, Reading and Speaking Weaknesses of Dyslexia. In Proceedings of the 2019 14th International Conference on Computer Science & Education (ICCSE), Toronto, ON, Canada, 19–21 August 2019; pp. 808–813. [CrossRef]
65. Brooks, B.M.; Rose, F.D.; Attree, E.A.; Elliot-Square, A. An evaluation of the efficacy of training people with learning disabilities in a virtual environment. *Disabil. Rehabil.* **2002**, *24*, 622–626. [CrossRef]
66. Redel-Macías, M.D.; Pinzi, S.; Martínez-Jiménez, M.P.; Dorado, G.; Dorado, M.P. Virtual Laboratory on Biomass for Energy Generation. *J. Clean. Prod.* **2016**, *112*, 3842–3851. [CrossRef]
67. Lantto, E.; Simpura, F.; Uusitalo, J.; Lukander, K.; Räsänen, T.; Teperi, A.-M. Evaluation of the Efficacy of a Virtual Reality-Based Safety Training and Human Factors Training Method: Study Protocol for a Randomised-Controlled Trial. *Inj. Prev.* **2020**, *26*, 360–369. [CrossRef]
68. Herrero, J.F.; Lorenzo, G. An Immersive Virtual Reality Educational Intervention on People with Autism Spectrum Disorders (ASD) for the Development of Communication Skills and Problem Solving. *Educ. Inf. Technol.* **2020**, *25*, 1689–1722. [CrossRef]
69. El Naqa, I.; Pedroli, E.; Padula, P.; Guala, A.; Meardi, M.T.; Riva, G.; Albani, G. A Psychometric Tool for a Virtual Reality Rehabilitation Approach for Dyslexia. *Comput. Math. Methods Med.* **2017**, *2017*, 7048676. [CrossRef]
70. Kalyvioti, K.; Mikropoulos, T.A. Memory Performance of Dyslexic Adults in Virtual Environments. *Procedia Comput. Sci.* **2012**, *14*, 410–418. [CrossRef]
71. Siti Suhaila, A.H.; Admodisastro, A.N.; Manshor, N.; Kamaruddin, A.; Abd Ghani, A.A. Dyslexia Adaptive Learning Model: Student Engagement Prediction Using Machine Learning Approach. In Proceedings of the International Conference on Soft Computing and Data Mining (SCDM 2018), Johor, Malaysia, 12 January 2018; pp. 372–384. [CrossRef]
72. Costantini, A.; Ceschi, A.; Sartori, R. Psychosocial Interventions for the Enhancement of Psychological Resources among Dyslexic Adults: A Systematic Review. *Sustainability* **2020**, *12*, 7994. [CrossRef]
73. Sprenger-Charolles, L.; Siegel, S.L.; Jiménez, J.E.; Ziegler, J.C. Prevalence and Reliability of Phonological, Surface, and Mixed Profiles in Dyslexia: A Review of Studies Conducted in Languages Varying in Orthographic Depth. *Sci. Stud. Read.* **2011**, *15*, 498–521. [CrossRef]
74. Moll, K.; Kunze, S.; Neuhoff, N.; Bruder, J.; Schulte-Körne, G. Specific Learning Disorder: Prevalence and Gender Differences. *PLoS ONE* **2014**, *9*, e103537. [CrossRef] [PubMed]
75. Scuola, Pubblicati I Dati Sugli Alunni Con Disturbi Specifici Dell'apprendimento. Available online: https://www.miur.gov.it/web/guest/-/scuola-pubblicati-i-dati-sugli-alunni-con-disturbi-specifici-dell-apprendimento#:~{}:text=Complessivamente%2C%20nel%202017%2F2018%2C,(disturbo%20nel%20calcolo%20matematico) (accessed on 14 February 2021).
76. Katz, J. The Three Block Model of Universal Design for Learning (UDL): Engaging students in inclusive education. *Can. J. Educ.* **2013**, *36*, 153–194.
77. Smeby, J. Disciplinary differences in university teaching. *Stud. High. Educ.* **1996**, *21*, 69–79. [CrossRef]

78. Ray, S. A Quick Review of Machine Learning Algorithms. In Proceedings of the 2019 International Conference on Machine Learning, Big Data, Cloud and Parallel Computing (COMITCon), Faridabad, India, 14–16 February 2019; pp. 35–39. [CrossRef]
79. Taborri, J.; Palermo, E.; Rossi, S. Automatic detection of faults in race walking: A comparative analysis of machine-learning algorithms fed with inertial sensor data. *Sensors* **2019**, *19*, 1461. [CrossRef]
80. Freire, A.P.; Petrie, H.; Power, C. Empirical results from an evaluation of the accessibility of websites by dyslexic users. In Proceedings of the Workshop on Accessible Design in the Digital World, Lisboa, Portugal, 5–9 September 2011; pp. 41–53.
81. Bergua, V.; Meillon, C.; Potvin, O.; Bouisson, J.; Le Goff, M.; Rouaud, O.; Ritchie, K.; Dartigues, J.-F.; Amieva, H. The STAI-Y trait scale: Psychometric properties and normative data from a large population-based study of elderly people. *Int. Psychogeriatr.* **2012**, 1–9. [CrossRef]
82. Luszczynska, A.; Scholz, U.; Schwarzer, R. The General Self-Efficacy Scale: Multicultural Validation Studies. *J. Psychol.* **2005**, *139*, 439–457. [CrossRef]
83. Sica, C.; Chiri, L.R.; Favilli, R.; Marchetti, I. *Test Q-PAD-Questionario per la Valutazione Della Psicopatologia in Adolescenza*; Erickson: Trento, Italy, 2011; ISBN 9788861378797.
84. Goldsmith, R.E. Dimensionality of the Rosenberg Self-Esteem Scale. *J. Soc. Behav. Personal.* **1986**, *1*, 253–264.
85. Mikropoulos, T.A.; Natsis, A. Educational Virtual Environments: A Ten-Year Review of Empirical Research (1999–2009). *Comput. Educ.* **2011**, *56*, 769–780. [CrossRef]
86. Mikropoulos, T. A Virtual Reality Test for the Identification of Memory Strengths of Dyslexic Students in Higher Education. *J. Univers. Comput. Sci.* **2013**, *19*, 2698–2721. [CrossRef]
87. Seymour, P.H.K.; Aro, M.; Erskine, J.M. Foundation literacy acquisition in European orthographies. *Br. J. Psychol.* **2003**, *94*, 143–174. [CrossRef] [PubMed]

applied sciences

MDPI

Article

Pre-Training Autoencoder for Lung Nodule Malignancy Assessment Using CT Images

Francisco Silva [1,2], Tania Pereira [1], Julieta Frade [1,2], José Mendes [1,2], Claudia Freitas [3,4], Venceslau Hespanhol [3,4], José Luis Costa [4,5,6], António Cunha [1,7] and Hélder P. Oliveira [1,8,*]

1 INESC TEC—Institute for Systems and Computer Engineering, Technology and Science, Porto 4200-465, Portugal; francisco.c.silva@inesctec.pt (F.S.); tania.pereira@inesctec.pt (T.P.); julietafrade97@gmail.com (J.F.); ee12293@fe.up.pt (J.M.); acunha@utad.pt (A.C.)
2 FEUP—Faculty of Engineering, University of Porto, Porto 4200-465, Portugal
3 CHUSJ—Department of Pulmonology, Centro Hospitalar e Universitário de São João, Porto 4200-319, Portugal; claudiaasfreitas@gmail.com (C.F.); hespanholv@gmail.com (V.H.);
4 FMUP—Faculty of Medicine, University of Porto, Porto 4200-319, Portugal; jcosta@ipatimup.pt
5 i3S—Institute for Research and Innovation in Health, University of Porto, Porto 4200-135, Portugal
6 IPATIMUP—Institute of Molecular Pathology and Immunology, University of Porto, Porto 4200-135, Portugal
7 UTAD—University of Trás-os-Montes and Alto Douro, Vila Real 5001-801, Portugal
8 FCUP—Faculty of Science, University of Porto, Porto 4169-007, Portugal
* Correspondence: helder.f.oliveira@inesctec.pt

Received: 28 September 2020; Accepted: 27 October 2020; Published: 5 November 2020

Abstract: Lung cancer late diagnosis has a large impact on the mortality rate numbers, leading to a very low five-year survival rate of 5%. This issue emphasises the importance of developing systems to support a diagnostic at earlier stages. Clinicians use Computed Tomography (CT) scans to assess the nodules and the likelihood of malignancy. Automatic solutions can help to make a faster and more accurate diagnosis, which is crucial for the early detection of lung cancer. Convolutional neural networks (CNN) based approaches have shown to provide a reliable feature extraction ability to detect the malignancy risk associated with pulmonary nodules. This type of approach requires a massive amount of data to model training, which usually represents a limitation in the biomedical field due to medical data privacy and security issues. Transfer learning (TL) methods have been widely explored in medical imaging applications, offering a solution to overcome problems related to the lack of training data publicly available. For the clinical annotations experts with a deep understanding of the complex physiological phenomena represented in the data are required, which represents a huge investment. In this direction, this work explored a TL method based on unsupervised learning achieved when training a Convolutional Autoencoder (CAE) using images in the same domain. For this, lung nodules from the Lung Image Database Consortium and Image Database Resource Initiative (LIDC-IDRI) were extracted and used to train a CAE. Then, the encoder part was transferred, and the malignancy risk was assessed in a binary classification—benign and malignant lung nodules, achieving an Area Under the Curve (AUC) value of 0.936. To evaluate the reliability of this TL approach, the same architecture was trained from scratch and achieved an AUC value of 0.928. The results reported in this comparison suggested that the feature learning achieved when reconstructing the input with an encoder-decoder based architecture can be considered an useful knowledge that might allow overcoming labelling constraints.

Keywords: transfer learning; autoencoder; lung cancer; malignancy assessment

1. Introduction

Lung cancer is on the top of cancer-related mortality numbers worldwide [1,2]. Only 16% of lung cancer cases are diagnosed as local stage tumors. In these cases, patients have a five-year survival rate of more than 50%; however, when diagnosed in an advanced stage, the chances of a five-year survival decrease to 5%. Thus, achieving an earlier diagnosis is critical to increase survival rate, and systems able to provide screening support might play an important role. As a non-invasive method, computed tomography (CT) images have shown the ability to provide valuable information on tumor status, rising opportunities to the development of computer-aided diagnoses (CAD) systems able to provide an automatic assessment of lung nodules malignancy risk to help the clinical decision. Considering the use of qualitative data, factors like the high interobserver variability associated with the visual assessment of relevant characteristics, and the amount of radiological data to be analyzed makes the development of completely automatic systems a more attractive approach.

Several methods based on convolutional neural networks have been proposed to investigate the ability to distinguish between malignant and benign pulmonary nodules, taking advantage of the ability to directly detect relevant patterns with abstract and complex imaging manifestations [3]. Several previous works proposed deep learning-based solutions for lung nodule malignancy classification using the Lung Image Database Consortium image collection (LIDC-IDRI) [4,5], which is a public dataset of thoracic CT scans with expert annotations, and the most commonly used to develop AI-based solutions for lung cancer. Shen et al. [6] proposed a hierarchical learning framework to capture the nodule heterogeneity by utilizing a Convolutional neural network (CNN) to extract features and a random forest classifier for the final classification with the highest accuracy of 0.868. Lu et al. [7] obtained an accuracy of 0.919 using a CNN to extract the features and a support vector machine (SVM) for the final classification. Yan et al. [8] compared the performance obtained with 2D and 3D CNN implementations, achieving a mean area under the curve (AUC) of 0.937 for 2D analysis and an AUC of 0.947 using 3D inputs. Song et al. [9] proposed a comparison between a CNN, a deep neural network (DNN), and a stacked autoencoder (SAE) for the classification of benign or malignant lung nodules, and the CNN showed the highest performance, with an accuracy of 0.842. Yutong et al. [10] developed an algorithm that uses a deep convolutional neural network to automatically learn the feature representation of nodules on a 2D analysis, fuses this information with other more common features (shape, texture), and obtained an AUC of 0.966. A similar approach was developed by Causey et al. [11] that combines the deep learning CNN features with radiomics features as input in a random forest classifier and obtained an accuracy of 0.990. The success of deep learning-based feature extraction is due to the ability of not only taking the information of different conventional semantic features (such as shape or margins) or radiomic features (texture or histogram-based properties) but also taking into account abstract features, where deeper levels provide more complex and abstract knowledge [12,13].

Despite offering successful approaches, deep learning models require a large number of training data to be able to generalize over unseen images, and the lack of publicly available data is often a problem in the majority of medical applications. One strategy to overcome this issue is by using transfer learning (TL), a learning method that consists of applying a network already trained for a different task. By taking advantage of general patterns learned in the first layers, this technique reduces the number of trainable parameters alongside the necessary dataset size. Only a few and most recent works have explored this approach for this classification task. Lindsay et al. [14] used a pre-trained 3D-CNN on the LIDC-IDRI dataset to identify nodules in CT scans. Three new untrained layers were added to the existing pre-trained network, a private dataset of 796 patients who underwent CT-guided lung biopsy were used to retrain and test the approach. The biopsy results were used as ground truth labels removing the subjectivity of the annotations by the radiologists that are present in the LIDC-IDRI dataset. The TL of the initial layers of the networks with the retrain on the new dataset achieved an AUC of 0.70. Nóbegra et al. [15,16] proposed an investigation with multiple ImageNet [17,18] pre-trained feature extractors and different classifiers tested on the LIDC-IDRI dataset. The highest

AUC of 0.931 was achieved with the ResNet-50 [19] architecture and a SVM with a radial basis kernel as a classifier. Zhang et al. [20] explored a pulmonary nodule classification method based on the pre-trained ResNet for feature extraction and classification of the pulmonary nodules in an end-to-end manner that was evaluated on LIDC-IDRI and achieved an AUC of 0.912. Shi et al. [21] developed an approach based on a pre-trained VGG-16 model in ImageNet combined with an SVM classifier for a false-positive reduction in pulmonary nodule detection on CT slices from the LIDC-IDRI dataset and obtained an overall accuracy of 0.915.

The autoencoder is a dimensionality reduction method that adds the opportunity to extract features using unlabelled data, which allows overcoming one of the biggest limitations in the medical data. Autoencoders extract representative patterns from the images using the image reconstruction mechanism [22]. Kumar et al. [23] proposed to use an unsupervised denoising autoencoder to extract deep features from 2D lung nodule slices from the LIDC-IDRI dataset, achieving a classification mean accuracy value of 0.750, using a decision tree as a classifier. Cheng et al. [24] used a stacked denoising auto-encoder (SDAE) and constructed a pre-training architecture to use as network initialization for the latter supervised training, which achieved an AUC of 0.984 on the LIDC-IDRI dataset.

The work presented in this paper addressed a binary lung nodule malignancy classification by a TL approach based on a trained Convolutional Autoencoder (CAE), using the LIDC-IDRI dataset. Taking advantage of the unsupervised feature learning ability of the CAE while reconstructing nodule 2D images, the relevance of the learned patterns was explored to prevent the overfitting occurrence. Thus, a first experiment was conducted to train the CAE, and then a classification model was developed to distinguish between benign and malignant lung nodules in a 2D perspective. The major contribution of this work relies on three main points: simple architecture, by taking advantage of the computational resources lower consumption; no annotations required on pre-training, allowing to use large available datasets without the huge investment on labeling all data; pre-trained architecture, without the need of explicit design and selection of problem-oriented features that can be used for other tasks.

2. Materials and Methods

This section presents the dataset used in this work and describes the pipeline implemented to predict the malignancy of the lung nodules.

2.1. CT Image Database

The LIDC-IDRI [4,5] is a lung cancer screening dataset which comprises thoracic CT scans for a total of 1010 patients, alongside with annotated nodules belonging to one of three classes: (a) nodule $\geq$ 3 mm; (b) nodule < 3 mm or (c) non-nodule $\geq$ 3 mm, made during a two-phase annotation process by four experienced radiologists. Regarding data acquisition, slice thickness ranged from 0.6 to 5.0 mm, with X-ray current ranged from 40 to 627 mA (mean: 222.1 mA) at 120–140 kVp.

Data Preparation

To standardize the dataset, all CT scans were resampled. The pixel spacing was set to [1.00, 1.00, 1.00] mm and each CT dimension was calculated to match this new spacing, obtaining the resampled image by interpolation. Additionally, each pixel intensity value, measured in the Hounsfield Units (HU) scale, was normalized using the *min-max* normalization method, and values under −1000 HU, which corresponds to air's radiodensity value, were transformed into 0 and values above 400 HU, representing hard tissues like bones, were transformed into 1. A linear transformation was computed to map all values in the middle into the [0, 1] intended range.

To assess the nodule location, the final binary mask that included the pixels with at least 50% consensus of the provided contours was used, and an image with size (80 × 80 × 80) voxels centered on the nodule was extracted for each considered example in this study.

From the 7371 detected nodules, only 2669 were classified as larger than 3 mm. For these examples, this database also provides nodule contours marked by each radiologist, as well as quantitative labels for a set of nodule related features. Although the feature learning task did not require any labels, this inclusion criterium was the one selected to facilitate the evaluation process. It was considered that a reconstructed image with a larger nodule would provide more clear information to evaluate the quality of the outputs by visualization.

Considering data inclusion in the classification task, the provided malignancy risk value assessed by the radiologists was analyzed. This value corresponded to an integer ranging from 1 to 5 with the following designations related to the malignancy degree: (1) *Highly Unlikely*, (2) *Moderately Unlikely*, (3) *Indeterminate*, (4) *Moderately Suspicious* and (5) *Highly Suspicious*. The interobserver agreement in rating malignancy was previously studied and a modest agreement was found [25,26]. For cases rated by two radiologists, they only agree in 43% of the cases. The percentage of agreement decreases for nodules annotated by more radiologists, with an agreement of 19% in the cases where the annotation was performed by three radiologists, and 12% of agreement in annotations from four radiologists. In addition, the analysis found in [5] regarding the percentage of nodules that were labelled by one, two, three or four radiologists, shows that almost 64% of the 2669 nodules were marked by a single clinician or by all of them, 29.1% and 34.8% respectively. When only assigned by one radiologist, the model only takes into account a single opinion, being susceptible to a possible error; on the other hand, if a nodule was labelled by all the four clinicians, the computed agreement rate of 12% tells that a lack of confidence in the resultant label would continue to exist. Having this analysis in mind and the fact that the subjectivity in the labelling process would not be eliminated if only the nodules annotated by all clinicians were included, this study includes the entire set of nodules marked by at least one radiologist.

To take into account the different annotations provided for the same example, each nodule's malignancy value was computed by averaging the provided annotations made by all radiologists, and a mean malignancy value ≤ 2.0 was considered as benign, and ≥ 4.0 as malignant, excluding the examples with resultant mean value outside the selected ranges. Although this database provides multi-class labeled nodules, this work aimed to perform a binary classification given the high inter-observer variability present in the available annotations. Considering the original range of values used to label each nodule, the arithmetic mean combined the different classifications with an equal contribution of each annotation [27]. This process helped to decrease the chances of using mislabelled information in the training or evaluation phases. Considering the explained inclusion criteria, only 1095 nodules were included in this classification task: 789 benign and 306 malignant.

The models implemented in this work were designed to receive two-dimensional inputs. Although this slice-level approach does not allow a complete nodule analysis, it increased the number of training examples by sampling different slices from the same nodule and also helped to achieve a better class balancing. Considering this, in the feature learning task, the slice over-sampling operation consisted of extracting middle slices from the axial, coronal and sagittal planes from the $(80 \times 80 \times 80)$ nodule centered image, as illustrated in Figure 1. In the classification task, this operation included the extraction of four additional random slices to balance positive and negative classes in the training set, using the cube symmetry planes to obtain these extra nodule perspectives.

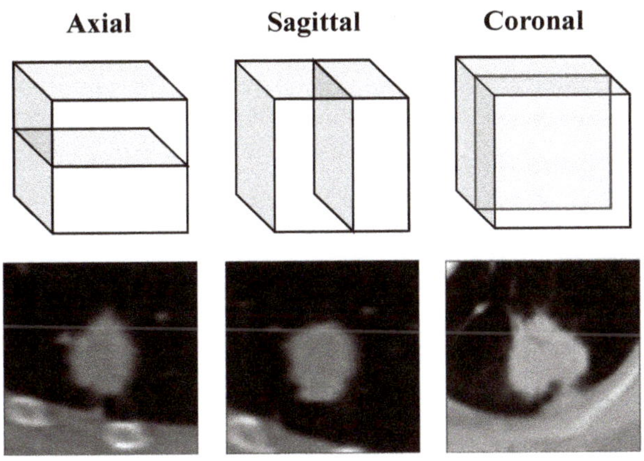

Figure 1. Lung nodule slice extraction example: middle slice from axial, sagittal and coronal planes from the same 3D nodule.

2.2. Methodology

The TL approach used in this work consisted of training a CAE to use the encoder layers as a feature extractor, followed by a classifier that uses those features to produce the final malignancy binary classification. Considering basic observation on CNN behavior, the first convolutional layers of the encoder learn generic features useful for many different tasks, but progressively, as the network goes deeper, more specific patterns are detected and more relevant information is learned regarding the target task [28]. In this approach, the relevance of the knowledge achieved while pre-training the encoder layers was explored in order to reduce the number of trainable parameters in the classification task. The pipeline proposed in this work is illustrated in Figure 2.

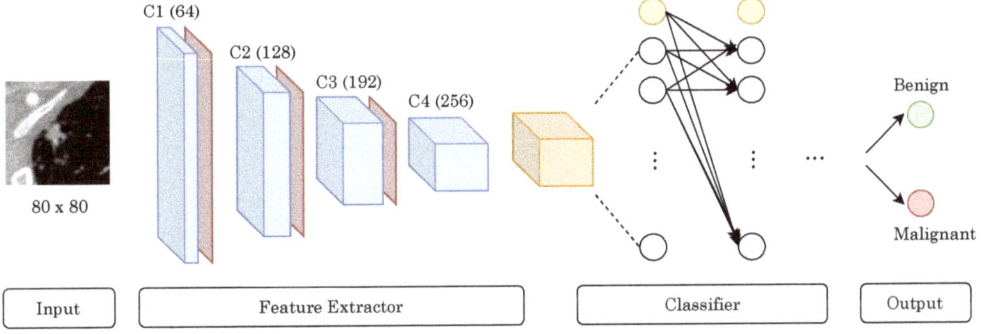

Figure 2. Pipeline developed for lung nodule malignancy classification composed by the feature extractor (CAE encoder) followed by a classifier.

In this classification task, a multi-stage training strategy was adopted, where initially only the classifier fully connected layers were trained. Given the fact that the encoder layers were pre-trained with images from the same domain, an initially good convergence was expected, but still far from a local minimum. Thus, to find the best classification performance, the convolutional layers were progressively unfrozen and retrained to detect more representative patterns related to nodule malignancy. By analyzing the learning curves, when learning stopped improving, a deeper convolutional layer was unfrozen, repeating this process until the overfitting occurrence. The major advantage of this layer-wise fine-tuning approach relies on the fact that immediately unfreezes the entire feature extractor at once, which was not expected to be a viable option given the small amount of training data. Due to large similarities between source and target domains, it was also considered an unnecessary approach.

In training, validation, and testing phases, completely independent subsets of the LIDC-IDRI cohort were used to avoid the occurrence of images from the same nodule in more than one of those sets.

2.2.1. Feature Extraction

Extending the Autoencoder, a CAE enables a dimensionality reduction while preserving the original spatial representation, which is an absolutely critical characteristic when analyzing multi-dimensional data [29]. The overall structure of a CAE comprises two main phases: (1) in the encoding phase, a convolutional encoder transforms the input data into a lower-dimensional feature space, followed by a decoding phase (2), where the compressed representation passes through decoding layers to achieve the original input image reconstruction. Having the input as the target, the CAE learns the best set of features that enables the input reconstruction while eliminating the need for labeled data. The application of this dimensionality reduction technique lies in the idea that the knowledge achieved while training to reconstruct the input data might provide a useful weight initialization if one intends to use the encoder part for a different task in a transfer learning approach.

In the proposed CAE architecture, represented in Figure 3, the encoding phase is composed of four convolutional layers with 3×3 kernels, and an increasing number of filters as the network goes deeper. All the convolutional layers are followed by a Rectified Linear Unit (ReLU) activation function and a *max-pooling* layer to reduce the output feature map by half. Giving an input tensor of size $C \times H \times W$, passing through the encoding layers results in a feature map of 256 filters with size $\frac{H}{8} \times \frac{W}{8}$. To reconstruct the original input, three *max-unpooling* layers were employed to double the input size before the first three convolution blocks. The last convolutional layer is followed by a sigmoid activation, ensuring that all output pixels belong to the $[0, 1]$ range of values.

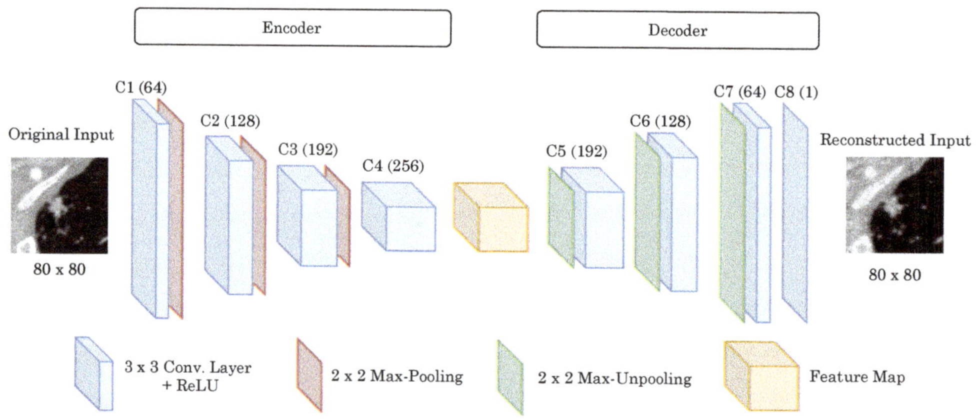

Figure 3. Proposed CAE architecture for lung nodule unsupervised feature learning by optimizing input reconstructions.

Table 1 depicts the considered values in the hyper-parameter search.

Table 1. Set of hyper-parameters values considered in the search for the CAE training.

Hyper-Parameter	Range Values
Learning Rate	0.0001, 0.001, 0.01, 0.1
Optimizer	SGD [1], Adam
Momentum	0.1, 0.5, 0.9

[1] Stochastic Gradient Descent.

2.2.2. Malignancy Classification

Given the end-to-end characteristic of this approach and the backpropagation based learning, a Multi-layer Perceptron (MLP) was used as a classifier to perform the intended predictions. An MLP is an artificial neural network composed of an input layer where all the input values are received, an output layer with a number of neurons depending on the classification task in hands, and in between a variable number of hidden layers. Since it is a fully-connected neural network, each neuron is connected to all neurons of the following layer. When approaching a classification task with TL techniques, the developed model will consist of the feature extraction pre-trained layers and a classifier stacked on top to be completely trained with the new target data (Figure 2). Each feature extracted by the convolutional block is used to feed the input fully-connected layer of the classifier. For regularization, a dropout layer was employed before this input layer to reduce the number of features taken into account at each training iteration, forcing the classifier to learn in a more robust manner. Additionally, the output of each neuron in the input and hidden layers passed through a ReLU activation, and a sigmoid activation was employed for the output neuron. The binary cross-entropy was used as the cost function to be minimized. The MLP is a backpropagation neural network, which works by approximating the non-linear relationship between the input and the output by adjusting the weight values internally. The ability of this neural network-based classifier to propagate the prediction error to the feature extractor layers allowed fine-tuning the higher-level layers of the convolutional encoder, where the most useful patterns are detected. The feature extractor is an encoder network pre-trained to reconstruct images in the same domain, not optimised to detect representative patterns related to nodule malignancy. To capture the most useful features for this classification task, the feature extractor needed to be fine-tuned, and this process is possible due to the backpropagation-based learning of MLP.

In this binary malignancy classification task, a more extensive search was employed to find the set of hyper-parameters that achieved the desired performance, with considered values depicted in Table 2.

Table 2. Set of hyper-parameters values used in the search for the malignancy classification model.

Hyper-Parameter	Range Values
Learning Rate	0.0001, 0.001, 0.01, 0.1
Batch-size	4, 8, 16, 32, 64
Momentum	0.1, 0.5, 0.9, 0.99
Weight decay	0.00001, 0.0001, 0.001, 0.01
Dropout	0.25, 0.5, 0.75
Hidden Layers	1, 2, 3
Hidden Neurons	32, 64, 128, 256, 512
Optimizer	SGD [1], Adam

[1] Stochastic Gradient Descent.

2.3. *Performance Metrics*

As common choice for learning, the Mean Squared Error (MSE) was used as the loss function, representing the averaged error of each output pixel value when compared to the same pixel in the original image [30,31]. Training stopped when no change in loss value was reported during 50 consecutive training iterations.

Considering the CAE training, the optimization criterium was based on the Mean Squared Error (MSE) value computed [32]. As a feature learning task, and even though a perfect reconstruction does not ensure the best set of learned features to be applied in the transfer learning approach under investigation, these experiments were conducted in order to optimize the input and its reconstruction similarities.

To evaluate the malignancy classification performance, the Receiver Operation Characteristic (ROC) curve was analyzed, as well as the AUC value. Additionally, Precision, Recall, and F-score metrics were also computed to measure the generalization ability of the developed models. To assess the reliability of the transfer learning approach adopted, the same architecture was also completely trained from scratch and evaluated.

3. Results

We evaluated the malignancy predictions in two training methods: using transfer learning and trained from the scratch. The results for the approach optimization and the classification performance achieved are presented in this section.

3.1. Hyper-Parameters Selection

The hyper-parameters selected for the CAE training in the nodule reconstruction task that achieved the best results were the following: mini-batches of four images with Stochastic Gradient Descent (SGD) as the optimizer, learning rate of 0.01, and momentum with the value of 0.9.

Considering the classification task, the hyper-parameters values that achieved the best performance on the test set are presented in Table 3. All the classification performance metrics were computed in a hold-out test set with 5-fold cross-validation applied to the training data, over five random train/test combinations to prevent some possible bias on results induced by a specific set of inputs.

Table 3. Hyper-parameters values that achieved best performance.

Hyper-Parameter	Value
Learning Rate	0.001
Batch-size	8
Momentum	0.9
Weight decay	0.0001
Dropout	0.5
Hidden Layers	1
Hidden Neurons	64
Optimizer	SGD [1]

[1] Stochastic Gradient Descent.

3.2. Malignancy Classification

The classifier training convergence was achieved after 200 iterations. After this first training stage, fine-tuning was applied by unfreezing the encoder's last convolutional layer; finally, after the second training convergence, gradients were updated for the last two convolutional layers for a final training. This strategy obtained the best results by preventing the model to overfit, which occurred when it was tried to add one more convolutional layer for retraining.

Mean values for each performance metric considered are presented in Table 4. By taking advantage of cross-validation data, the decision threshold was tuned for F-score maximization by the recall. This threshold optimization was employed by evaluating the precision-recall trade-off which, given the context, represents the cost of a missing malignant nodule (false negative) over a false suspicious of a benign tumor (false positive). Thus, as a missing malignant tumor should be a more penalized error, recall was maximized over precision in this classification task. With the TL approach, the mean value for the classification threshold was 0.413 ± 0.101, which clearly illustrates the necessity of choosing a value under 0.5, the default threshold value, to allow to increase the confidence in non-malignant predictions.

Figure 4 shows the ROC curve for the binary classification between benign and malignant cases for our proposed method using the TL approach, with a maximum mean AUC of 0.936 ± 0.009. The small gray shading in the figure shows the consistency of the results and the independence of the performance with the subsets selected for training and testing, given the small variation on the AUC obtained.

Table 4. Lung nodule malignancy classification results.

	Performance Metrics (Mean ± Standard Deviation)			
Training Method	**AUC**	**Precision**	**Recall**	**F-Score**
Transfer Learning	0.936 ± 0.009	0.794 ± 0.026	0.848 ± 0.035	0.817 ± 0.020
Trained from scratch	0.928 ± 0.027	0.842 ± 0.035	0.789 ± 0.069	0.808 ± 0.022

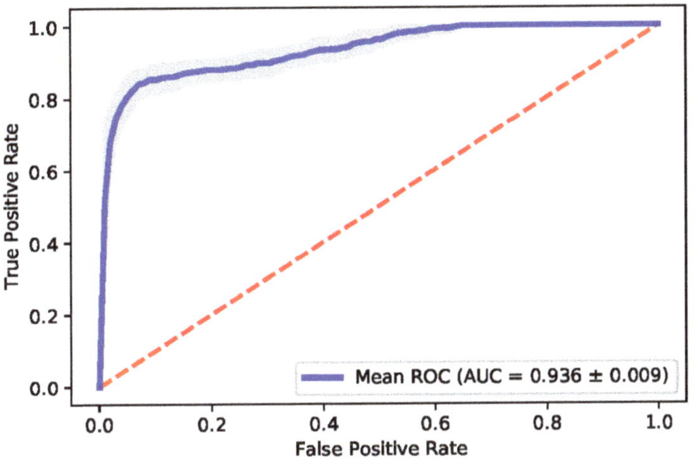

Figure 4. Averaged ROC curve for lung nodule malignancy classification using the transfer learning approach. The ROC curve is computed for each iteration, the arithmetic average is then calculated and represented by the blue line with a standard deviation represented by the gray shading area. The red dashed-line represents an at-chance classifier ROC curve.

To better understand the relationship between the success of a prediction and the output reconstruction of the correspondent nodule, Figure 5 shows one example for each case. Well and wrongly classified examples are depicted, alongside with the correspondent MSE value obtained by analysis of the CAE output. The reported MSE mean value for the successfully classified examples was 0.00162 ± 0.00088, in contrast with the value 0.00337 ± 0.00249, correspondent to wrong predictions. It was possible to verify that, in general, a correct prediction was associated with a reconstructed image with a lower pixel error, which might demonstrate the impact of the low-level features learning in the detection of more relevant patterns for classification.

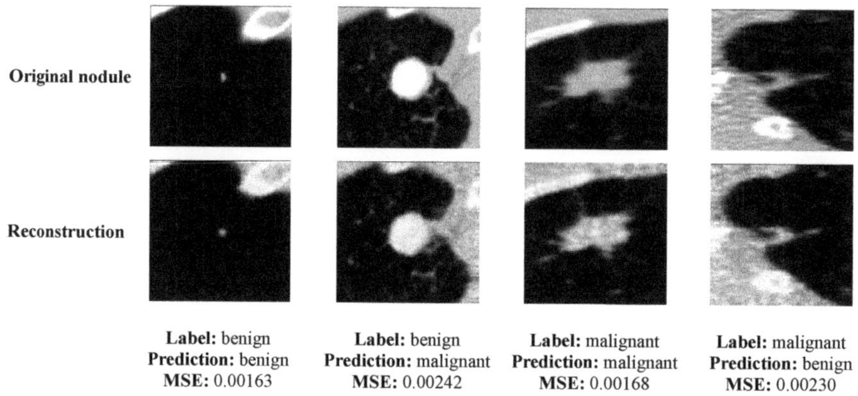

Figure 5. Testing examples, representing benign and malignant nodules, well and wrongly classified. The MSE value reported for each example represents the mean squared pixel error between the original image and its reconstruction.

4. Discussion

In the study, we proposed a deep structured algorithm to automatically extract features based on a convolutional autoencoder and an end-to-end learning classification network to predict the malignancy risk of nodules in CT images using TL techniques. A baseline experiment was also implemented, where the proposed architecture was trained from the scratch, in order to assess the capability to use the proposed TL strategy for this classification task. The results achieved suggest that the transfer learning approach was able to perform as well as the network trained from scratch, which means that the unsupervised learning ability of the proposed CAE represented a useful knowledge for the classification task, helping the network to avoid overfitting.

Considering the previous works, it is not possible to make a fair and direct comparison of the performance results, since the selection of the data and the criteria for final labelling of the nodules were different, which impacts the final performance. The results obtained in the current work did not overcome the performance obtained in some studies but achieve the same high level (above 0.9 of accuracy) using an approach that is not dependent on the massive label data and without needing feature engineering. In fact, the presented approach has several advantages that makes this contribution relevant in the medical image classification: (1) end-to-end approach (automatic feature extraction avoiding the *ad hoc* feature engineering); (2) unsupervised learning (allowing to use massive datasets without annotations to train the feature extractor) and (3) transfer learning (use the knowledge acquired with an unlabelled dataset to use in a different but related problem).

Additionally, studies using TL were usually based on pre-trained ImageNet models, since this is a massive annotated dataset. In a comparison with this more conventional transfer learning strategy, an important advantage provided by the proposed approach relies on the similarities between source and target domains, which is often a problem when applying a model trained on natural images in a medical imaging task like CT scans. This domain gap might compromise the feature extraction since the pre-trained model might not be able to detect the relevant features for the classification task, which might cause an impact on the required fine-tuning depth, alongside with the dataset size required for this deeper fine-tuning. The approach presented in this study allows the use of public CT datasets to train the feature extractor, ensuring that the model can capture the relevant features for the final problem.

One of the limitations of this work is the binary output characteristic. The classification performed by the clinicians considers multiple classes since the malignancy evaluation in a clinical environment is not limited by a binary benign/malignant result. Different classes might give additional and useful information to help with diagnosis. As future work, a multi-class solution will be developed. However, a very clear annotated dataset must be labeled in order to decrease the label noise introduced by the inter-observer variability of the annotations. With a fine-grained classification, it is expected a higher disagreement between annotators, due to the higher difficult to distinguish between multiple classes. Several scientific societies recommended guidelines for nodule management [33]. The first assessment of the nodule is based on patient history and imaging studies. Further invasive investigation with biopsies is performed for the ones evaluated with a higher risk of malignancy. The fact that the malignancy label was based on subjective ratings by radiologists and not in an objective result such as a biopsy for all cases on the dataset, represents a limitation that will limit the performance that can be achieved by the developed models [14].

Some improvements might also be important to note to address other limitations in this work. Lung nodules are 3D elements, and with 2D or even 2.5D analysis, a large portion of useful information might be lost, which makes these perspectives sub-optimal ways of approaching this classification task. However, besides consuming more computational resources, a 3D approach does not allow a slice oversampling operation as employed in this study, which might be a problem given the lower amount of available training data, as well as the class imbalance naturally present. In the proposed feature learning task, the CAE is trained to minimize a cost function based on the pixel-wise error between the input and the correspondent reconstruction. The use of this loss might lead to blurred areas in the reconstructed images, meaning that the high-frequency components of the original image were not clearly learned by the encoder. These areas often correspond to edges or other detailed shapes, and clear learning of these low-level features might play an important role in the detection of relevant patterns related to the lung nodule malignancy [34].

Finally, the proposed approach is still limited on the explainable level since it is an end-to-end solution, which represents a black box for the clinicians. The importance of interpretability in machine learning is increasing due to the need to trust the final classification and understand the information used by the models for the prediction. The novel AI-based solutions should be transparent and understandable in order to generate scientific knowledge [35]. As future work, there is a need to create trustful models that allow the clinicians to understand which features contribute to the malignancy prediction [35]. However, this work showed important results to prove that this approach can have several advantages compared to other machine learning and deep learning solutions, maintaining the performance level for lung nodule classification.

5. Conclusions

We developed and applied an approach based on two steps: features extraction and classification, to help the diagnosis of lung nodules in CT images. This work was motivated by the need to explore options to overcome the lack of annotated biomedical data, which have been limiting the development of robust AI-based solutions in the medical field. In conclusion, this work showed that feature learning achieved when reconstructing the input with an encoder-decoder based architecture can be considered as useful knowledge in a transfer learning approach. This approach allows the use of data to learn without labeling constraints, which is one of the biggest limitations when using medical data, since the annotation is an expensive and extremely complex process.

Author Contributions: F.S., T.P., A.C. and H.P.O. conceived the scientific idea, C.F. and V.H. gave the pneumology insights about the malignancy risk assessment, and J.L.C. gave the molecular biology insights. F.S. conducted all the experiments. F.S., T.P., J.F., J.M., A.C. and H.P.O. contributed to the critical analysis of the results. F.S. and T.P. drafted the manuscript. All authors provided critical feedback and contributed to the final manuscript. All authors have read and agreed to the published version of the manuscript.

Funding: This work is financed by National Funds through the Portuguese funding agency, FCT—Fundação para a Ciência e a Tecnologia within project UIDB/50014/2020.

Acknowledgments: We acknowledged the National Cancer Institute and the Foundation for the National Institutes of Health for the free publicly available LIDC-IDRI Database used in this work.

Conflicts of Interest: The authors declare no conflict of interest.

References

1. World Health Organisation. *Latest Global Cancer Data: Cancer Burden Rises to 18.1 Million New Cases and 9.6 Million Cancer Deaths in 2018*; International Agency for Research on Cancer: Lyon, France, 2018.
2. American Cancer Society. *Facts & Figures 2019*; Technical Report; American Cancer Society: Atlanta, GA, USA, 2019.
3. Riquelme, D.; Akhloufi, M.A. Deep Learning for Lung Cancer Nodules Detection and Classification in CT Scans. *AI* **2020**, *1*, 28–67. [CrossRef]
4. Armato, S.G., III; McLennan, G.; Bidaut, L; McNitt-Gray, M.F.; Meyer, C.R.; Reeves, A.P.; Zhao, B.; Aberle, D.R.; Henschke, C.I.; Hoffman, E.A.; et al. Data From LIDC-IDRI. The Cancer Imaging Archive: Little Rock, AR, USA, 2015. [CrossRef]
5. Armato, S.G., III; McLennan, G.; Bidaut, L.; McNitt-Gray, M.F.; Meyer, C.R.; Reeves, A.P.; Zhao, B.; Aberle, D.R.; Henschke, C.I.; Hoffman, E.A.; et al. The Lung Image Database Consortium (LIDC) and Image Database Resource Initiative (IDRI): A completed reference database of lung nodules on CT scans. *Med. Phys.* **2011**, *38*, 915–931. [CrossRef] [PubMed]
6. Shen, W.; Zhou, M.; Yang, F.; Yang, C.; Tian, J. *Multi-Scale Convolutional Neural Networks for Lung Nodule Classification*; Lecture Notes in Computer Science (Including Subseries Lecture Notes in Artificial Intelligence and Lecture Notes in Bioinformatics); Springer: Cham, Switzerland, 2015. [CrossRef]
7. Liu, L.; Liu, Y.; Zhao, H. Benign and malignant solitary pulmonary nodules classification based on CNN and SVM. In Proceedings of the ACM International Conference Proceeding Series, Singapore, 23–25 April 2018. [CrossRef]
8. Yan, X.; Pang, J.; Qi, H.; Zhu, Y.; Bai, C.; Geng, X.; Liu, M.; Terzopoulos, D.; Ding, X. Classification of lung nodule malignancy risk on computed tomography images using convolutional neural network: A comparison between 2d and 3d strategies. In Proceedings of the Asian Conference on Computer Vision, Taipei, Taiwan, 20–24 November 2016; pp. 91–101. [CrossRef]
9. Song, Q.Z.; Zhao, L.; Luo, X.K.; Dou, X.C. Using Deep Learning for Classification of Lung Nodules on Computed Tomography Images. *J. Healthc. Eng.* **2017**, *2017*, 8314740. [CrossRef] [PubMed]
10. Xie, Y.; Zhang, J.; Xia, Y.; Fulham, M.; Zhang, Y. Fusing texture, shape and deep model-learned information at decision level for automated classification of lung nodules on chest CT. *Inf. Fusion* **2018**, *42*, 102–110. [CrossRef]
11. Causey, J.L.; Zhang, J.; Ma, S.; Jiang, B.; Qualls, J.A.; Politte, D.G.; Prior, F.; Zhang, S.; Huang, X. Highly accurate model for prediction of lung nodule malignancy with CT scans. *Sci. Rep.* **2018**, *8*, 9286. [CrossRef] [PubMed]
12. Rizzo, S.; Botta, F.; Raimondi, S.; Origgi, D.; Fanciullo, C.; Morganti, A.G.; Bellomi, M. Radiomics: The facts and the challenges of image analysis. *Eur. Radiol. Exp.* **2018**, *2*, 36. [CrossRef] [PubMed]
13. Soleymani, S.; Dabouei, A.; Kazemi, H.; Dawson, J.; Nasrabadi, N.M. Multi-Level Feature Abstraction from Convolutional Neural Networks for Multimodal Biometric Identification. In Proceedings of the International Conference on Pattern Recognition, Beijing, China, 20–24 August 2018. [CrossRef]
14. Lindsay, W.; Wang, J.; Sachs, N.; Barbosa, E.; Gee, J. *Transfer Learning Approach to Predict Biopsy-Confirmed Malignancy of Lung Nodules from Imaging Data: A Pilot Study*; Lecture Notes in Computer Science (Including Subseries Lecture Notes in Artificial Intelligence and Lecture Notes in Bioinformatics); Springer: Cham, Switzerland, 2018. [CrossRef]
15. Da Nóbrega, R.V.M.; Peixoto, S.A.; Da Silva, S.P.P.; Filho, P.P.R. Lung Nodule Classification via Deep Transfer Learning in CT Lung Images. In Proceedings of the International Symposium on Computer-Based Medical Systems (CBMS), Karlstad, Sweden, 18–21 June 2018; pp. 244–249. [CrossRef]
16. da Nóbrega, R.V.M.; Rebouças Filho, P.P.; Rodrigues, M.B.; da Silva, S.P.; Dourado Júnior, C.M.; de Albuquerque, V.H.C. Lung nodule malignancy classification in chest computed tomography images using transfer learning and convolutional neural networks. *Neural Comput. Appl.* **2020**, *32*, 11065–11082. [CrossRef]
17. ImageNet. Available online: http://www.image-net.org/ (accessed on 27 January 2020).
18. Russakovsky, O.; Deng, J.; Su, H.; Krause, J.; Satheesh, S.; Ma, S.; Huang, Z.; Karpathy, A.; Khosla, A.; Bernstein, M.; et al. ImageNet Large Scale Visual Recognition Challenge. *Int. J. Comput. Vis.* **2015**, *115*, 211–252. [CrossRef]

19. He, K.; Zhang, X.; Ren, S.; Sun, J. Deep Residual Learning for Image Recognition. In Proceedings of the 2016 IEEE Conference on Computer Vision and Pattern Recognition, Las Vegas, NV, USA, 27–30 June 2016; pp. 770–778.
20. Zhang, Y.; Zhang, J.; Zhao, L.; Wei, X.; Zhang, Q. Classification of Benign and Malignant Pulmonary Nodules Based on Deep Learning. In Proceedings of the 2018 5th International Conference on Information Science and Control Engineering (ICISCE), Zhengzhou, China, 20–22 July 2018. [CrossRef]
21. Shi, Z.; Hao, H.; Zhao, M.; Feng, Y.; He, L.; Wang, Y.; Suzuki, K. A deep CNN based transfer learning method for false positive reduction. *Multimed. Tools Appl.* **2019**, *78*, 1017–1033. [CrossRef]
22. Cavallari, G.; Ribeiro, L.; Ponti, M. Unsupervised Representation Learning Using Convolutional and Stacked Auto-Encoders: A Domain and Cross-Domain Feature Space Analysis. In Proceedings of the 31st Conference on Graphics, Patterns and Images, (SIBGRAPI), Parana, Brazil, 29 October–1 November 2018. [CrossRef]
23. Kumar, D.; Wong, A.; Clausi, D.A. Lung Nodule Classification Using Deep Features in CT Images. In Proceedings of the 12th Conference on Computer and Robot Vision, Halifax, NS, Canada, 3–5 June 2015; pp. 133–138. [CrossRef]
24. Cheng, J.Z.; Ni, D.; Chou, Y.H.; Qin, J.; Tiu, C.M.; Chang, Y.C.; Huang, C.S.; Shen, D.; Chen, C.M. Computer-Aided Diagnosis with Deep Learning Architecture: Applications to Breast Lesions in US Images and Pulmonary Nodules in CT Scans. *Sci. Rep.* **2016**, *6*, 24454. [CrossRef] [PubMed]
25. Wiemker, R.; Bergtholdt, M.; Dharaiya, E.; Kabus, S.; Lee, M.C. Agreement of CAD features with expert observer ratings for characterization of pulmonary nodules in CT using the LIDC-IDRI database. *SPIE Med. Imaging* **2009**, *7260*, 72600. [CrossRef]
26. Lin, H.; Huang, C.; Wang, W.; Luo, J.; Yang, X.; Liu, Y. Measuring Interobserver Disagreement in Rating Diagnostic Characteristics of Pulmonary Nodule Using the Lung Imaging Database Consortium and Image Database Resource Initiative. *Acad. Radiol.* **2017**, *24*, 401–410. [CrossRef] [PubMed]
27. Nibali, A.; He, Z.; Wollersheim, D. Pulmonary nodule classification with deep residual networks. *Int. J. Comput. Assist. Radiol. Surg.* **2017**, *12*, 1799–1808. [CrossRef] [PubMed]
28. Albawi, S.; Mohammed, T.A.; Al-Zawi, S. Understanding of a convolutional neural network. In Proceedings of the 2017 International Conference on Engineering and Technology (ICET), Antalya, Turkey, 21–23 August 2017; pp. 1–6. [CrossRef]
29. Wang, Y.; Yao, H.; Zhao, S.; Zheng, Y. Dimensionality reduction strategy based on auto-encoder. In Proceedings of the ACM International Conference Proceeding Series, Zhangjiajie, China, 19–21 August 2015. [CrossRef]
30. Pihlgren, G.G.; Sandin, F.; Liwicki, M. Improving Image Autoencoder Embeddings with Perceptual Loss. *arXiv* **2020**, arXiv:cs.CV/2001.03444.
31. Alain, G.; Bengio, Y. What Regularized Auto-Encoders Learn from the Data Generating Distribution. *arXiv* **2012**, arXiv:cs.LG/1211.4246.
32. Zhao, H.; Gallo, O.; Frosio, I.; Kautz, J. Loss Functions for Image Restoration with Neural Networks. *IEEE Trans. Comput. Imaging* **2016**, 3, 47–57. [CrossRef]
33. Loverdos, K.; Fotiadis, A.; Kontogianni, C.; Iliopoulou, M.; Gaga, M. Lung nodules: A comprehensive review on current approach and management. *Ann. Thorac. Med.* **2019**, *14*, 226–238. [CrossRef] [PubMed]
34. Ichimura, N. Spatial Frequency Loss for Learning Convolutional Autoencoders. *arXiv* **2018**, arXiv:cs.CV/1806.02336.
35. Carvalho, D.V.; Pereira, E.M.; Cardoso, J.S. Machine learning interpretability: A survey on methods and metrics. *Electronics* **2019**, *8*, 832. [CrossRef]

applied sciences

MDPI

Article

Real Driving Cycle-Based State of Charge Prediction for EV Batteries Using Deep Learning Methods

Seokjoon Hong, Hoyeon Hwang, Daniel Kim, Shengmin Cui and Inwhee Joe *

Department of Computer Science, Hanyang University, Seoul 04763, Korea; daniel379@hanyang.ac.kr (S.H.); kd3122@hanyang.ac.kr (H.H.); danielkim96@hanyang.ac.kr (D.K.); shengmincui@hanyang.ac.kr (S.C.)
* Correspondence: iwjoe@hanyang.ac.kr; Tel.: +82-02-2220-1088

Abstract: An accurate prediction of the State of Charge (SOC) of an Electric Vehicle (EV) battery is important when determining the driving range of an EV. However, the majority of the studies in this field have either been focused on the standard driving cycle (SDC) or the internal parameters of the battery itself to predict the SOC results. Due to the significant difference between the real driving cycle (RDC) and SDC, a proper method of predicting the SOC results with RDCs is required. In this paper, RDCs and deep learning methods are used to accurately estimate the SOC of an EV battery. RDC data for an actual driving route have been directly collected by an On-Board Diagnostics (OBD)-II dongle connected to the author's vehicle. The Global Positioning System (GPS) data of the traffic lights en route are used to segment each instance of the driving cycles where the Dynamic Time Warping (DTW) algorithm is adopted, to obtain the most similar patterns among the driving cycles. Finally, the acceleration values are predicted from deep learning models, and the SOC trajectory for the next trip will be obtained by a Functional Mock-Up Interface (FMI)-based EV simulation environment where the predicted accelerations are fed into the simulation model by each time step. As a result of the experiments, it was confirmed that the Temporal Attention Long–Short-Term Memory (TA-LSTM) model predicts the SOC more accurately than others.

Keywords: electric vehicle; real driving cycle; recurrent neural network; simulation; state of charge; temporal attention

Citation: Hong, S.; Hwang, H.; Kim, D.; Cui, S.; Joe, I. Real Driving Cycle-Based State of Charge Prediction for EV Batteries Using Deep Learning Methods. *Appl. Sci.* **2021**, *11*, 11285. https://doi.org/10.3390/app112311285

Academic Editors: João M. F. Rodrigues, Pedro J. S. Cardoso and Marta Chinnici

Received: 23 October 2021
Accepted: 22 November 2021
Published: 29 November 2021

1. Introduction

Since Electric Vehicles (EVs) do not emit CO_2, they have a great potential to prevent air pollution [1]. In addition, it is attractive for drivers to use EVs, because the cost of charging the battery of an electric vehicle is much cheaper than the cost of refueling [2]. In recent years, the performance of electric vehicle batteries has greatly improved and, consequently, there are EVs capable of driving more than 500 km when fully charged. However, there are few battery-charging stations compared to conventional gas stations, and batteries take a long time to charge [3]. Therefore, when driving an electric vehicle, it is very important to predict the driving distance through the state-of-charge (SOC) of the battery.

Accurate battery SOC measurement is an important feature in the BMS of electric vehicles, which can be implemented with microcontrollers (MCUs). Recently, as pre-trained neural network models have been installed and used in automotive MCUs, it has become possible to apply AI to the Battery Management System (BMS) [4]. In addition, among recent studies, a method to more accurately measure SOC using a cloud data center that provides high-capacity and high-performance calculations for big-data-based, data-driven Deep Learning (DL), and interworking with the vehicle's BMS, was also proposed [5].

Future applications for fully Connected and Autonomous Vehicles (CAVs) include an AR-based navigation system, a video-conferencing application, real-time image-processing and inferencing solutions implemented by a neural network model equipped on board [6,7]. Since these applications require a high level of computation power and low power consumption, various hardware implementations such as a Graphics Processing Unit (GPU), field-programmable gate array (FPGA), and Application-Specific Integrated Circuits (ASIC) are employed. Specifically, a Tesla Full Self-Driving (FSD) computer, built on ASIC, meets 50 Tera floating point operations per second (TFLOPS) of the Neural Network model, and uses 100 W or less power consumption during computation [8]. To benefit from these state-of-the-art embedded systems adapted to the full CAVs, it is important to both estimate the current SOC and predict the future SOC to aid the drivers in decision-making.

It is important to accurately measure the current SOC, but it is also necessary to predict the future SOC based on the current driving data. An accurate prediction of the SOC can help determine the possible driving distance and charging time. There are two main methods to predict the SOC of an EV battery. The first one is predicting the SOC through simulation based on the vehicle dynamics. This approach was implemented by [9], which used standard driving cycles (SDC) as input of the simulation model. The second method consists of predicting the SOC through Machine Learning (ML) techniques using the battery data collected while driving the electric vehicle [10]; for example, using information from the battery itself (voltage, current, temperature) and ML algorithms to predict the future SOC.

Recently, there have been studies to predict SOC through complex mathematical formulas, considering the actual driving cycle of electric vehicles, but ML was not used.

In this paper, we propose an algorithm that accurately predicts the SOC of an EV battery by acquiring actual driving information through OBD-II, dividing and preprocessing this driving information by section based on GPS information and DTW, training a neural network with the dataset and running simulations using Functional Mock-Up Units (FMU).

2. Related Work

This section classifies the related studies into two major categories: modeling and simulation, and SOC estimation and prediction in the EV domain. Since there have been significant research efforts in these fields, we are focused on work that has been conducted rather recently, and discuss the limitations of these works.

First of all, in the field of modeling and simulation, extensive research has been carried out to improve the accuracy of EV energy consumption estimation with high-fidelity simulation models [11–17]. By using simulation environments such as MATLAB/Simulink and DACCOSIM, each physical component of an EV is mathematically modeled to form a subsystem block, which is composed of primitive blocks. The simulation is performed by the subsystem blocks mapping the input signals to the output values, which are passed to the subsequent blocks over time during the simulation. Recently, distributed co-simulation based on Functional Mock-up Interface (FMI), which provides a standardized interface to ease the sharing of different models [18], has been widely used. Techniques to reduce the simulation time while maintaining the simulation accuracy have also been developed. S. Hong proposed a redundancy reduction algorithm (RRA) to increase the simulation speed by adaptively adjusting the simulation step size, increasing it when a specific pattern of the cycle is detected as repeated, and decreasing it when a zero-crossing point is detected [19]. However, this is mainly applicable to the SDCs, especially the New European Driving Cycle (NEDC), where many repetitive parts are known ahead of the simulation.

These studies heavily rely on the SDCs to validate the methods. However, as mentioned above, SDCs have inherent limitations, as they are not real driving cycles (RDCs). Even though a variety of SDCs have been developed to emulate urban, rural, and highway environments, they do not reflect personalized factors, i.e., individual drivers' habits, which are highly complicated as they dynamically change over time [20,21]. Moreover, they can hardly incorporate factors such as traffic conditions, traffic signals, and time of driving, which are unique to each driving instance [22]. Therefore, in order to provide more personalized simulation results, both the mathematically defined models and data-driven methods that learn the drivers' patterns from the previous RDCs are required.

Next, there have been a myriad of studies in the area of SOC estimation and prediction. In the application of the pure EV, accurate prediction of the battery SOC consumption is highly important, since this would directly lead to an accurate prediction of the EV range [23]. According to the literature [24–26], methodological approaches to the SOC estimation can be divided into two: ML-based and non-ML-based methods. In brief, Coulomb counting and open-circuit voltage (OCV) methods have been used as conventional methods, while, recently, ML models consisting of deep neural networks (DNNs) have been intensely applied [24–26]. In the ML-based data-driven approach, the SOC values are predicted by automatically adjusting parameters to minimize the resultant error by learning from the pattern of the input features such as current, voltage, and temperature [27–29]. In this way, the models are fitted to the specific real-world measurement data, which enables predictions with enhanced accuracy under similar conditions.

However, few studies have been conducted on the prediction of future SOC trends through simulation. Most of the research has been focused on the internal parameters of the battery management system (BMS) [30] to estimate the SOC value. That is, research on the relationship between the EV subsystems in an integrated view to improve the performance of the simulation, as well as reducing the overhead of the tedious model-based simulation, has been insufficient. If an individual driver's OBD data can be repeatedly collected on the road, one can simulate the SOC patterns for the next cycle for the same segment of the route with the velocity and acceleration data as inputs to an EV model. In this way, an approximate trajectory of the SOC consumption can be obtained, which can, in turn, help the driver be aware of the possible range and adjust her driving behavior from the perspective of an SOC.

3. Overall Procedure for Real Driving Cycle-Based SOC Estimation

This paper proposes a method to predict SOC using a real-world driving cycle. The proposed SOC prediction method using real-world driving cycle is as shown in the Figure 1.

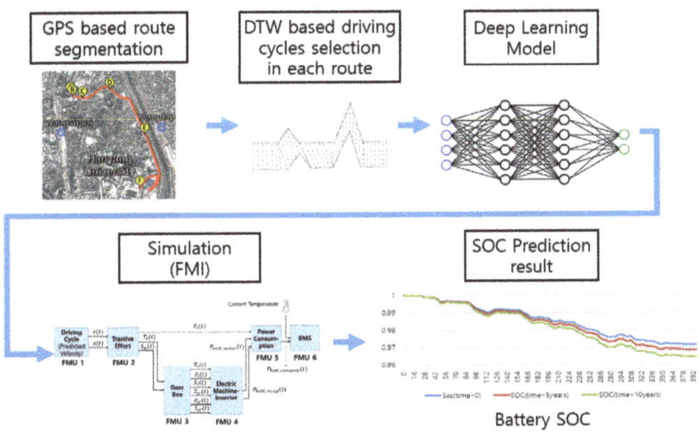

Figure 1. Procedure of Real Driving cycle based SOC estimation.

The tested driving routes run from a fixed site to the other, i.e., from the homeplace of one of the authors to a parking lot in the Hanyang University campus. The route consists of several intersections between the start point and the destination point. The driving data were collected onboard while the author was driving, using an application named Torque Pro running on an Android device connected to a Bluetooth dongle, which supports the OBD-II specifications. The OBD data were comprised of various second-by-second temporal data, such as GPS latitude, longitude, velocities, etc., and were saved in a Comma-Separated-Value (CSV) format. The route is highlighted, as shown in Figure 2. The route was extracted and drawn on the map using the Google API with the GPS latitude and longitude information from the OBD data.

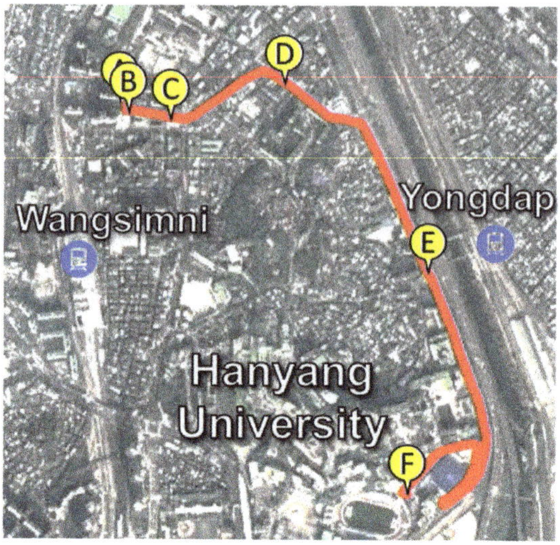

Figure 2. Real Driving Cycle with Google MAP.

Since the overall driving cycles are highly dependent on the operations of en-route traffic lights, to enhance the accuracy of the predicted driving cycles, we first divided each of the end-to-end routes into sub-routes by the traffic lights (as can be seen from Figure 2 as circular markers). The segments obtained from different instances of the route were then compared to one another. To find the most similar intervals across all of the instances of the route, a Dynamic Time Warping (DTW) algorithm was employed. After grouping the segments of the route by their similarity, different Machine Learning (ML) algorithms were used to learn from them and predict future driving cycles.

To validate the performance of the prediction of the SoC consumption, we generated the eventual SoC values by an Electric Vehicle (EV) simulation model, implemented based on Functional Mock-up Units (FMU). The FMU-based EV simulation model reads as inputs the velocities, calculates the electric current from power consumption and regeneration by acceleration and deceleration, respectively, and writes as outputs the resultant SoC values to a file at each second. As expected, the test results showed that the SoC values could be predicted with high accuracy, as long as the driving cycle was predicted with high accuracy.

The paper is structured as follows: first, we proposed the preprocessing method using the DTW algorithm to group the driving intervals by similarity. Secondly, we proposed a deep learning algorithm that accurately predicts the speed after learning from similar driving cycles. The speed was also predicted using the existing Seasonal Autoregressive Integrated Moving Average (SARIMA) model. Finally, to evaluate the models, the predicted speeds by both algoritms were used as input to a simulation model, which computes the SOC of the driving cycle.

4. Methodology

4.1. Dynamic Time Warping (DTW) for Similarity of Real Driving Cycles

First proposed by [31] in the speech recognition field, the DTW is an algorithm that measures similarity between time series, which may vary in length. The technique performs non-linear warpings so that the time series are stretched or shrinked in order to find the optimal alignment between them. It has a wide range of applications, such as data mining, gesture recognition, robotics, speech-processing, manufacturing and medicine [32].

We propose a method to find similar driving routes based on the DTW algorithm, using the acceleration of the vehicle as input. The reason why acceleration is used is that the SOC is closely related to power consumption, which, in turn, is closely related to the acceleration of the vehicle. There were more than 30 driving-cycle-related data collected and, due to the large noise and uncertainty existent in the dataset, only similar segments were used for the prediction. Therefore, the goal of the algorithm is to find the most similar acceleration time series segments to use them as input to the ML models.

The algorithm that was used can be described in the following manner:

- First of all, the *base* time series which we want to use to find other similar time series must be defined. For this present work, we defined the route shown in the map of the Figure 2, represented by the time series in the Figure 3, as the base route to which the remaining 30 time series will be compared.
- Secondly, the base time series may further be split into an arbitrarily number of base segments. Considering the Figures 2 and 3, the markers from A to F and the dotted black vertical lines represent situations in which the velocity of the car reached zero or, in other words, the start and the end points of each of the four segments (A to C, C to D, D to E, E to F). Note that the point B was not taken into account, because we regarded it as noise due to its proximity to point A, and that the each of the base segments are shown in greater detail in the subfigures of Figure 4.
- Thirdly, for each of the four *base* segments, we set the lower bound and the upper bound times for the similar patter- finder algorithm by subtracting and adding an ϵ arbitrary small amount of time. For example, the second *base* segment was extracted from second 69 to 135 of the *base* time series. If $\epsilon = 50$, then the lower-bound and upper-bound times for a similar pattern-finder algorithm will be, respectively, 19 s and 185 s, as shown in Table 1.
- With the lower and upper bounds of the algorithm defined, we split the period between the lower and upper bounds into smaller search periods, spaced by an arbitrary time interval. Intuitively, the lower the interval is, the higher the accuracy of the algorithm. These smaller search periods are shown in Table 1 for an interval of five seconds between each period.
- Next, we looped through the search periods.
- Inside the search-periods-loop, we loop through the remaining 30 time series, which will be compared to the *base* time series.
- Next, we extract the corresponding search period of the time series to be used in the comparison.
- Next, we obtain the distance between the extracted time series and the *base* segment by DTW.
- Finally, after the two loops are finished, we rank the all the extracted time series according to their cost, computed by DTW, and obtain the 10 periods that have the lowest cost, which will form the training and test datasets for the machine learning models.

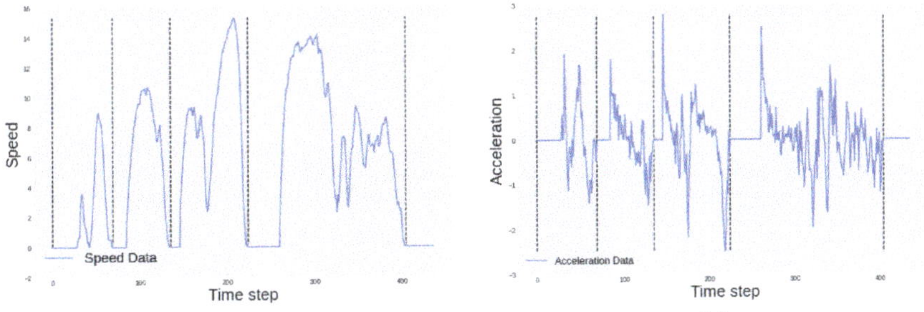

(a) (b)

Figure 3. Speed and acceleration over time of the base time series. (**a**) Speed × time of the base time series. (**b**) Acceleration × time of the base time series.

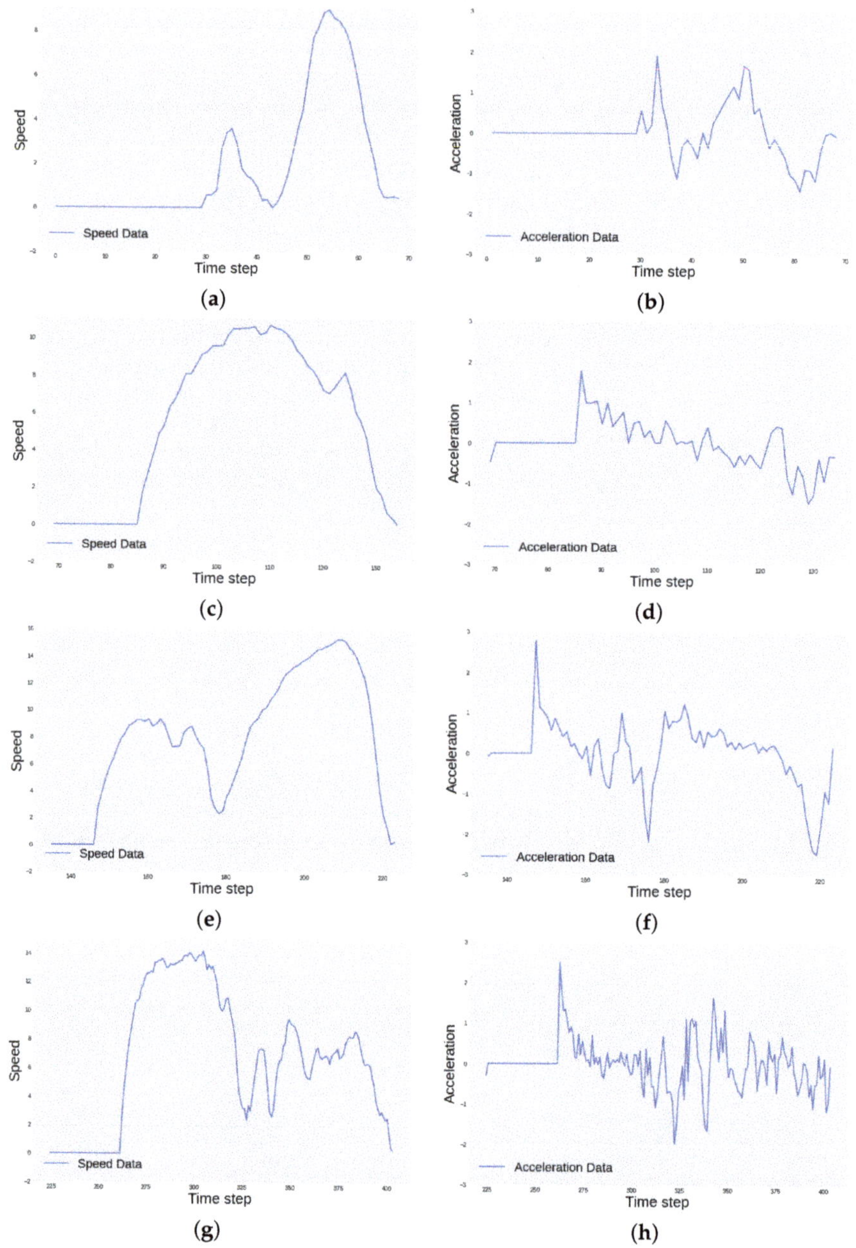

Figure 4. Speed and acceleration over time of the four base segments. (**a**) Speed × Time (A to C segment); (**b**) Acceleration × Time (A to C segment); (**c**) Speed × Time (C to D segment); (**d**) Acceleration × Time (C to D segment); (**e**) Speed × Time (D to E segment); (**f**) Acceleration × Time (D to E segment); (**g**) Speed × Time (E to F segment); (**h**) Acceleration × Time (E to F segment).

Table 1. Table showing the start and end time of the segments and the search periods.

	First Base Segment	Second Base Segment	Third Base Segment	Fourth Base Segment
Start Point (s)	0	69	135	224
End Point (s)	69	135	224	404
Search Start Point (s)	0	19	85	174
Search End Point (s)	119	185	274	454
examles of Search periods for interval = 5 s (start, end)	(0, 69), (5, 74), ... (45, 114), (50, 119)	(19, 85), (24, 90), ... (114, 180), (119, 185)	(85, 174), (90, 179), ... (180, 269), (185, 274)	(174, 354), (179, 359), ... (269, 449), (274, 454)

A summary of the algorithm is shown in Algorithm 1.

Algorithm 1 Similar time series finding algorithm

```
Set base_time_series
Split base_time_series into base_segments
for bs in base_segments do
    Set lower_bound and upper_bounds
    Set interval
    Compute search_periods
    Initialize empty list costs
    for period in search_periods do
        for ts in time_series_list do
            Extract period from ts
            cost ← DTW(extracted_ts_period, bs)
            Append cost to costs
        end for
    end for
    get_n_series_least_cost(costs, num_series)
end for
```

4.2. Deep Learning-Based SOC Prediction

In this section, our intention is to predict the future speed from the past speed of the EV; hence, we consider this task as a univariate time series prediction task. Given the observed time series data $\mathbf{x} = [x_1, x_2, \ldots, x_T] \in \mathbb{R}^T$, the task is to predict the future value $x_{T+1} \in \mathbb{R}$. Formally, we intend to predict $\hat{x}_{T+1} \in \mathbb{R}$ through a function f, as follows:

$$\hat{x}_{T+1} = f(x_1, x_2, \ldots, x_T), \tag{1}$$

where $f(\cdot)$ is a linear or nonlinear function that needs to be learned.

Our main contribution is presenting a model for predicting speed, which is based on an LSTM with a temporal attention mechanism. The architecture of the prediction model is shown in Figure 5. First, we use an LSTM layer to encode the information from the input sequence into a feature representation. The final prediction is then made by utilizing the temporal attention mechanism over the output features of the LSTM layer.

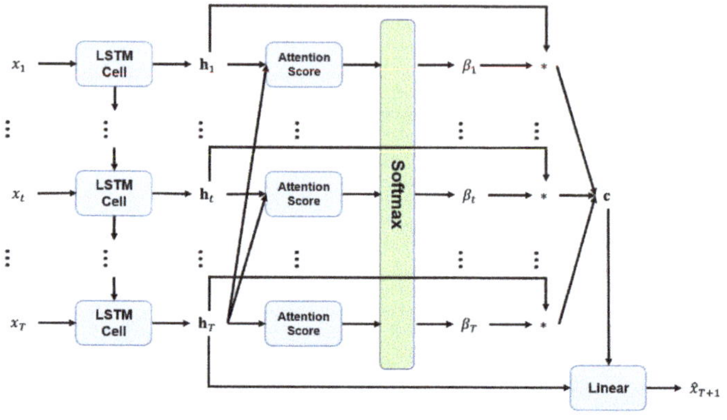

Figure 5. Architecture of TA-LSTM.

Recurrent Neural Network (RNN) is one of the family of neural networks specialized in processing sequential data. If a general feed-forward neural network approach is used to process fixed-length data, the network has separate parameters for each input feature, so it needs to learn them separately at each time position. In contrast, an RNN shares weights across multiple time steps. This sharing of weights is important, as it allows for the generalization of unseen sequences and the sharing of statistical strength across time steps. Long short-term memory (LSTM) and gated recurrent unit (GRU), which are variants of RNN, are commonly used for handling sequential data such as language modeling [33,34] and time series prediction [35,36]. LSTM outperforms RNN and GRU for many sequence prediction tasks because of its gating systems, so we chose LSTM for this task. The forward process of an RNN cell is defined as:

$$\mathbf{f}_t = \sigma(\mathbf{W}_f[x_t, \mathbf{h}_{t-1}] + \mathbf{b}_f) \tag{2}$$

$$\mathbf{i}_t = \sigma(\mathbf{W}_i[x_t, \mathbf{h}_{t-1}] + \mathbf{b}_i) \tag{3}$$

$$\tilde{\mathbf{c}}_t = \tanh(\mathbf{W}_c[x_t, \mathbf{h}_{t-1}] + \mathbf{b}_c) \tag{4}$$

$$\mathbf{c}_t = \mathbf{f}_t \cdot \mathbf{c}_{t-1} + \mathbf{i}_t \cdot \tilde{\mathbf{c}}_t \tag{5}$$

$$\mathbf{o}_t = \sigma(\mathbf{W_o}[x_t, \mathbf{h}_{t-1}] + \mathbf{b}_o) \tag{6}$$

$$\mathbf{h}_t = \mathbf{o}_t \cdot \tanh(c_t) \tag{7}$$

where $\mathbf{W}_f, \mathbf{W}_i, \mathbf{W}_c, \mathbf{W}_o \in \mathbb{R}^{m \times (m+1)}$ are weight matrices, and $\mathbf{b}_f, \mathbf{b}_i, \mathbf{b}_c, \mathbf{b}_o \in \mathbb{R}^m$ are bias to be learned. $\mathbf{f}_t$ is the forget gate, which determines how much previous information is forgotten, $\mathbf{i}_t$ is the input gate, which determines how much new information is added, and $\mathbf{o}_t$ is the output gate, which, along with the current cell state $\mathbf{c}_t$, determines the output of this cell.

Recently, researchers have proposed multiple attention mechanisms for time series tasks and achieved better results than LSTM and GRU [37–39]. Inspired by the dynamic spatial–temporal attention mechanism [39], we employ the temporal attention mechanism to predict the acceleration of EVs in this work, since our task is univariate time series forecasting, which does not require consideration of spatial attention. Therefore, the temporal attention mechanism assigns a weight to each hidden state by correlating the output at each time step with the output at the last time step. Intuitively, for the prediction of the next time step, the current state is very important, but we cannot ignore the state of previous time steps. A general RNN-based prediction model either uses a fully connected layer to connect the hidden states of all time steps, giving a relatively accurate view of the hidden states of all time steps, or uses the last hidden state for prediction. However, we believe that temporal attention chooses the balanced option of adaptively attributing all time steps an attention score according to the relevance of different time steps to the state of the last time step, and finally connecting the fused context vector with the last hidden state for prediction, which, in turn, enhances the contribution of the current state to the prediction of the next time step.

First, the attention score is computed based on the relevance between the current state and the last state.

$$\text{Attention Score}(\mathbf{h}_t, \mathbf{h}_T) = \mathbf{h}_t^T \mathbf{W}_s \mathbf{h}_T, \quad 1 \leq t \leq T \tag{8}$$

where $\mathbf{W}_s \in \mathbb{R}^{m \times m}$ is learnable weights, which can be trained jointly with the LSTM layer to adaptively learn the correlation of hidden state of each time step with the last time step.

Then, for each time step, the attention score is converted into probabilistic form using the Softmax function, and the attention weight for this timepoint is obtained by

$$\beta_t = \frac{\exp(\text{Attention Score}(\mathbf{h}_t, \mathbf{h}_T))}{\sum_{j=1}^{T} \exp(\text{Attention Score}(\mathbf{h}_j, \mathbf{h}_T))} \tag{9}$$

where the attention weight $\beta_t \in \mathbb{R}$ demonstrates the importance of hidden state h_t for prediction and the Softmax function is applied to ensure all β sum to 1.

Next, the context vector is obtained by aggregating the hidden states of the RNN layer:

$$\mathbf{c} = \sum_{t=1}^{T} \beta_t \mathbf{h}_t \tag{10}$$

where c is a weighted sum of all hidden states and can be considered as an adaptive selection of relevant hidden states among all time steps.

Finally, the prediction $\hat{x}_{T+1}$ can be obtained through the linear combination of context vector and the last hidden state:

$$\hat{x}_{T+1} = \mathbf{W}_l^T [\mathbf{c}, \mathbf{h}_T] + b_o \tag{11}$$

where $\mathbf{W}_l \in \mathbb{R}^{2m}$ and $b_o \in \mathbb{R}$ are learnable parameters.

The proposed model is differentiable, so the learnable parameters can be updated by back propagation. The mean squared error (MSE) is applied as loss function:

$$Loss(x_{T+1}, \hat{x}_{T+1}) = \frac{1}{N} \sum_{i=1}^{N} (x_{T+1}^i - \hat{x}_{T+1}^i)^2 \tag{12}$$

where N is the number of samples.

5. Experimental Results

Before applying the proposed method to the RDCs, we checked the prediction results of an SDC using Seasonal Auto Regressive Integrated Moving Average (SARIMA), a conventional time-series forecasting model. This model is an extended version of ARIMA, supporting the modeling of the seasonal components of the time-series data. We used NEDC as the SDC and predicted the velocity after training the model. The mean absolute percentage error (MAPE) is used as criteria for evaluating the performance.

$$MAPE = \frac{1}{N} \sum_{t=1}^{N} \left| \frac{x_t - \hat{x}_t}{x_t} \right| \tag{13}$$

where x_t and $\hat{x}_t$ are ground truth and predicted value, respectively.

As can be seen from Table 2, the SDC can be predicted accurately enough without using the DL model. However, the accuracy for the RDC is much lower than the SDC. Therefore, we will use the proposed DL method to improve the predictions for the RDC.

Table 2. MAPE velocity results of SARIMA for the SDC (NEDC) and RDC.

Model	SDC (NEDC)	RDC (Four Road Segments)			
		AtoC	CtoD	DtoE	EtoF
SARIMA	0.014	0.06	0.06	0.12	0.35

To validate our proposed model using RDC, we collected real EV driving data through OBD II. Our dataset of EV speed was obtained from actual driving and divided into four road segments: location A to location C, location C to location D, location D to location E, and location E to location F. Each route has 11 driving cycles, and we took the seen driving cycles of each route as the training set and the rest as the test set. We first verified the performance of our speed predictions. by analyzing the performance of our model using the grid search method with different combinations of hyperparameters. The purpose was to find the best combination of hyperparameters and then verify the effect of each component on the results. Once we obtained the best model, we predicted the speed/acceleration along with the SOC values and compared them with some commonly used methods.

5.1. Speed/Acceleration Prediction

5.1.1. Performance with Different Hyperparameters

First, we integrated all the training sets in the dataset and adopted a five-fold cross validation (CV) method to find the best combination of hyperparameters. We implemented our model using the pytorch library and trained and tested it on an Nvidia GTX 1080 Ti GPU. Adam optimizer was used to train the models. The hyperparameters we need to consider here are the time step length T and the hidden state size m of the LSTM layer. For the proposed model, we set T as varying among $[5, 10, 15]$ and m as varying among $[16, 32, 64, 128]$. The average RMSE results for the 5-fold CV are tabulated in Table 3. First, we describe the effect of the time step length T. It is obvious from Table 3 that the average RMSE results for $T = 10$ and $T = 15$ are notably larger than for the case of $T = 5$. Therefore, for speed prediction, the best choice of time step length is 5. Then, for the case $T = 5$, the average RMSE value is the smallest when the hidden state size m equals to 32. In deep learning, too few parameters can lead to underfitting problems while too many parameters can lead to overfitting problems. Therefore, we believe that $m = 16$ belongs to underfitting, $m = 64$ and $m = 128$ belong to overfitting. Thus, the optimal combination of hyperparameters for our model for this dataset is $\{T = 5, m = 32\}$.

Table 3. RMSE results of 5-fold CV.

Hyperparameters		Average RMSE
T	m	
5	16	0.4908
	32	0.4880
	64	0.4926
	128	0.4926
10	16	0.4988
	32	0.4998
	64	0.5013
	128	0.4982
15	16	0.5007
	32	0.5055
	64	0.5025
	128	0.5059

5.1.2. Performance with Different Models

First, to verify that the LSTM in TA-LSTM is the preferred encoding layer, we compared LSTM and other RNN variants on four road segments of the dataset. RNN and GRU are commonly used for time series forecasting, and all three models follow the five-fold CV approach introduced in the previous section to identify the best combination of hyperparameters. All three models obtained the minimum average RMSE at $\{T = 5, m = 32\}$. The first seven driving cycles of each road segment were used as the training set and the remaining four driving cycles were used as the test set, and each road segment was trained and tested separately. In addition, 20% of the data in the training set were randomly selected as the validation set during the training process, and the best-performing model in the validation set was saved for testing. The best test results for each model after training multiple times are shown in Table 4, as can be seen in Table 4, LSTM performs the best in all four road sections. Therefore, we choose LSTM as the coding layer in this task.

Table 4. RMSE results of different models.

Models	AtoC	CtoD	DtoE	EtoF
RNN	0.4485	0.3702	0.4549	0.4924
GRU	0.4369	0.3689	0.4572	0.4873
LSTM	0.4240	0.3556	0.4524	0.4864

Then, to verify the effectiveness of temporal attention, we compared the LSTM with TA-LSTM. The training process and the allocation of training and test sets are the same as in the previous experiment. The results of the comparison are depicted in Figure 6. The comparison shows that the addition of temporal attention yields better results.

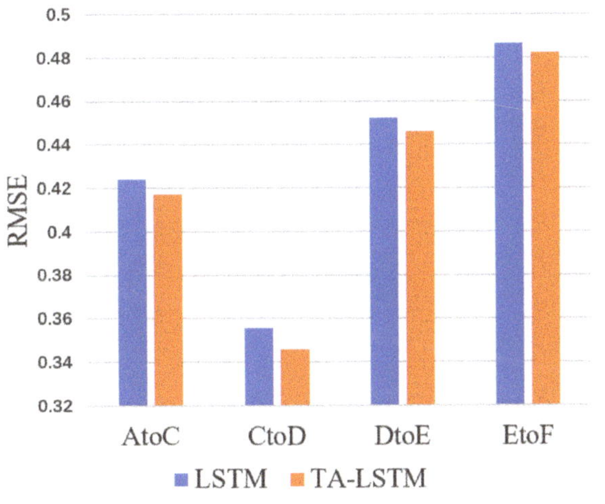

Figure 6. The effect of temporal attention on RMSE results.

Since temporal attention can be combined with various variants of RNN, we compared the performance of different encoding layers combined with temporal attention. From Tables 4 and 5, it can be seen that adding temporal attention to all three different encoding layers can improve the prediction accuracy. TA-LSTM has the best performance. Therefore, we chose TA-LSTM for the task of predicting speed. The comparison between the predicted and true values of each driving cycle for each road section is shown in Figures 7–10. As can be observed from these figures, our model can fit the real data very well in most cases.

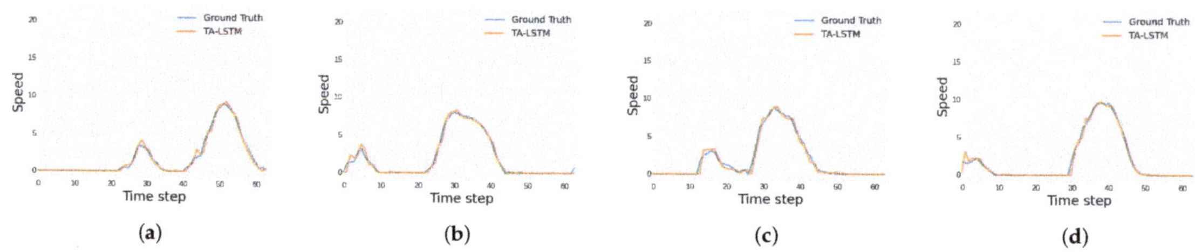

Figure 7. Speed prediction of route A to C: (**a**) cycle 8; (**b**) cycle 9; (**c**) cycle 10; (**d**) cycle 11.

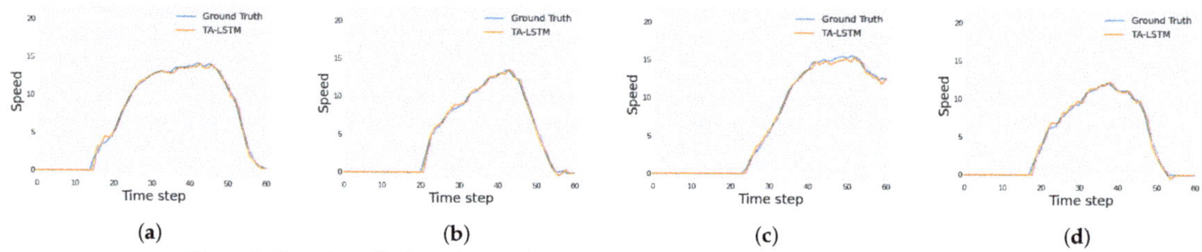

Figure 8. Speed prediction of route C to D: (**a**) cycle 8; (b) cycle 9; (**c**) cycle 10; (**d**) cycle 11.

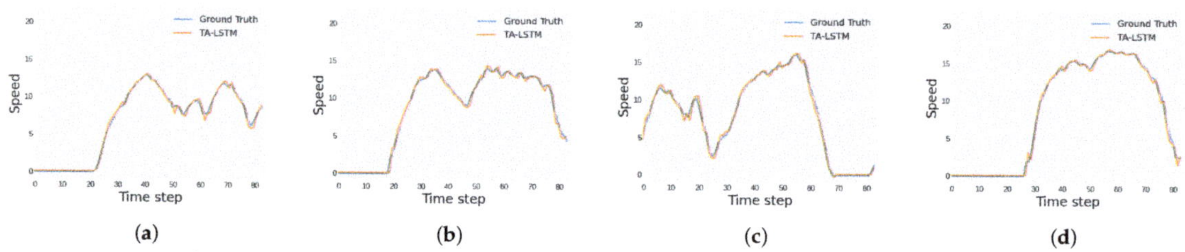

Figure 9. Speed prediction of route D to E: (**a**) cycle 8; (**b**) cycle 9; (**c**) cycle 10; (**d**) cycle 11.

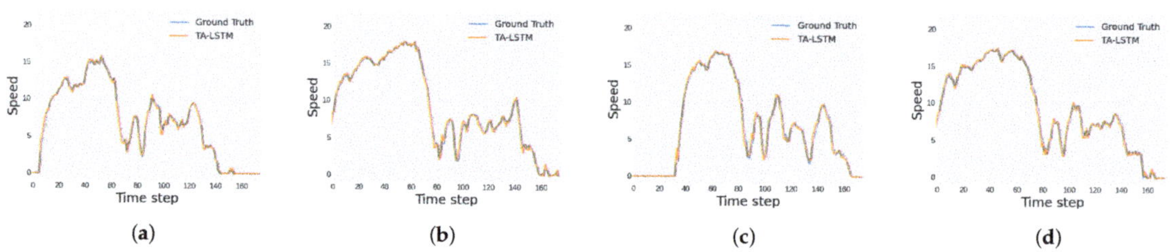

Figure 10. Speed prediction of route E to F: (**a**) cycle 8; (**b**) cycle 9; (**c**) cycle 10; (**d**) cycle 11.

Table 5. RMSE results of temporal attention with different models.

Models	AtoC	CtoD	DtoE	EtoF
TA-RNN	0.4412	0.3614	0.4515	0.4870
TA-GRU	0.4295	0.3586	0.4527	0.4832
TA-LSTM	0.4171	0.3456	0.4461	0.4821

5.2. SOC Prediction Using FMI-Based EV Simulation

After predicting the speed of the EV, we subsequently utilized the predicted results and our simulator to predict the SOC.

We used FMI-based EV simulation with the previously predicted velocity and acceleration as inputs to predict the EV's SOC. Our EV Simulation model is shown in Figure 11. As can be seen in the figure, given the inputs, the FMU2 to FMU6 are responsible for computing the SOC using the Formulas (14)–(38), whose variables are detailed in Table 6. Additionally, the lookup2d function in Formulas (29)–(31) performs the same operation as lookup2d in MATLAB/Simulink. The necessary efficiency values of the electric machine were obtained through the efficiency curves given in [9].

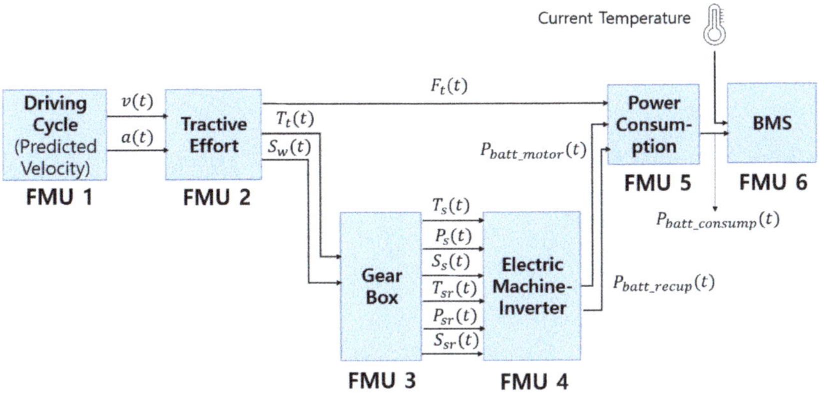

Figure 11. FMI-based model for EV power consumption.

Table 6. Parameters for FMU-based EV simulation (Default setting).

Parameters	Description	Values
m	Mass of the vehicle (kg)	1000
r_w	Wheel radius (m)	0.2736
g	Gravity acceleration (m/s^2)	9.81
ρ	Air density	1.2
A	Front area of vehicle (m^2)	2.36
α	Angle of driving surface (rad)	0
μ_{rr}	Rolling resistance coefficient	0.015
C_d	Aerodynamic drag coefficient	0.3
η_g	Gearbox efficiency	0.98
G	Gearbox ratio	8.59
C_0	Initial capacity (C)	720,000
R_{bi}	Internal resistance (Ω)	0.008
E_{b0}	Open-circuit voltage (V)	53.6
K_{ct}	Linear t dependency of the capacity C	6.084436×10^{-10}
t_{bu}	battery usage time (s)	0
T_{ref}	Reference temperature (°C)	20 °C
T_{curr}	Ambient temperature (°C)	20 °C
$alphaC$	Linear temperature coefficient of capacity (K^{-1})	0

$$F_t = F_{rr} + F_{ad} + F_{hc} + F_{la} + F_{wa} \tag{14}$$

$$F_{rr} = \mu_{rr} m g \tag{15}$$

$$F_{ad} = \frac{1}{2} \rho A C_d v^2 \tag{16}$$

$$F_{hc} = mg\sin(\alpha) \tag{17}$$

$$F_{la} = ma \tag{18}$$

$$F_{wa} = 0.05 \times F_{la} \tag{19}$$

$$T_t = F_t \times r_w \tag{20}$$

$$P_t = F_t \times v \tag{21}$$

$$\omega_w = \frac{v}{r_w} \tag{22}$$

$$S_w = \frac{30}{\pi} \times \omega_w \tag{23}$$

$$T_s = \frac{T_t}{\eta_g \times G}(P_t > 0) \tag{24}$$

$$T_{sr} = -\eta_g \times \frac{T_t}{G}(P_t < 0) \tag{25}$$

$$S_s = S_{sr} = G \times S_w \tag{26}$$

$$P_s = T_s \times S_s \times \frac{\pi}{30} \tag{27}$$

$$P_{sr} = T_{sr} \times S_s \times \frac{\pi}{30} \tag{28}$$

$$\text{Efficiency}(\eta) = \text{lookup2d}(\text{Torque}(\text{Nm}), \text{Speed}(\text{rpm})) \tag{29}$$

$$\eta_m = \text{lookup2d}(T_s, S_s) \tag{30}$$

$$\eta_r = \text{lookup2d}(T_{sr}, S_{sr}) \tag{31}$$

$$P_{bm} = \frac{P_s}{\eta_m} \tag{32}$$

$$P_{br} = \eta_r \times P_{sr} \tag{33}$$

$$P_{bc}(t) = \begin{cases} P_{bm}(t) + P_{aux}, & F_t(t) > 0 \\ P_{br}(t) + P_{aux}, & F_t(t) < 0 \end{cases} \tag{34}$$

$$I_B(t) = \frac{E_{B0}}{2 \times R_{Bi}} - \sqrt{\left(\frac{E_{B0}}{2 \times R_{Bi}}\right)^2 - \frac{P_{bc}}{R_{Bi}}} \tag{35}$$

$$Q(t) = \int_0^t I_B dt \tag{36}$$

$$C = C_0 * (1 - K_{Ct} * t_{bu}) * \left(1 + alphaC * \left(T_{curr} - T_{ref}\right)\right) \tag{37}$$

$$SOC(t) = \frac{C - Q(t)}{C} \tag{38}$$

To verify the effectiveness of our speed prediction model for the final SOC prediction, we compared the proposed model results with the speed and SOC prediction obtained with the Seasonal Autoregressive Integrated Moving Average (SARIMA) model. The final SOC prediction results are shown in Table 7. As can be seen from the table, the RMSE of TA-LSTM drops 56.69%, 84.97%, 82.34%, and 91.62% compared to SARIMA in the AtoC, CtoD, DtoE, and EtoF sections, respectively.

Table 7. RMSE ($\times 10^{-2}$) results of SOC prediction.

Models	AtoC	CtoD	DtoE	EtoF
SARIMA	0.0820	0.0998	0.1512	0.3938
TA-LSTM	0.0356	0.0150	0.0267	0.0330

The predicted SOC results made by TA-LSTM and SARIMA for the driving cycle of each road section are shown in Table 8 and Figures 12–15. From Table 8, we can see that TA-LSTM performed better than SARIMA in every driving cycle. In addition, the RMSE results of TA-LSTM for all driving cycles in the CtoD and DtoE sections are less than 0.0004. Although the RMSE for the AtoC and EtoF section is slightly worse than the first three sections, the RMSE results are less than 0.0006. The comparison between the real SOC and the predicted values shows that the predicted results of our model are very close to the real values. This indicates that our proposed method is able to perform SOC prediction properly.

Table 8. RMSE ($\times 10^{-2}$) results of SOC prediction of each driving cycle.

Cycle	AtoC		CtoD		DtoE		EtoF	
	SARIMA	TA-LSTM	SARIMA	TA-LSTM	SARIMA	TA-LSTM	SARIMA	TA-LSTM
8	0.0892	0.0161	0.1288	0.0170	0.1236	0.0195	0.2505	0.0176
9	0.0792	0.0499	0.0411	0.0139	0.1179	0.0237	0.5170	0.0252
10	0.0604	0.0368	0.1429	0.0163	0.1408	0.0381	0.2615	0.0288
11	0.0950	0.0310	0.0339	0.0122	0.2060	0.0214	0.4711	0.0509

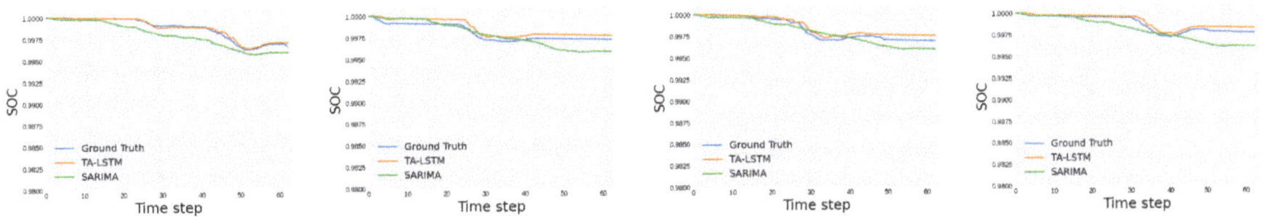

(a) (b) (c) (d)

Figure 12. SOC prediction of route A to C: (**a**) cycle 8; (**b**) cycle 9; (**c**) cycle 10; (**d**) cycle 11.

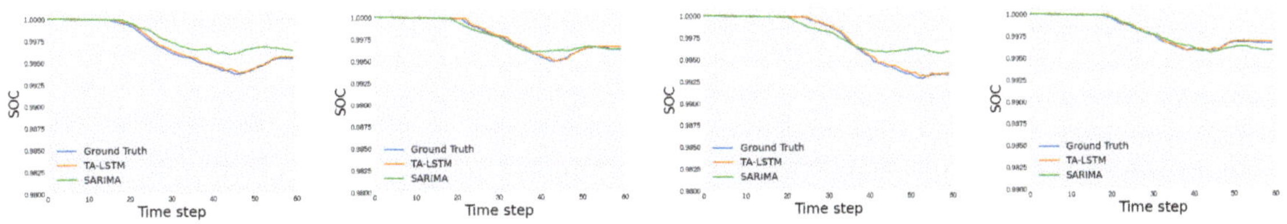

(a) (b) (c) (d)

Figure 13. SOC prediction of route C to D: (**a**) cycle 8; (**b**) cycle 9; (**c**) cycle 10; (**d**) cycle 11.

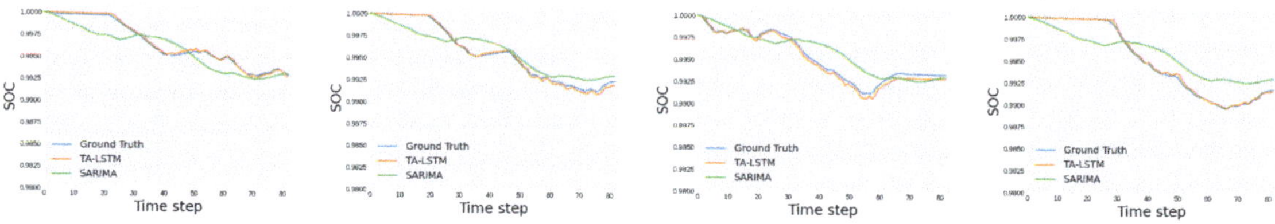

(**a**) (**b**) (**c**) (**d**)

Figure 14. SOC prediction of route D to E: (**a**) cycle 8; (**b**) cycle 9; (**c**) cycle 10; (**d**) cycle 11.

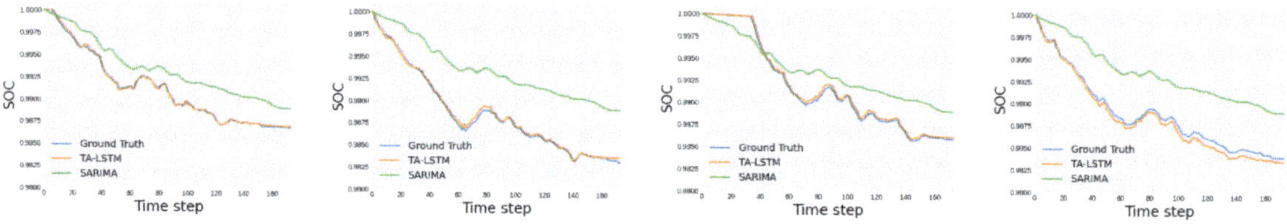

(**a**) (**b**) (**c**) (**d**)

Figure 15. SOC prediction of route E to F: (**a**) cycle 8; (**b**) cycle 9; (**c**) cycle 10; (**d**) cycle 11.

In addition, in order to simulate the battery model by reflecting more realistic situations, we checked the change in SOC by varying the ambient temperature and battery usage time. The real driving cycle, the parameters of the battery model, and the ambient temperature and battery operating time values for the simulation are shown in Table 9.

Table 9. Parameters for the battery model simulation.

Parameters	Description	Values
Driving cycle	Real driving cycle	Cycle 8
K_{ct}	Linear t dependency of the capacity C	6.084436×10^{-10}
t_{bu}	battery usage time (s)	0, 157,698,305 s (5 years), 315,396,610 s (10 years)
T_{ref}	Reference temperature (°C)	20 °C
T_{curr}	Ambient temperature (°C)	0, 20, 30 °C
$alphaC$	Linear temperature coefficient of capacity (K^{-1})	0.03 (0)

From the results in Figure 16, it can be confirmed that the battery SOC decreased faster because the battery capacity decreases as the temperature decreases.

From the results of Figure 17, it can be confirmed that the battery SOC decreased more quickly because the battery capacity decreases as the battery usage time increases.

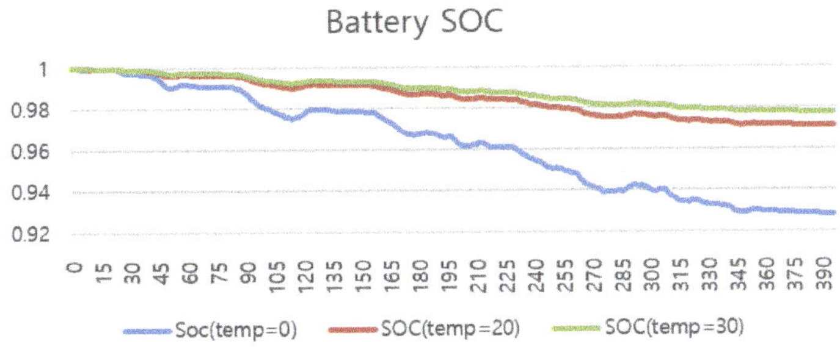

Figure 16. Temperature effect on the battery SOC.

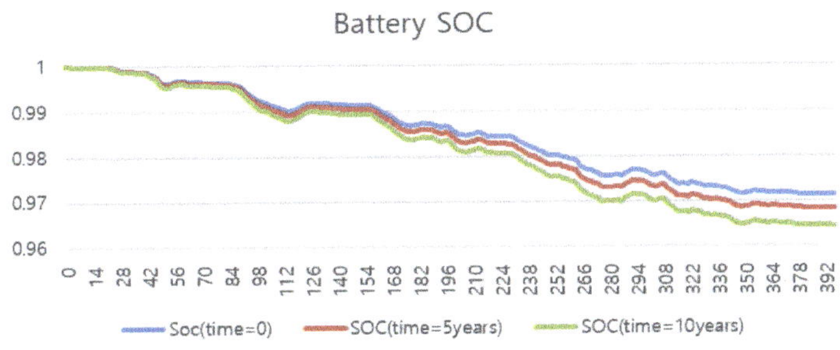

Figure 17. Battery aging effect on the battery SOC.

6. Conclusions

In our paper, we proposed a method for predicting the real driving-cycle-based SOC for EV batteries using deep-learning-based algorithms. For the RDC data, the authors used the data collected through the OBD II standard while driving from home to university in an actual vehicle. The proposed method first classifies detailed routes based on the section with traffic lights using GPS information, and finds the most similar patterns through DTW among the driving cycles of each section. Next, we predicted these sections by learning speed using TA-LSTM based on temporal attention, and predicted SOC based on this predicted speed. From the experimental results, it was confirmed that the case of using TA-LSTM could most accurately predict the speed, and that this could be very close to the actual values when used as a basis to predict the SOC. Therefore, if the proposed method is applied to the Real Driving Cycle, in which the same section is repeatedly driven, an accurate SOC prediction can be expected.

All the algorithms we propose can be implemented as an in-vehicle embedded system. It has been confirmed that DL models can be pre-trained and executed in an MCU. FMU simulation for the battery SOC can also be executed in an embedded system, such as an MCU or a sensor node [4,40]. If a high-capacity memory is required due to an increase in the demands of storing a massive amount of data or improving the calculation speed, a prediction can be performed in conjunction with a cloud data center or an edge server through the vehicular network, such as the Vehicle to Infrastructure (V2I) network.

The idea of the paper is to propose an algorithm that accurately predicts the SOC value of the next time step. However, predicting the SOC after multiple time steps may also be useful to the driver. Therefore, in a future work, the algorithm could be adapted so that it may predict, for example, the SOC after the completion of the entire driving cycle.

Author Contributions: Methodology, S.H., H.H. and D.K.; Project administration, S.H.; Software, S.H., H.H., D.K. and S.C.; Supervision, I.J.; Writing–original draft, S.H. and H.H. All authors have read and agreed to the published version of the manuscript.

Funding: This work was supported partly by the Institute for Information & Communications Technology Promotion (IITP) grant funded by the Korea government (MSIP) (No. 2020-0-00107, Development of the technology to automate the recommendations for big data analytic models that define data characteristics and problems), and partly by Basic Science Research Program through the National Research Foundation of Korea (NRF) funded by the Ministry of Education (NRF-2019R1I1A1A01058964).

Conflicts of Interest: The authors declare no conflict of interest.

Abbreviations

ASIC	Application Specific Integrated Circuits
BMS	Battery Management System
CAVs	Connected and Autonomous Vehicles
CSV	Comma Separated Value
CV	Cross Validation
DL	Deep Learning
DNNs	Deep Neural Networks
DTW	Dynamic Time Warping
EV	Electric Vehicle
FGPA	Field-programmable gate array
FMI	Functional Mock-up Interface
FMU	Functional Mock-up Units
FSD	Full Self-Driving
GPS	Global Positioning System
GPU	Graphics Processing Unit
GRU	Gated recurrent unit
MAPE	Mean Absolute Percentage Error
MCU	Microcontroller
ML	Machine Learning
MSE	Mean Squared Error
NEDC	New European Driving Cycle
OBD	On Board Diagnostics
OCV	Open circuit voltage
RDC	Real Driving Cycle
RMSE	Root Mean Squared Error
RNN	Recurrent Neural Network
RRA	Redundancy reduction algorithm
SARIMA	Seasonal Autoregressive Integrated Moving Average
SDC	Standard Driving Cycle
SOC	State of Charge
TA-LSTM	Temporal Attention Long Short-Term Memory
TFLOPS	Tera floating point operations per second
V2I	Vehicle to Infrastructure
Variables	
α	Angle of the driving surface, rad
η_g	Gearbox efficiency
η_m	Efficiency of power consumed (motoring mode)
η_r	Efficiency of power generated (regenerative braking mode, W
α_c	Linear temperature coefficient of the capacity C, K^{-1}
μ_{rr}	Coefficient of rolling resistance
ρ	Density of the air
A	Frontal area of the vehicle, m^2
a	Aceleration of the vehicle, m/s^2
C	Battery capacity, Ah
C_d	Aerodynamic drag coefficient
E_{b0}	Open-circuit voltage of the battery, V
F_t	Traction force of the vehicle, N
F_{ad}	Aerodynamic drag, N

F_{hc}	Hill climbing force, N
F_{la}	Force required to give linear acceleration, N
F_{rr}	Rolling resistance force of the wheels, N
F_{wa}	Force required to give angular acceleration to the rotating motor, N
G	Gear ratio of differential
g	Gravity acceleration, m/s^2
I_B	Battery current
m	Vehicle mass, kg
P_s	Shaft power of electric machine (motoring mode), W
P_t	Traction power, W
P_{aux}	Power consumed by auxiliary loads, W
P_{bc}	Total power consumed in the EV, W
P_{bm}	Power consumed by electric machine (motoring mode), W
P_{br}	Power consumed by electric machine (regenerative braking mode), W
P_{sr}	Shaft power of electric machine (regenerative braking mode), W
Q	Total charge of the battery
r_w	Wheel radius, m
R_{Bi}	Internal resistance of the battery, Ω
S_s	Shaft angular velocity of electric machine (motoring mode), rpm
S_W	Angular velocity of the wheels, rpm
S_{sr}	Shaft angular velocity of electric machine (regenerative braking mode), rpm
T_s	Shaft torque of electric machine (motoring mode), Nm
T_t	Traction torque, Nm
T_{ref}	Reference temperature, K
T_{sr}	Shaft torque of electric machine (regenerative braking mode, Nm)
v	Velocity of the vehicle, m/s
w_W	Angular velocity of the wheels, rad/s

References

1. Cao, J.; Emadi, A. A New Battery/UltraCapacitor Hybrid Energy Storage System for Electric, Hybrid, and Plug-In Hybrid Electric Vehicles. *IEEE Trans. Power Electron.* **2012**, *27*, 122–132. [CrossRef]
2. Sivak, M.; Schoettle, B. Relative Costs of Driving Electric and Gasoline Vehicles in the Individual US States, University of Michigan. Report No. SWT-2018-1. 2018. Available online: http://websites.umich.edu/~umtriswt/PDF/SWT-2018-1.pdf (accessed on 19 November 2021).
3. Moghaddam, Z.; Ahmad, I.; Habibi, D.; Phung, Q.V. Smart Charging Strategy for Electric Vehicle Charging Stations. *IEEE Trans. Transp. Electrif.* **2018**, *4*, 76–88. [CrossRef]
4. STMicroelectronics. Artificial Intelligence (AI) Plugin for Automotive SPC5 MCUs. Available online: https://www.st.com/en/development-tools/spc5-studio-AI.html (accessed on 19 November 2021).
5. Li, S.; He, H.; Li, J.; Wang, H. Big data driven Deep Learning algorithm based Lithium-ion battery SoC estimation method: A hybrid mode of C-BMS and V-BMS. In Proceedings of the Applied Energy Symposium: MIT A+B, Boston, MA, USA, 22–24 December 2019.
6. Lu, S.; Shi, W. The Emergence of Vehicle Computing. *IEEE Internet Comput.* **2021**, *25*, 18–22. [CrossRef]
7. Liu, L.; Lu, S.; Zhong, R.; Wu, B.; Yao, Y.; Zhang, Q.; Shi, W. Computing Systems for Autonomous Driving: State of the Art and Challenges. *IEEE Internet Things J.* **2021**, *8*, 6469–6486. [CrossRef]
8. Findelair, A. Tomorrow's Car Silicon Brain, How Is It Made? Towards Data Science. Available online: https://towardsdatascience.com/tomorrows-car-silicon-brain-how-is-it-made-9090e1f06c9d (accessed on 23 March 2021).
9. Bhatt, A. Planning and application of Electric Vehicle with MATLAB®/Simulink®. In Proceedings of the 2016 IEEE International Conference on Power Electronics, Drives and Energy Systems (PEDES), Trivandrum, India, 14–17 December 2016; pp. 1–6.
10. Wei, M.; Ye, M.; Li, J.B.; Wang, Q.; Xu, X. State of Charge Estimation of Lithium-Ion Batteries Using LSTM and NARX Neural Networks. *IEEE Access* **2020**, *8*, 189236–189245. [CrossRef]
11. Amrhein, M.; Krein, P.T. Dynamic simulation for analysis of hybrid electric vehicle system and subsystem interactions, including power electronics. *IEEE Trans. Veh. Technol.* **2005**, *54*, 825–836. [CrossRef]
12. Lv, C.; Zhang, J.; Li, Y.; Yuan, Y. Mechanism analysis and evaluation methodology of regenerative braking contribution to energy efficiency improvement of electrified vehicles. *Energy Convers. Manag.* **2015**, *92*, 469–482. [CrossRef]
13. Fiori, C.; Ahn, K.; Rakha, H.A. Power-based electric vehicle energy consumption model: Model development and validation. *Appl. Energy* **2016**, *168*, 257–268. [CrossRef]
14. Genikomsakis, K.N.; Mitrentsis, G. A computationally efficient simulation model for estimating energy consumption of electric vehicles in the context of route planning applications. *Transp. Res. Part D Transp. Environ.* **2017**, *50*, 98–118. [CrossRef]

15. Liu, K.; Wang, J.; Yamamoto, T.; Morikawa, T. Exploring the interactive effects of ambient temperature and vehicle auxiliary loads on electric vehicle energy consumption. *Appl. Energy* **2018**, *227*, 324–331. [CrossRef]
16. Iora, P.; Tribioli, L. Effect of Ambient Temperature on Electric Vehicles' Energy Consumption and Range: Model Definition and Sensitivity Analysis Based on Nissan Leaf Data. *World Electr. Veh. J.* **2019**, *10*, 2. [CrossRef]
17. Luin, B.; Petelin, S.; Al-Mansour, F. Microsimulation of electric vehicle energy consumption. *Energy* **2019**, *174*, 24–32. [CrossRef]
18. Blochwitz, T.; Otter, M.; Akesson, J.; Arnold, M.; Clauß, C.; Elmqvist, H.; Friedrich, M.; Junghanns, A.; Mauß, J.; Neumerkel, D.; et al. Functional Mockup Interface 2.0: The Standard for Tool independent Exchange of Simulation Models. In Proceedings of the 9th International MODELICA Conference, Munich, Germany, 3–5 September 2012; pp. 173–184.
19. Hong, S.; Lim, D.; Joe, I.; Kim, W. F-DCS: FMI-Based Distributed CPS Simulation Framework with a Redundancy Reduction Algorithm. *Sensors* **2020**, *20*, 252. [CrossRef] [PubMed]
20. Bär, T.; Nienhüser, D.; Kohlhaas R.; Zöllner, J.M. Probabilistic driving style determination by means of a situation based analysis of the vehicle data. In Proceedings of the 14th International IEEE Conference on Intelligent Transportation Systems (ITSC), Washington, DC, USA, 5–7 October 2011; pp. 1698–1703. [CrossRef]
21. Ellison, A.B.; Greaves, S.P.; Bliemer, M.C.J. Driver behaviour profiles for road safety analysis. *Accid. Anal. Prev.* **2015**, *76*, 118–132. [CrossRef]
22. Wu, X.; He, X.; Yu, G.; Harmandayan, A.; Wang, Y. Energy-Optimal Speed Control for Electric Vehicles on Signalized Arterials. *IEEE Trans. Intell. Transp. Syst.* **2015**, *16*, 2786–2796. [CrossRef]
23. Varga, B.O.; Sagoian, A.; Mariasiu, F. Prediction of Electric Vehicle Range: A Comprehensive Review of Current Issues and Challenges. *Energies* **2019**, *12*, 946. [CrossRef]
24. How, D.N.T.; Hannan, M.A.; Hossain Lipu, M.S.; Ker, P.J. State of Charge Estimation for Lithium-Ion Batteries Using Model-Based and Data-Driven Methods: A Review. *IEEE Access* **2019**, *7*, 136116–136136. [CrossRef]
25. Xiong, R.; Cao, J.; Yu, Q.; He, H.; Sun, F. Critical Review on the Battery State of Charge Estimation Methods for Electric Vehicles. *IEEE Access* **2018**, *6*, 1832–1843. [CrossRef]
26. Zhang, R.; Xia, B.; Li, B.; Cao, L.; Lai, Y.; Zheng, W.; Wang, H.; Wang, W. State of the Art of Lithium-Ion Battery SOC Estimation for Electrical Vehicles. *Energies* **2018**, *11*, 1820. [CrossRef]
27. Yang, F.; Song, X.; Xu, F.; Tsui, K. State-of-Charge Estimation of Lithium-Ion Batteries via Long Short-Term Memory Network. *IEEE Access* **2019**, *7*, 53792–53799. [CrossRef]
28. Chemali, E.; Kollmeyer, P.J.; Preindl, M.; Emadi, A. State-of-charge estimation of Li-ion batteries using deep neural networks: A machine learning approach. *J. Power Sources* **2018**, *400*, 242–255. [CrossRef]
29. Babaeiyazdi, I.; Rezaei-Zare, A.; Shokrzadeh, S. State of charge prediction of EV Li-ion batteries using EIS: A machine learning approach. *Energy* **2021**, *223*, 120116. [CrossRef]
30. Xing, Y.; Ma, E.W.M.; Tsui, K.L.; Pecht, M. Battery Management Systems in Electric and Hybrid Vehicles. *Energies* **2011**, *4*, 1840–1857. [CrossRef]
31. Sakoe, H.; Chiba, S. Dynamic programming algorithm optimization for spoken word recognition. *IEEE Trans. Acoust. Speech Signal Process.* **1978**, *26*, 43–49. [CrossRef]
32. Keogh, E.J.; Pazzani, M.J. Derivative Dynamic Time Warping. In Proceedings of the 2001 SIAM International Conference on Data Mining, Chicago, IL, USA, 5–7 April 2001.
33. Sundermeyer, M.; Schluter, R.; Ney, H. LSTM Neural Networks for Language Modeling. In Proceedings of the Thirteenth Annual Conference of the International Speech COMMUNICATION association, Portland, OR, USA, 9–13 September 2012.
34. Sutskever, I.; Martens, J.; Hinton, G. Generating Text with Recurrent Neural Networks. In Proceedings of the 28th International Conference on Machine Learning, Bellevue, WA, USA, 28 June–2 July 2011.
35. Fu, R.; Zhang, Z.; Li, L. Using LSTM and GRU neural network methods for traffic flow prediction. In Proceedings of the 31st Youth Academic Annual Conference of Chinese Association of Automation (YAC), Wuhan, China, 11 November 2016.
36. Cui, S.; Joe, I. Collision prediction for a low power wide area network using deep learning methods. *J. Commun. Netw.* **2020**, *22*, 205–214. [CrossRef]
37. Qin, Y.; Song, D.; Cheng, H.; Cheng, W.; Jiang, G.; Cottrell, G.W. A dual-stage attention-based recurrent neural network for time series prediction. In Proceedings of the Twenty-Sixth International Joint Conference on Artificial Intelligence, Melbourne, Australia, 19–25 August 2017.
38. Fan, C.; Zhang, Y.; Pan, Y.; Li, X.; Zhang, C.; Yuan, R.; Wu, D.; Wang, W.; Pei, J.; Huang, H. Multi-horizon time series forecasting with temporal attention learning. In Proceedings of the 25th ACM SIGKDD International Conference on Knowledge Discovery and Data Mining, New York, NY, USA, 25 July 2019.
39. Cui, S.; Joe, I. A Dynamic Spatial-Temporal Attention-Based GRU Model With Healthy Features for State-of-Health Estimation of Lithium-Ion Batteries. *IEEE Access* **2021**, *9*, 27374–27388. [CrossRef]
40. Bertsch, C.; Neudorfer, J.; Ahle, E.; Armugham, S.; Ramachandran, K.; Thuy, A. FMI for Physical Models on Automotive Embedded Targets. In Proceedings of the 11th International Modelica Conference, Versailles, France, 21–23 September 2015. [CrossRef]

applied sciences

MDPI

Article

Long Short-Term Memory Network-Based Metaheuristic for Effective Electric Energy Consumption Prediction

Simran Kaur Hora [1], Rachana Poongodan [2], Rocío Pérez de Prado [3,*], Marcin Wozniak [4] and Parameshachari Bidare Divakarachari [5]

1 Department of Information Technology, Chameli Devi Group of Institutions, Indore 452020, India; simrankaur.hora@cdgi.edu.in
2 Department of Computer Science and Engineering, New Horizon College of Engineering, Bangalore 560103, India; dr.rachanap@newhorizonindia.edu
3 Department of Telecommunication Engineering, University of Jaén, 23700 Linares (Jaén), Spain
4 Faculty of Applied Mathematics, Silesian University of Technology, 44-100 Gliwice, Poland; marcin.wozniak@polsl.pl
5 Department of Telecommunication Engineering, GSSS Institute of Engineering and Technology for Women, Mysuru 570016, India; paramesh@gsss.edu.in
* Correspondence: rperez@ujaen.es

Abstract: The Electric Energy Consumption Prediction (EECP) is a complex and important process in an intelligent energy management system and its importance has been increasing rapidly due to technological developments and human population growth. A reliable and accurate model for EECP is considered a key factor for an appropriate energy management policy. In recent periods, many artificial intelligence-based models have been developed to perform different simulation functions, engineering techniques, and optimal energy forecasting in order to predict future energy demands on the basis of historical data. In this article, a new metaheuristic based on a Long Short-Term Memory (LSTM) network model is proposed for an effective EECP. After collecting data sequences from the Individual Household Electric Power Consumption (IHEPC) dataset and Appliances Load Prediction (AEP) dataset, data refinement is accomplished using min-max and standard transformation methods. Then, the LSTM network with Butterfly Optimization Algorithm (BOA) is developed for EECP. In this article, the BOA is used to select optimal hyperparametric values which precisely describe the EEC patterns and discover the time series dynamics in the energy domain. This extensive experiment conducted on the IHEPC and AEP datasets shows that the proposed model obtains a minimum error rate relative to the existing models.

Keywords: butterfly optimization algorithm; electric energy consumption prediction; long short-term memory network; time series analysis; transformation methods

Citation: Hora, S.K.; Poongodan, R.; de Prado, R.P.; Wozniak, M.; Divakarachari, P.B. Long Short-Term Memory Network-Based Metaheuristic for Effective Electric Energy Consumption Prediction. *Appl. Sci.* **2021**, *11*, 11263. https://doi.org/10.3390/app112311263

Academic Editors: João M. F. Rodrigues, Pedro J. S. Cardoso and Marta Chinnici

Received: 20 October 2021
Accepted: 23 November 2021
Published: 27 November 2021

1. Introduction

In recent decades, the demand for electricity has been rising on a global scale due to the massive growth of electronic markets [1], the development of electrical vehicles [2], the use of heavy machinery equipment (e.g., line excavators, pile boring machines) [3], technological advancements, and rapid population growth [4,5]. As a result, accurate electric load forecasting has greater importance in the field of power system planning [6]. An underestimation reduces the reliability of the power system, while overestimation wastes energy resources and effectively enhances operational costs [7]. Therefore, a precise electric load forecasting system is necessary for power systems, the electrical load series being affected by several influencing factors [8]. Currently, several electrical load forecasting models are being developed. The models fall into two categories: multi-factor forecasting models and time series forecasting models [9]. The time series forecasting models are quicker and easier in EECP compared to the multi-factor forecasting models. Numerous non-objective factors and electric load series are affected in practical applications, and it is difficult to control these with multi-factor forecasting models [10–12]. Hence, the multi-factor forecasting models simply evaluate the relations between forecasting variables and influencing factors [13–15]. In this research, a novel metaheuristic based on an LSTM model is developed to generate a more effective EECP. The main contributions are specified below:

- Input data sequences are collected from IHEPC and AEP datasets, and data refinement is accomplished using min-max along with standard transformation methods in order to eliminate redundant, missing, and outlier variables.
- Next, the EECP is generated using the proposed metaheuristic based on the LSTM model. The proposed model superiorly handles the irregular tendencies of energy consumption relative to other deep learning models and conventional LSTM networks.
- The effectiveness of the proposed metaheuristic based on the LSTM model is evaluated in terms of mean squared error (MSE), root MSE (RMSE), mean absolute error (MAE) and mean absolute percentage error (MAPE) on both IHEPC and AEP datasets.

This article is structured as follows. Previous existing research studies on the topic of EECP are reviewed in Section 2. The mathematical explanations of the proposed metaheuristic based on the LSTM model and a quantitative study including experimental results are specified in Sections 3 and 4, respectively. Finally, the conclusion of this work is stated in Section 5.

2. Related Works

In this section, previous works in the area are reviewed in order to justify the contribution of the proposal and the selection of strategies considered for comparison in the experimental section.

Le et al. [16] combined a Bidirectional Long Short-Term Memory (Bi-LSTM) network and a Convolutional Neural Network (CNN) to forecast household EEC. Firstly, the CNN was employed to extract the discriminative feature values from the IHEPC dataset and then the Bi-LSTM network was used to make predictions. Ishaq et al. [17] introduced a new ensemble-based deep learning model to forecast and predict energy consumption and demands. Initially, data pre-processing was performed using transformation, normalization, and cleaning techniques, and then the pre-processed data were fed into the ensemble model, the CNN and Bi-LSTM network extracting discriminative feature values. In this work, an active learning concept was created on the basis of the moving window to improve and ensure the prediction performance of the presented model. In the resulting phase, the effectiveness of the presented model was tested on a Korean commercial building dataset in light of MAPE, RMSE, MAE, and MSE values. Lin et al. [18] integrated an Extreme Learning Machine (ELM) and Variational Mode Decomposition (VMD) techniques for electrical load forecasting. Firstly, the VMD technique was employed to transform the collected electric load series into components with dissimilar frequencies, which helps to eliminate fluctuation properties and enhances the overall accuracy of prediction. Finally, EEC forecasting was carried out utilizing ELM with a differential evolution algorithm.

Xu et al. [19] combined a Deep Belief Network (DBN) and linear regression techniques to predict time series data. In this study, the linear regression technique captures the non-linear and linear behaviors of the time series data. Initially, the linear regression technique was used to obtain the residuals between input and predicted data, and then the residuals were fed into the DBN for the final forecasting. In the time series forecasting, the DBN significantly extracts the features between self-organization properties and layers. Maldonado et al. [20] applied Support Vector Regression (SVR) to the time series data for electric load forecasting. The SVR technique successfully modelled the nonlinear relation between the target variables and the exogenous covariates. Wan et al. [21] developed a new multivariate temporal convolutional network for time series prediction that has been extensively used in applications such as transportation, finance, aerology, and power/energy. In the time series data forecasting, the presented convolution network superiorly enhanced the results of EECP. Further, this study concentrates on the trade-off between prediction accuracy and implementation complexity. Bouktif et al. [22] combined a genetic algorithm and a Particle Swarm Optimization (PSO) algorithm to select optimal hyperparameters in LSTM for an effective EECP.

Qiu et al. [23] introduced an oblique random forest classifier for time series forecasting. In the developed classification technique, every node of the decision tree was replaced by the optimal feature-based orthogonal classifier. Additionally, the least square classification technique was used to perform feature partition. The efficiency of the oblique random forest classifier was investigated using five electricity load time series datasets and eight general time series datasets. Further, Kuo and Huang [24] presented a new deep learning network for short term energy load forecasting. The obtained results showed that the deep-energy model was robust and had a strong generalization ability in data series forecasting. Similarly, Qiu et al. [25] combined DBN and empirical mode decomposition for electricity load demand forecasting. Initially, the acquired data series were decomposed into several Intrinsic Mode Functions (IMFs). Further, the DBN was applied to model each of the extracted IMFs for accurate prediction. Pham et al. [26] implemented a random forest classifier to forecast household short-term energy consumption. The effectiveness of the random forest classifier was tested on five one-year datasets. The evaluation outcome showed that the presented random forest classifier obtained better predictive accuracy by means of MAE.

Galicia et al. [27] introduced an ensemble classifier by combining random forest, gradient boosted trees and decision trees to forecast big data time series. The evaluation results showed that the developed ensemble classifier performed well in time series data prediction compared to other models and individual ensemble models. Khairalla et al. [28] presented a new stacking multi-learning ensemble model for forecasting time series data. The presented model includes three main techniques—SVR, linear regression, and a back-propagation neural network—and the presented ensemble model comprises four major steps: integration, pruning, generation, and ensemble prediction tasks. Jallal et al. [29] introduced a hybrid model that integrates a firefly algorithm and an Adaptive Neuro Fuzzy Inference System (ANFIS) classifier for EECP, though the improved search space diversification in the presented model enhances its predictive accuracy. Bandara et al. [30] introduced a new LSTM Multi-Seasonal Net (LSTM-MSNet) for time series forecasting with multiple seasonal patterns. The evaluation outcome showed that the presented LSTM-MSNet model achieved better computational time and prediction accuracy compared to existing systems. Abbasimehr and Paki [31] combined multi-head attention and LSTM networks to predict the time series data precisely. Sajjad et al. [32] initially used min-max and standard transformation techniques to eliminate outlier, redundant, and null values from the IHEPC and AEP datasets. Then, EECP was accomplished using CNN with a Gated Recurrent Units (GRUs) model. The experimental evaluation showed that the presented model obtained a significant performance in EECP by means of MAE, RMSE, and MSE.

Khan et al. [33] combined a Bi-LSTM network and dilated CNN to predict power consumption in local energy systems. As can be seen in the resulting phase, the presented model effectively predicts multiple step power consumption that includes monthly, weekly, daily, and hourly outputs. Khan et al. [34] has integrated multilayer bi-directional GRU and CNNs for household electricity consumption prediction. The effectiveness of the presented model was evaluated in terms of MAE, RMSE, and MSE on the IHEPC and AEP datasets.

Nowadays, artificial intelligence techniques are applied more in the application of EECP because of its reliability and high performance results. The artificial intelligence-based techniques, such as CNN, GRU, multi head attentions, ANFIS, and the ensemble schemes, are extensively applied for energy forecasting and time series issues. The GRU technique obtained a better outcome in EECP related to conventional techniques, but it is ineffective in handling long-term time series data sequences, and it is also historically dependent. In addition, the aforementioned techniques failed in long-term consequence forecasting and includes vanishing gradient issues [35]. To overcome the above stated concerns, a new metaheuristic based on the LSTM network is proposed in this article to predict and handle the short-term and long-term dependencies in energy forecasting.

3. Proposal

The proposed metaheuristics based on the LSTM network includes three major phases in EECP, namely, data collection (AEP and IHEPC datasets), data refinement (min-max transformation and standard transformation methods) and consumption prediction (using metaheuristics based on the LSTM network). The flow-diagram of the proposed model is specified in Figure 1.

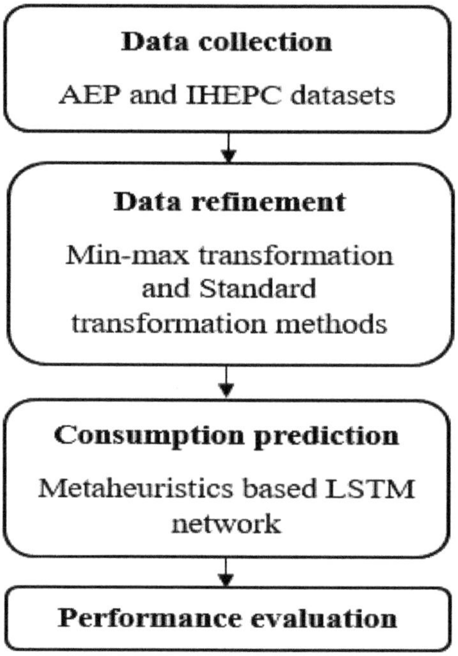

Figure 1. Flow-diagram of the proposed model.

3.1. Dataset Description

In the household EECP application, the effectiveness of the proposed metaheuristics-based LSTM network is validated with AEP and IHEPC datasets. The AEP dataset contains 29 parameters related to appliances' energy consumption, lights, and weather information (pressure, temperature, dew point, humidity, and wind speed), which are statistically depicted in Table 1. The AEP dataset includes data for four and half months of a residential house at a ten-minute resolution. In the AEP dataset, the data are recorded from the outdoor and indoor environments using a wireless sensor network, the outdoor data being acquired from a near-by airport [36]. The residential house contains one outdoor temperature sensor, nine indoor temperature sensors, and nine humidity sensors; one sensor is placed in the outdoor environment and seven humidity sensors are placed in the indoor environment. The humidity, outdoor pressure, temperature, dew point, and visibility are recorded at the near-by airport. The statistical information about the AEP dataset is depicted in Table 1.

Table 1. Statistical information about the AEP dataset.

Attributes	Information	Units
Dew point	Outside dew point recorded from Chievres Weather Station (CWS)	C
Visibility	Outside visibility recorded from CWS	Km
Wind speed	Outside wind speed recorded from CWS	m/s
Rho	Outside humidity recorded from CWS	%
Pressure	Outside pressure recorded from CWS	Mm Hg
To	Outside temperature recorded from CWS	C
RH1	Humidity of parents' room	%
T1	Temperature of parents' room	C
RH2	Humidity of teenager's room	%
T2	Temperature of teenager's room	C
RH3	Humidity of ironing room	%
T3	Temperature of ironing room	C
RH4	Outside humidity of building	%
T4	Outside temperature of building	C
RH5	Humidity of bathroom	%
T5	Temperature of bathroom	C
RH6	Humidity of office room	%
T6	Temperature of office room	C
RH7	Humidity of laundry room	%
T7	Temperature of laundry room	C
RH8	Humidity of living room	%
T8	Temperature of living room	C
RH9	Humidity of kitchen	%
T9	Temperature of kitchen	C
Light	Total energy consumption by lights	Watt-hour (Wh)
Appliances	Total energy consumption by appliances	Wh

In addition, the IHEPC dataset comprises of 2,075,259 instances, which are recorded from a residential house in France for five years (From December 2006 to November 2010) [37]. The IHEPC dataset includes nine attributes like voltage, minute, global intensity, month, global active power, year, global reactive power, day and hour. Three more variables are acquired from energy sensors: sub metering 1, 2, and 3 with proper meaning. The statistical information about IHEPC dataset is represented Table 2. The data samples of AEP and IHEPC datasets are graphically presented in the Figure 2.

Table 2. Statistical information about IHEPC dataset.

Attributes	Information
Sub metering 1	Energy utilized in kitchen (Wh)
Sub metering 2	Energy utilized in laundry room (Wh)
Sub metering 3	Energy utilized by water heater (Wh)
Date	dd/mm/yyyy
Time	hh:mm:ss
Voltage	Minute averaged voltage of household (volt)
Global reactive voltage	Minute averaged global reactive voltage of household (kilowatt (kW))
Global active voltage	Minute averaged global active voltage of household (kW)
Global intensity	Minute averaged global intensity of household (ampere)

(a)
(b)

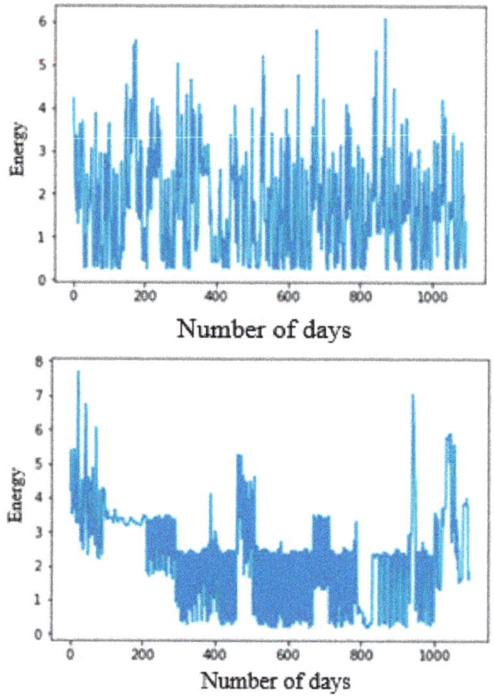

Figure 2. Data samples of (**a**) the AEP dataset and (**b**) the IHEPC dataset.

3.2. Data Refinement

After the acquisition of data from the AEP and IHEPC datasets, data refinement is performed to eliminate missing and outlier variables and to normalize the acquired data. In the AEP dataset, a standard transformation technique is employed for converting the acquired data into a particular range. In the AEP dataset, the feature vectors range lies between 0 and 800, and by using the standard transformation technique the feature vectors range is transformed into −4 and −6. The mathematical expression of the standard transformation technique is defined in Equation (1):

$$T_{standard} = (X - U)/S \tag{1}$$

where S indicates standard deviation, X denotes actual acquired data, and U represents the mean. In addition, the IHEPC dataset comprises redundant, outlier, and null values, so a min-max scalar is applied to eliminate non-significant values and to bring the feature vectors into a particular range of values. In the IHEPC dataset, the feature vectors range lies between 0 to 250, and by using the min-max transformation technique, the feature vectors range is transformed into −2 and −3. The mathematical expression of the min-max transformation technique is defined in Equation (2):

$$T_{min-max} = \frac{X - X_{min}}{X_{max} - X_{min}} \tag{2}$$

where X_{max} and X_{min} indicates maximum and minimum values of the IHEPC dataset. A total of 2890 and 25,980 missing values are eliminated in the AEP and IHEPC datasets utilizing the pre-processing techniques. The refined data samples of AEP dataset and IHEPC dataset are presented in Figure 3.

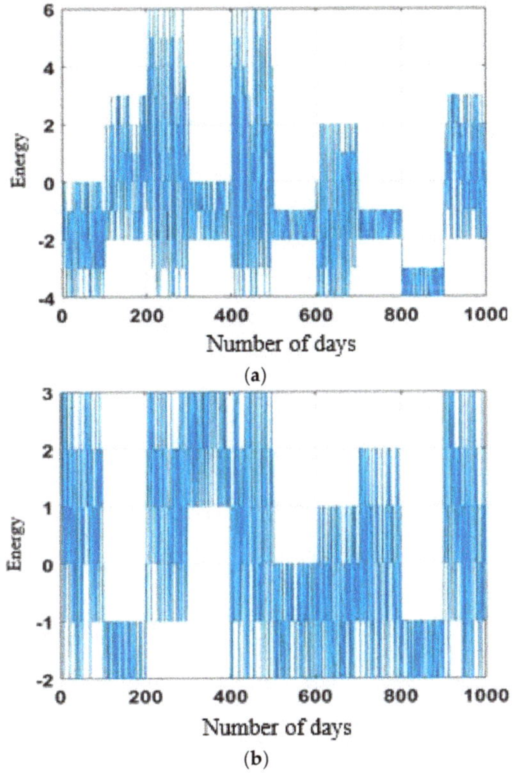

Figure 3. Refined data samples of (**a**) the AEP dataset and (**b**) the IHEPC dataset.

3.3. Energy Consumption Prediction

After refining the acquired data, the EECP is accomplished using the metaheuristics-based LSTM network. The LSTM network is an extension of a Recurrent Neural Network (RNN). The RNN has numerous problems, such as short-term memory and vanishing gradient issues, when it processes large data sequences [38]. In addition, the RNN is inappropriate for larger data sequences because it removes the important information from the input data. In the RNN model, the gradient updates the weights during back propagation, where sometimes it reduced highly and the initial layers get low gradient and stops further learning. To tackle these issues, the LSTM network was developed by Hochreiter [39]. The LSTM network overcomes the issues of RNNs by replacing hidden layers with memory cells for modelling long-term dependencies. The LSTM network includes dissimilar gates, such as a forget gate, input gate, and output gate, along with activation functions for learning time-based relations. The LSTM network and the individual LSTM unit are graphically depicted in Figures 4 and 5.

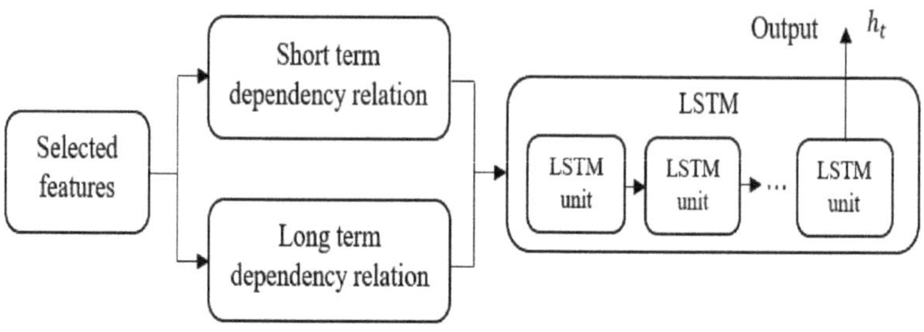

Figure 4. Graphical presentation of the LSTM network.

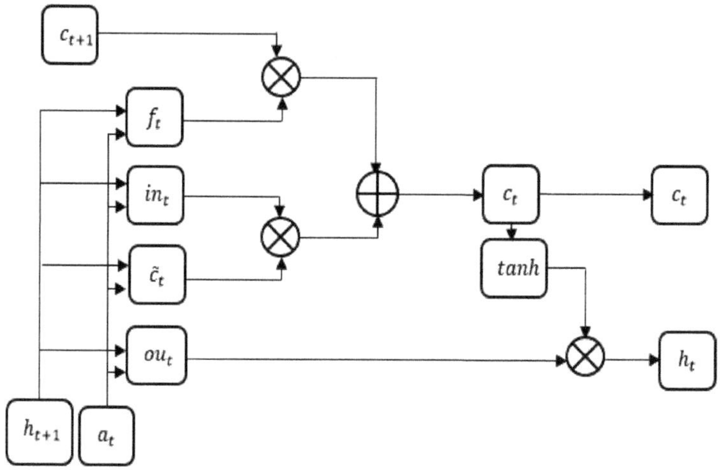

Figure 5. Graphical presentation of the LSTM unit.

The mathematical expressions of the input gate in_t, forget gate f_t, cell c_t, and output gate ou_t are defined in Equations (3)–(6):

$$in_t = \sigma(W_{inh}h_{t-1} + W_{ina}a_t + b_{in}) \tag{3}$$

$$f_t = \sigma(W_{fh}h_{t-1} + W_{fa}a_t + b_f) \tag{4}$$

$$c_t = f_t \odot c_{t-1} + in_t \odot \tan\text{h}(W_{ch}h_{t-1} + W_{ca}a_t + b_c) \tag{5}$$

$$ou_t = \sigma(W_{ouh}h_{t-1} + W_{oua}a_t + b_{ou}) \tag{6}$$

where t represents different time steps, $a_t = A[t, .] \in \mathbb{R}^F$ represents temporal quasi-periodic feature vectors, $tanh$ (.) denotes a hyperbolic tangent function, $\sigma(.)$ states a sigmoid function, and W and b work coefficients. The output of the LSTM unit h_{t-1} is mathematically specified in Equation (7), and it is graphically presented in Figure 5:

$$h_t = ou_t \odot \tanh(c_t) \tag{7}$$

The cell state $\{c_t | t = 1, 2, ..T|\}$ learns the necessary information from a_t on the basis of the dependency relationship during the training and testing mechanism. Finally, the extracted feature vectors are specified by the last LSTM unit output h_T. The hyperparametric values selected using BOA for the LSTM network are listed as follows: the number of sequences are 2 and 3, the sequence length is from 7 to 12, the minimum batch size is 20, the learning rate is 0.001, the number of the LSTM unit is 55, the maximum epoch is 120, and the gradient threshold value is 1. The BOA is a popular metaheuristic algorithm, which mimics the butterfly's behavior in foraging and mating. Biologically, butterflies are well adapted for foraging, possessing sense receptors that allow them to detect the presence of food. The sense receptors are known as chemoreceptors and are dispersed over several of the butterfly's body parts, such as the antennae, palps, legs, etc. In the BOA, the butterfly is assumed as a search agent to perform optimization and the sensing process depends on three parameters such as sensory modality, power exponent and stimulus intensity. If the butterfly is incapable of sensing the fragrance, then it moves randomly in the local search space [40].

Whereas, the sensory modality is in the form of light, sound, temperature, pressure, smell, etc. and it is processed by the stimulus. In the BOA, the magnitude of the physical stimulus is denoted as M and it is associated with the fitness of butterfly with greater fragrance value in the local search space. In the BOA, the searching phenomenon depends on two important issues: formulation of fragrance q and variations of physical stimulus M. For simplicity purpose, the stimulus intensity M is related with encoded-objective-function. Hence, q is relative and is sensed by other butterflies in the local search space. In the BOA, the fragrance is considered as a function of the stimulus, which is mathematically defined in Equation (8):

$$q_i = zl^d \tag{8}$$

where z denotes the sensory modality, l the perceived magnitude of fragrance, M the stimulus intensity and d indicates the power exponent. The BOA consists of two essential phases: a local and a global search phase. In the global search phase, the butterfly identifies the fitness solution that is determined in Equation (9):

$$x_i^{t+1} = x_i^t + \left(levy\ (\lambda) \times g^* - x_i^t\right) \times q_i \tag{9}$$

where x_i^t indicates the vector x_i of the ith butterfly, t represents iteration, g^* the present best solution, q_i states fragrance of the butterfly and $levy\ (\lambda)$ denotes a random number that ranges between 0 and 1. The general formula for calculating the local search phase is given in Equation (10):

$$x_i^{t+1} = x_i^t + \left(levy\ (\lambda) \times x_k^t - x_i^t\right) \times q_i \tag{10}$$

where x_k^t and x_i^t are the kth/ith butterflies from the solution. If x_k^t and x_i^t belongs to the same flight, Equation (10) performs a local random walk. The flowchart of the BOA is depicted in Figure 6.

In this scenario, the iteration phase is continued until the stopping criteria is not matched. The pseudocode of the BOA is represented as follows (Algorithm 1):

Algorithm 1 Pseudocode of BOA

Objective function $q(x)$, x_i $(i = 1, 2, \ldots .n)$
Initialize butterfly population
In the initial population, best solution is identified
Determine the probability of switch P
While stopping criteria is not encountered **do**
For every butterfly **do**
Draw *rand*
Find butterfly *fragrance* utilizing Equation (8)
If $rand < P$ **then**
Accomplish global search utilizing Equation (9)
Else
Accomplish local search utilizing Equation (10)
End if
Calculate the new solutions
Update the best solutions
End for
Identify the present better solution
End while
Output: Better solution is obtained

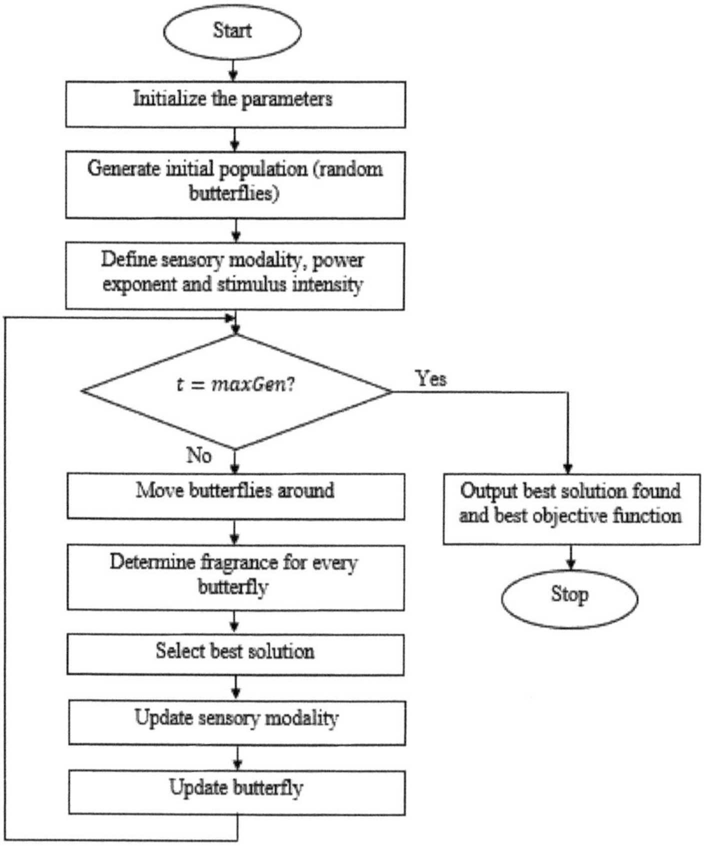

Figure 6. Flowchart of the BOA.

4. Experimental Results

In the EECP application, the proposed metaheuristic based on the LSTM network is simulated using a Python software environment on a computer with 64 GB random access memory, a TITAN graphics processing unit with Intel core i7 processor and Ubuntu operating system. The effectiveness of the proposed metaheuristic based on the LSTM network in EECP is validated by comparing its performance with benchmark models, such as a Bi-LSTM with CNN [16], an ensemble-based deep learning model [17], a CNN with GRU model [32], a Bi-LSTM with dilated CNN [33], and multilayer bi-directional GRU with CNN [34] on the AEP and IHEPC datasets. In this research, the experiment is conducted using four performance measures, MAPE, MAE, RMSE, and MSE, for time series data prediction. The MAPE is used to estimate the prediction accuracy of the proposed metaheuristic based on the LSTM network. The MAPE performance measure represents accuracy in percentage, as stated in Equation (11):

$$\text{MAPE} = \frac{1}{n}\sum_{1}^{n}|\frac{y-\hat{y}}{y}| \tag{11}$$

The MAE is used to estimate the average magnitude of the error between actual and predicted values by ignoring their direction. The MSE is used to determine the mean disparity between actual and predicted values. The mathematical expressions of MAE and MSE are stated in the Equations (12) and (13). Correspondingly, the RMSE is used to find the dissimilarity between the actual and predicted values, and then the mean of the square errors is computed. Lastly, the square root of the mean values is calculated, where the mathematical expression of RMSE is stated in Equation (14):

$$\text{MAE} = \frac{1}{n}\sum_{1}^{n}|y-\hat{y}| \tag{12}$$

$$\text{MSE} = \frac{1}{n}\sum_{1}^{n}(y-\hat{y})^2 \tag{13}$$

$$\text{RMSE} = \sqrt{\frac{1}{n}\sum_{1}^{n}(y-\hat{y})^2} \tag{14}$$

where n represents the number of instances, y the actual value and $\hat{y}$ the prediction value.

4.1. Quantitative Study on AEP Dataset

In this scenario, an extensive experiment is conducted on the AEP dataset to evaluate and validate the proposed metaheuristic based on the LSTM network's effectiveness and robustness for real-world issues. The refined AEP dataset is split into a 20:80% ratio for the proposed model's testing and training. The proposed metaheuristic based on the LSTM network utilizes 20% of data during testing and 80% of data during training. As seen in Table 3, the proposed metaheuristic based on the LSTM network obtained results closely related to the native properties of energy and the actual consumed energy level. By inspecting Table 3, the proposed model achieved effective results compared to other existing models, such as linear regression, CNN, SVR, the LSTM network and the Bi-LSTM network in light of MAPE, MAE, RMSE and MSE. Hence, the irregular tendencies of energy consumption are easily and effectively handled by the proposed metaheuristic based on the LSTM network. Hence, the proposed model attained a minimum MAPE of 0.09, an MAE of 0.07, an RMSE of 0.13, and an MSE of 0.05. In addition to this, the proposed model reduces prediction time by almost 30% compared to other models for the AEP dataset. A graphical presentation of the experimental models for the AEP dataset is depicted in Figure 7.

Table 3. Performance of the experimental models on the AEP dataset.

Models	MAPE	MAE	RMSE	MSE	Predicting Time (s)
Linear regression	1.54	1.99	2.12	1.24	38
CNN	0.34	0.40	1.02	0.49	29
SVR	0.72	0.88	1.92	0.82	49
LSTM network	0.28	0.17	0.92	0.29	21
Bi-LSTM network	0.24	0.13	0.78	0.14	18
Metaheuristic based on the LSTM network	0.09	0.07	0.13	0.05	12

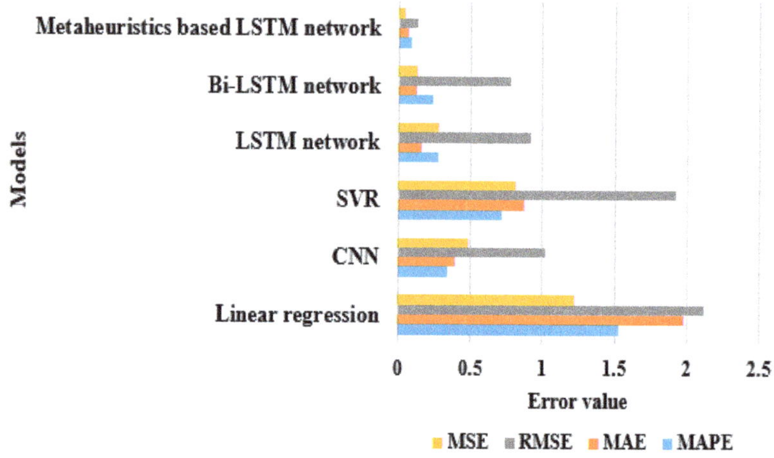

Figure 7. Graphical presentation of the experimental models for the AEP dataset.

In Table 4, the hyperparameter selection in the LSTM network is carried out with dissimilar optimization techniques, such as BOA, Grey Wolf Optimizer (GWO), Particle Swarm Optimizer (PSO), Genetic Algorithm (GA), Ant Colony Optimizer (ACO), and Artificial Bee Colony (ABC), and the performance validation is done using four metrics, namely, MAPE, MAE, RMSE, and MSE on the AEP dataset. As evident from Table 4, the combination of the LSTM network with BOA obtained an MAPE of 0.09, an MAE of 0.07, an RMSE of 0.13, and an MSE of 0.05, which are minimal compared to other optimization techniques. Due to naive selection of the hyperparametric values and the noisy electric data, the LSTM network obtained unacceptable forecasting results. An optimal LSTM network configuration is therefore needed to discover the time series dynamics in the energy domain and to describe the electric consumption pattern precisely. In this article, a metaheuristic-based BOA is applied to identify the optimal hyperparametric values of the LSTM network in the EEC domain. The BOA effectively learns the hyper parameters of the LSTM network to forecast energy consumption. Graphical presentation of dissimilar hyperparameter optimizers in the LSTM network on the AEP dataset is depicted in Figure 8.

Table 4. Performance of the dissimilar hyperparameter optimizers in the LSTM network on the AEP dataset.

	LSTM Network				
Optimizers	MAPE	MAE	RMSE	MSE	Predicting Time (s)
GWO	0.23	0.18	0.56	0.15	30
PSO	0.19	0.09	0.31	0.11	43
GA	0.26	0.13	0.82	0.19	64
ACO	0.22	0.17	0.44	0.13	29
ABC	0.12	0.11	0.37	0.08	25
BOA	0.09	0.07	0.13	0.05	12

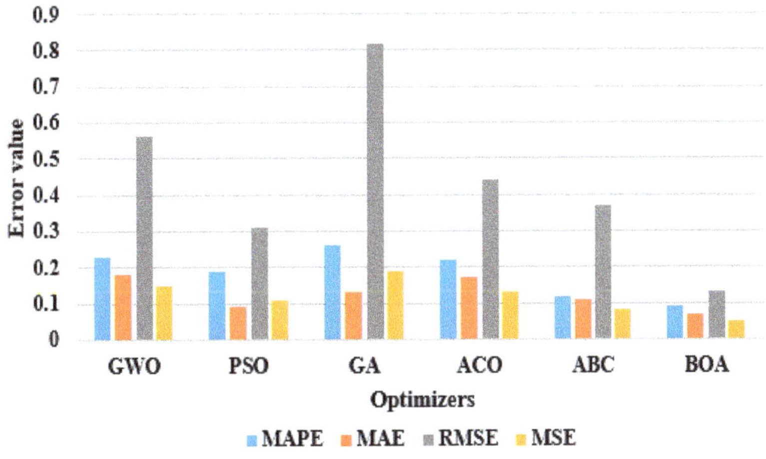

Figure 8. Graphical presentation of the dissimilar hyperparameter optimizers in the LSTM network on the AEP dataset.

4.2. Quantitative Study on IHEPC Dataset

Table 5 represents the extensive experiment conducted on the IHEPC dataset to evaluate the efficiency of the proposed metaheuristic based on the LSTM network by means of MAPE, MAE, RMSE, and MSE. The proposed metaheuristic based on the LSTM network obtained a minimum MAPE of 0.05, an MAE of 0.04, an RMSE of 0.16, and an MSE of 0.04, which are effective compared to other experimental models, such as linear regression, CNN, SVR, the LSTM network, and the Bi-LSTM network on the IHEPC database. In addition, the prediction time of metaheuristic based on LSTM network is 25% minimum compared to other experimental models. In this research, the metaheuristic based on the LSTM network superiorly handles the complex time series patterns and moderates the error value at every interval related to the other experimental models. Graphical presentation of the experimental models for the IHEPC dataset is depicted in Figure 9.

Table 5. Performance of the experimental models on IHEPC dataset.

Models	MAPE	MAE	RMSE	MSE	Predicting Time (s)
Linear regression	0.82	0.62	0.90	0.23	34
CNN	0.13	0.14	0.29	0.17	22
SVR	0.47	0.26	0.82	0.38	29
LSTM network	0.11	0.12	0.31	0.19	19
Bi-LSTM network	0.09	0.12	0.19	0.11	18.20
Metaheuristic based on the LSTM network	0.05	0.04	0.16	0.04	13

The LSTM network with BOA achieved better results in energy forecasting compared to other optimizers in light of MAPE, MAE, RMSE, and MSE. As seen in Table 6, the BOA reduced the error value in energy forecasting by almost 20–50%, and the prediction time by 25% compared to other hyperparameter optimizers in the LSTM network for the IHEPC dataset. The experimental result shows that the metaheuristic-based BOA model obtained a successful solution, and it effectively reduces computational complexity in determining the optimal parameters in the context of EECP. Graphical presentation of dissimilar hyperparameter optimizers in the LSTM network on the IHEPC dataset is depicted in Figure 10. Additionally, the fitness comparison of different optimizers by varying the iteration number is graphically presented in Figure 11.

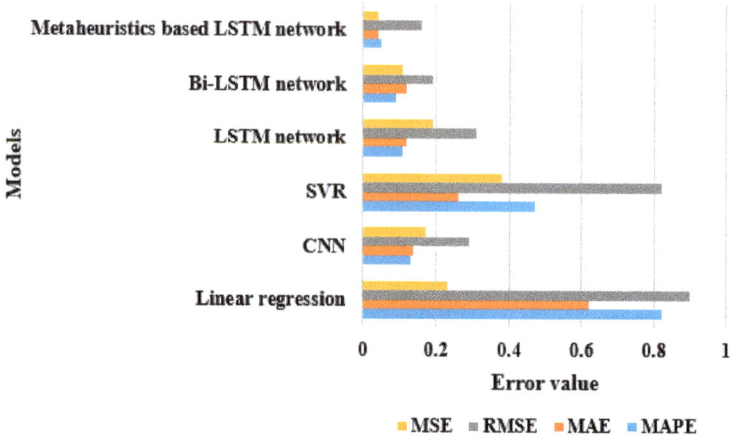

Figure 9. Graphical presentation of the experimental models for the IHEPC dataset.

Table 6. Performance of the dissimilar hyperparameter optimizers in the LSTM network on the IHEPC dataset.

			LSTM Network		
Optimizers	MAPE	MAE	RMSE	MSE	Predicting Time (s)
GWO	0.12	0.23	0.29	0.11	18
PSO	0.18	0.12	0.23	0.17	18.20
GA	0.09	0.07	0.20	0.07	17
ACO	0.12	0.21	0.20	0.11	16
ABC	0.14	0.17	0.18	0.12	14
BOA	0.05	0.04	0.16	0.04	13

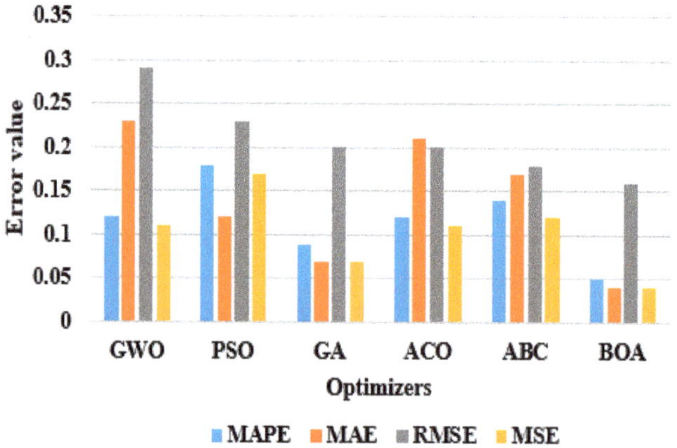

Figure 10. Graphical presentation of the dissimilar hyperparameter optimizers in the LSTM network on the IHEPC dataset.

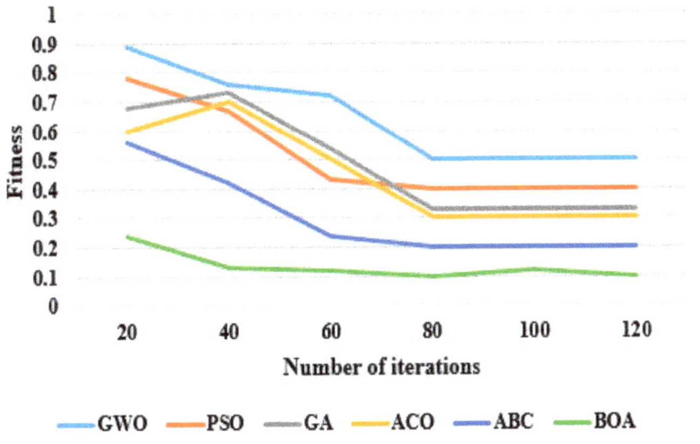

Figure 11. Fitness comparison of different optimizers achieved by varying the iteration number on the IHEPC dataset.

The prediction performance of the metaheuristic-based BOA model for the AEP and IHEPC datasets are graphically presented in Figures 12 and 13. Through an examination of these graphs, the proposed metaheuristic-based BOA model was shown to generate effective prediction results in the EECP domain.

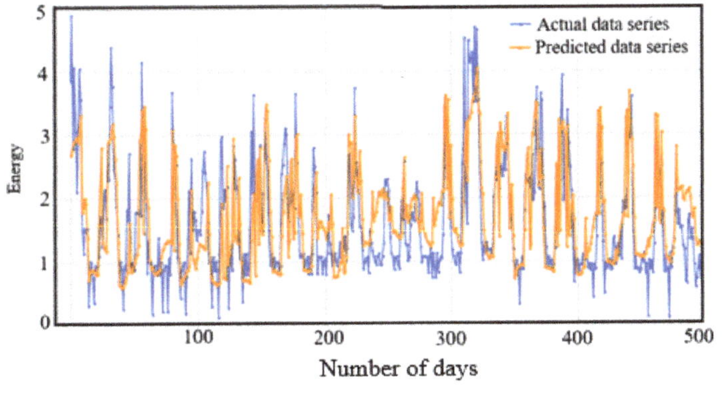

Figure 12. Prediction performance of the proposed model for the AEP dataset.

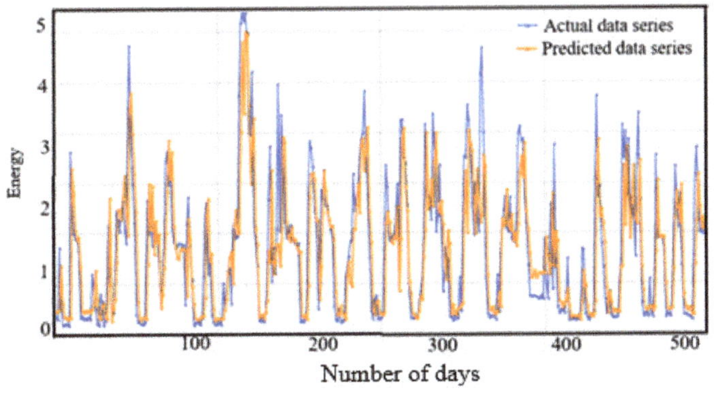

Figure 13. Prediction performance of the proposed model for the IHEPC dataset.

4.3. Comparative Study

In this scenario, the comparative investigation of the metaheuristic-based LSTM network and the existing models is detailed in Table 7 and Figure 14. T. Le et al. [16] integrated a Bi-LSTM network and a CNN for household EECP. Initially, the discriminative feature values were extracted from the IHEPC dataset using a CNN model, then the EECP was accomplished with the Bi-LSTM network. Extensive experimentation showed that the presented Bi-LSTM and the CNN model obtained an MAPE of 21.28, an MAE of 0.18, an RMSE of 0.22, and an MSE of 0.05 for the IHEPC dataset. M. Ishaq et al. [17] implemented an ensemble-based deep learning model to predict household energy consumption. In the resulting phase, the presented model performance was tested on the IHEPC dataset by means of MAPE, RMSE, MAE, and MSE. The presented ensemble-based deep learning model obtained an MAPE of 0.78, an MAE of 0.31, an RMSE of 0.35, and an MSE of 0.21 on the IHEPC dataset. M. Sajjad et al. [32] combined CNN with GRUs for an effective household EECP. Experimental evaluations showed that the presented model attained MAE values of 0.33 and 0.24, RMSE values of 0.47 and 0.31, and MSE values of 0.22 and 0.09 for both the IHEPC and AEP datasets.

Table 7. Statistical comparison of the proposed model with the existing models for the AEP and IHEPC datasets.

Models	Dataset	MAPE	MAE	RMSE	MSE
Bi-LSTM with CNN [16]	IHEPC	21.28	0.18	0.22	0.05
Ensemble-based deep learning model [17]	IHEPC	0.78	0.31	0.35	0.21
CNN with GRU model [32]	IHEPC	-	0.33	0.47	0.22
	AEP	-	0.24	0.31	0.09
Bi-LSTM with dilated CNN [33]	IHEPC	0.86	0.66	0.74	0.54
Multilayer bidirectional GRU with CNN [34]	IHEPC	-	0.29	0.42	0.18
	AEP	-	0.23	0.29	0.10
Metaheuristic based on the LSTM network	IHEPC	0.05	0.04	0.16	0.04
	AEP	0.09	0.07	0.13	0.05

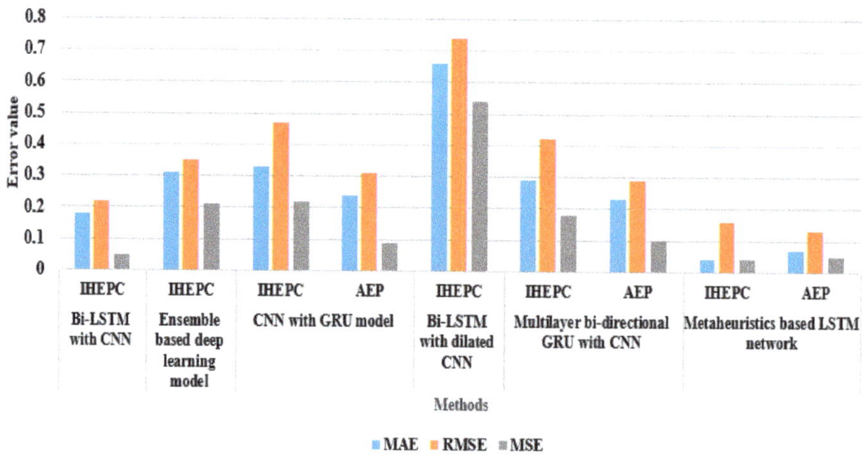

Figure 14. Comparison of the proposed model with the existing models.

Similarly, N. Khan et al. [33] integrated a Bi-LSTM network with a dilated CNN for predicting power consumption in the local energy system. Experimental evaluation showed that the presented model achieved an MAPE of 0.86, an MAE of 0.66, an RMSE of 0.74, and an MSE of 0.54 on the IHEPC dataset. Z.A. Khan et al. [34] combined a multilayer bidirectional GRU with a CNN for household electricity consumption prediction. The experimental investigation showed that the presented model achieved MAE values of 0.29

and 0.23, RMSE values of 0.42 and 0.29, and MSE values of 0.18 and 0.10 for the IHEPC and AEP datasets. As compared to the prior models, the metaheuristic based on the LSTM network achieved a good performance in EECP and also obtained a minimum error value for the IHEPC and AEP datasets. Hence, the obtained experimental results show that the metaheuristic based on the LSTM network significantly handles long and short time series data sequences to achieve better EECP with low computational complexity.

5. Conclusions

In this article, a new metaheuristic based on the LSTM model is proposed for effective household EECP. The metaheuristic based on the LSTM model comprises three modules, namely, data collection, data refinement, and consumption prediction. After collecting the data sequences from the IHEPC and AEP datasets, standard and min-max transformation methods are used for eliminating the missing, redundant, and outlier variables, and for normalizing the acquired data sequences. The refined data are fed into the metaheuristic-based LSTM model to extract hybrid discriminative features for EECP. In the LSTM network, the BOA selects the optimal hyperparameters, which improves the classifier's running time, and reduces system complexity. The effectiveness of the proposed model was tested on the IHEPC and AEP datasets in terms of MAPE, MAE, RMSE, and MSE, and the obtained results were compared with existing models, such as a Bi-LSTM with CNN, ensemble-based deep learning model, a CNN with a GRU model, a multilayer bidirectional GRU with a CNN, and a Bi-LSTM with a dilated CNN. As seen in the comparative analysis, the proposed metaheuristic based on the LSTM model obtained an MAPE of 0.05 and 0.09, an MAE of 0.04 and 0.07, an RMSE of 0.16 and 0.13, and an MSE of 0.04 and 0.05 for the IHEPC and AEP datasets, and these results were better than those generated by the comparative models. As a future extension of the present work, many non-linear exogenous data structures, such as monetary factors and climatic changes, will be explored in order to investigate power consumption.

Author Contributions: Investigation, resources, data curation, writing—original draft preparation, writing—review and editing, and visualization, S.K.H. and R.P. Conceptualization, software, validation, formal analysis, methodology, supervision, project administration, and funding acquisition relating to the version of the work to be published, R.P.d.P., M.W. and P.B.D. All authors have read and agreed to the published version of the manuscript.

Funding: Authors acknowledge contributions to this research from the Rector of the Silesian University of Technology, Gliwice, Poland.

Institutional Review Board Statement: Not applicable.

Informed Consent Statement: Not applicable.

Data Availability Statement: Data available in a publicly accessible repository that does not issue DOIs. Publicly available datasets were analyzed in this study. This data can be found here: [AEP Dataset. Available online: https://www.kaggle.com/loveall/appliances-energy-prediction (accessed on 12 September 2021). IHEPC Dataset. Available online: https://archive.ics.uci.edu/ml/datasets/Individual+household+electric+power+consumption (accessed on 12 September 2021)].

Conflicts of Interest: The authors declare no conflict of interest.

Abbreviations

ABC	Artificial Bee Colony
ACO	Ant Colony Optimizer
AEP	Appliances Load Prediction
ANFIS	Adaptive Neuro Fuzzy Inference System
Bi-LSTM	Bidirectional Long Short-Term Memory network
BOA	Butterfly Optimization Algorithm
CNN	Convolutional Neural Network
CWS	Chievres Weather Station
DBN	Deep Belief Network
EECP	Electric Energy Consumption Prediction
ELM	Extreme Learning Machine
GA	Genetic Algorithm
GRUs	Gated Recurrent Units
GWO	Grey Wolf Optimizer
IHEPC	Individual Household Electric Power Consumption
IMFs	Intrinsic Mode Functions
kW	kilowatt
LSTM	Long Short-Term Memory network
MAE	Mean Absolute Error
MAPE	Mean Absolute Percentage Error
MSE	Mean Square Error
PSO	Particle Swarm Optimization
RMSE	Root Mean Square Error
SVR	Support Vector Regression
VMD	Variational Mode Decomposition
Wh	Watt hour

References

1. Bandara, K.; Hewamalage, H.; Liu, Y.H.; Kang, Y.; Bergmeir, C. Improving the accuracy of global forecasting models using time series data augmentation. *Pattern Recognit.* **2021**, *120*, 108148. [CrossRef]
2. Gonzalez-Vidal, A.; Jimenez, F.; Gomez-Skarmeta, A.F. A methodology for energy multivariate time series forecasting in smart buildings based on feature selection. *Energy Build.* **2019**, *196*, 71–82. [CrossRef]
3. Chou, J.S.; Tran, D.S. Forecasting energy consumption time series using machine learning techniques based on usage patterns of residential householders. *Energy* **2018**, *165*, 709–726. [CrossRef]
4. Lara-Benítez, P.; Carranza-García, M.; Luna-Romera, J.M.; Riquelme, J.C. Temporal convolutional networks applied to energy-related time series forecasting. *Appl. Sci.* **2020**, *10*, 2322. [CrossRef]
5. Rodríguez-Rivero, C.; Pucheta, J.; Laboret, S.; Sauchelli, V.; Patiño, D. Energy associated tuning method for short-term series forecasting by complete and incomplete datasets. *J. Artif. Intell. Soft Comput. Res.* **2017**, *7*, 5–16. [CrossRef]
6. Di Piazza, A.; Di Piazza, M.C.; La Tona, G.; Luna, M. An artificial neural network-based forecasting model of energy-related time series for electrical grid management. *Math. Comput. Simul.* **2021**, *184*, 294–305. [CrossRef]
7. Coelho, I.M.; Coelho, V.N.; Luz, E.J.D.S.; Ochi, L.S.; Guimarães, F.G.; Rios, E. A GPU deep learning metaheuristic based model for time series forecasting. *Appl. Energy* **2017**, *201*, 412–418. [CrossRef]
8. Le, T.; Vo, M.T.; Kieu, T.; Hwang, E.; Rho, S.; Baik, S.W. Multiple electric energy consumption forecasting using a cluster-based strategy for transfer learning in smart building. *Sensors* **2020**, *20*, 2668. [CrossRef]
9. Choi, J.Y.; Lee, B. Combining LSTM network ensemble via adaptive weighting for improved time series forecasting. *Math. Probl. Eng.* **2018**, *2018*, 2470171. [CrossRef]
10. Ahmad, T.; Chen, H. Potential of three variant machine-learning models for forecasting district level medium-term and long-term energy demand in smart grid environment. *Energy* **2018**, *160*, 1008–1020. [CrossRef]
11. AlKandari, M.; Ahmad, I. Solar power generation forecasting using ensemble approach based on deep learning and statistical methods. *Appl. Comput. Inf.* **2020**. [CrossRef]
12. Talavera-Llames, R.; Pérez-Chacón, R.; Troncoso, A.; Martínez-Álvarez, F. MV-kWNN: A novel multivariate and multi-output weighted nearest neighbours algorithm for big data time series forecasting. *Neurocomputing* **2019**, *353*, 56–73. [CrossRef]
13. Ahmad, T.; Chen, H. Utility companies strategy for short-term energy demand forecasting using machine learning based models. *Sustain. Cities Soc.* **2018**, *39*, 401–417. [CrossRef]
14. Xiao, J.; Li, Y.; Xie, L.; Liu, D.; Huang, J. A hybrid model based on selective ensemble for energy consumption forecasting in China. *Energy* **2018**, *159*, 534–546. [CrossRef]

15. Gajowniczek, K.; Ząbkowski, T. Two-stage electricity demand modeling using machine learning algorithms. *Energies* **2017**, *10*, 1547. [CrossRef]
16. Le, T.; Vo, M.T.; Vo, B.; Hwang, E.; Rho, S.; Baik, S.W. Improving electric energy consumption prediction using CNN and Bi-LSTM. *Appl. Sci.* **2019**, *9*, 4237. [CrossRef]
17. Ishaq, M.; Kwon, S. Short-Term Energy Forecasting Framework Using an Ensemble Deep Learning Approach. *IEEE Access* **2021**, *9*, 94262–94271.
18. Lin, Y.; Luo, H.; Wang, D.; Guo, H.; Zhu, K. An ensemble model based on machine learning methods and data preprocessing for short-term electric load forecasting. *Energies* **2017**, *10*, 1186. [CrossRef]
19. Xu, W.; Peng, H.; Zeng, X.; Zhou, F.; Tian, X.; Peng, X. A hybrid modelling method for time series forecasting based on a linear regression model and deep learning. *Appl. Intell.* **2019**, *49*, 3002–3015. [CrossRef]
20. Maldonado, S.; Gonzalez, A.; Crone, S. Automatic time series analysis for electric load forecasting via support vector regression. *Appl. Soft Comput.* **2019**, *83*, 105616. [CrossRef]
21. Wan, R.; Mei, S.; Wang, J.; Liu, M.; Yang, F. Multivariate temporal convolutional network: A deep neural networks approach for multivariate time series forecasting. *Electronics* **2019**, *8*, 876. [CrossRef]
22. Bouktif, S.; Fiaz, A.; Ouni, A.; Serhani, M.A. Multi-sequence LSTM-RNN deep learning and metaheuristics for electric load forecasting. *Energies* **2020**, *13*, 391. [CrossRef]
23. Qiu, X.; Zhang, L.; Suganthan, P.N.; Amaratunga, G.A. Oblique random forest ensemble via least square estimation for time series forecasting. *Inf. Sci.* **2017**, *420*, 249–262. [CrossRef]
24. Kuo, P.H.; Huang, C.J. A high precision artificial neural networks model for short-term energy load forecasting. *Energies* **2018**, *11*, 213. [CrossRef]
25. Qiu, X.; Ren, Y.; Suganthan, P.N.; Amaratunga, G.A. Empirical mode decomposition based ensemble deep learning for load demand time series forecasting. *Appl. Soft Comput.* **2017**, *54*, 246–255. [CrossRef]
26. Pham, A.D.; Ngo, N.T.; Truong, T.T.H.; Huynh, N.T.; Truong, N.S. Predicting energy consumption in multiple buildings using machine learning for improving energy efficiency and sustainability. *J. Clean. Prod.* **2020**, *260*, 121082. [CrossRef]
27. Galicia, A.; Talavera-Llames, R.; Troncoso, A.; Koprinska, I.; Martínez-Álvarez, F. Multi-step forecasting for big data time series based on ensemble learning. *Knowl. Based Syst.* **2019**, *163*, 830–841. [CrossRef]
28. Khairalla, M.A.; Ning, X.; Al-Jallad, N.T.; El-Faroug, M.O. Short-term forecasting for energy consumption through stacking heterogeneous ensemble learning model. *Energies* **2018**, *11*, 1605. [CrossRef]
29. Jallal, M.A.; Gonzalez-Vidal, A.; Skarmeta, A.F.; Chabaa, S.; Zeroual, A. A hybrid neuro-fuzzy inference system-based algorithm for time series forecasting applied to energy consumption prediction. *Appl. Energy* **2020**, *268*, 114977. [CrossRef]
30. Bandara, K.; Bergmeir, C.; Hewamalage, H. LSTM-MSNet: Leveraging forecasts on sets of related time series with multiple seasonal patterns. *IEEE Trans. Neural Netw. Learn. Syst.* **2020**, *32*, 1586–1599. [CrossRef]
31. Abbasimehr, H.; Paki, R. Improving time series forecasting using LSTM and attention models. *J. Ambient Intell. Hum. Comput.* **2021**, 1–19. [CrossRef]
32. Sajjad, M.; Khan, Z.A.; Ullah, A.; Hussain, T.; Ullah, W.; Lee, M.Y.; Baik, S.W. A novel CNN-GRU-based hybrid approach for short-term residential load forecasting. *IEEE Access* **2020**, *8*, 143759–143768. [CrossRef]
33. Khan, N.; Haq, I.U.; Khan, S.U.; Rho, S.; Lee, M.Y.; Baik, S.W. DB-Net: A novel dilated CNN based multi-step forecasting model for power consumption in integrated local energy systems. *Int. J. Electr. Power Energy Syst.* **2021**, *133*, 107023. [CrossRef]
34. Khan, Z.A.; Ullah, A.; Ullah, W.; Rho, S.; Lee, M.; Baik, S.W. Electrical Energy Prediction in Residential Buildings for Short-Term Horizons Using Hybrid Deep Learning Strategy. *Appl. Sci.* **2020**, *10*, 8634. [CrossRef]
35. Mocanu, E.; Nguyen, P.H.; Gibescu, M.; Kling, W.L. Deep learning for estimating building energy consumption. *Sustain. Energy Grids Netw.* **2016**, *6*, 91–99. [CrossRef]
36. Candanedo, L.M.; Feldheim, V.; Deramaix, D. Data driven prediction models of energy use of appliances in a low-energy house. *Energy Build.* **2017**, *140*, 81–97. [CrossRef]
37. Kim, T.Y.; Cho, S.B. Predicting residential energy consumption using CNN-LSTM neural networks. *Energy* **2019**, *182*, 72–81. [CrossRef]
38. Ullah, A.; Muhammad, K.; Hussain, T.; Baik, S.W. Conflux LSTMs network: A novel approach for multi-view action recognition. *Neurocomputing* **2021**, *435*, 321–329. [CrossRef]
39. Hochreiter, S.; Schmidhuber, J. Long short-term memory. *Neural Comput.* **1997**, *9*, 1735–1780. [CrossRef] [PubMed]
40. Arora, S.; Singh, S. Butterfly optimization algorithm: A novel approach for global optimization. *Soft Comput.* **2019**, *23*, 715–734. [CrossRef]

applied sciences

MDPI

Article

Towards Facial Biometrics for ID Document Validation in Mobile Devices

Iurii Medvedev [1,*], Farhad Shadmand [1], Leandro Cruz [1,2] and Nuno Gonçalves [1,3]

1 Institute of Systems and Robotics, University of Coimbra, R. Silvio Lima, 3030-194 Coimbra, Portugal; farhad.shadmand@isr.uc.pt (F.S.); l.cruz@psenterprise.com (L.C.); nunogon@deec.uc.pt (N.G.)
2 Siemens Process Systems Engineering, London W6 7HA , UK
3 Portuguese Mint and Official Printing Office (INCM), 1000-042 Lisbon, Portugal
* Correspondence: iurii.medvedev@isr.uc.pt

Abstract: Various modern security systems follow a tendency to simplify the usage of the existing biometric recognition solutions and embed them into ubiquitous portable devices. In this work, we continue the investigation and development of our method for securing identification documents. The original facial biometric template, which is extracted from the trusted frontal face image, is stored on the identification document in a secured personalized machine-readable code. Such document is protected from face photo manipulation and may be validated with an offline mobile application. We apply automatic methods of compressing the developed face descriptors to make the biometric validation system more suitable for mobile applications. As an additional contribution, we introduce several print-capture datasets that may be used for training and evaluating similar systems for mobile identification and travel documents validation.

Keywords: artificial neural networks; biometrics; document handling; face recognition

Citation: Medvedev, I.; Shadmand, F.; Cruz, L.; Gonçalves, N. Towards Facial Biometrics for ID Document Validation in Mobile Devices. *Appl. Sci.* **2021**, *11*, 6134. https://doi.org/10.3390/app11136134

Academic Editor: Takayoshi Kobayashi

Received: 13 May 2021
Accepted: 28 June 2021
Published: 1 July 2021

1. Introduction

Document security has been an important issue since the appearance of the first documents and banknotes. Physical documents are still the first and ultimate authentication method and that is why their protection against spoofing attacks is important. Since the face image is one of the most important biometric components of ID and travel documents, its security is a prominent concern for official issuing organizations [1].

Face recognition techniques have been drawing a lot of attention in the last decade and, particularly, with the development of deep learning tools such as convolutional neural networks, they achieve outstanding accuracy levels.

Facial recognition technology impacts the overall security, allowing to automate ID document validation. The pipeline of this process usually follows the differential scenario, which implies that, during the verification procedure, the trusted reference is available. As a source of this reference, various systems may use the face image, a template that is previously stored and secured, or the trusted live captured image of a person from the border control camera.

The scenario of 1-1 authentication (verification) is a form of identity validation of a tested individual. At this scale, it is not required to store the remote biometric template/samples database, which eliminates the risks related to identity database theft or fraud at the point of control. The limitation of such a scenario provides face recognition systems with a number of benefits. They can be convenient and safe in applications, such as accessing the security area of personal devices or proving the identity during automated border control when trusted live capture is compared with the face image printed in the passport.

The particular case of 1-1 authentication is the so-called match-on-document scenario, which assumes that the trusted and secured biometric template is stored on the documents themselves. This strategy allows performing document validation in an offline mode to reduce the information security risks when storing or accessing databases of face images and templates are not allowed. The approach we discuss in this paper is directed to the applications for this scenario.

Various face recognition systems usually follow a 1-N authentication scenario which has some peculiarities but which is indeed less relevant to this work.

Face recognition systems are threatened by presentation attacks (spoofing attacks) of different kinds. In general, they usually intend to disguise the real subject identity or deceive the system to be impersonated as a different identity [2]. Such impostor attacks can lead to the gaining of illegitimate access by the fraudulent user and pose significant threads for security fields in companies and government sectors [3].

In applications of facial biometric recognition technologies in security systems for ID documents, several aspects are relevant. These solutions tend to become embedded in portable devices, such as smartphones. The specific tendency of certain solutions is to accomplish the offline validation of documents, to avoid risk related to compromising the network connectivity at any level. In this case, the minutia information is extracted from sources of biometric data (faces, fingerprints, iris, etc.), without storing or accessing databases of face images/templates.

The face recognition task for document security is usually constrained by considering frontal face images (in accordance with the ISO/IEC 19794-5 [4]). Since our method is focused on protecting ID and travel documents, we follow that statement. However, it is worth noting that, conceptually, instead of the frontal face image, any other source of biometric data (e.g., fingerprint pattern or iris) may be used in the design of such a validation system.

The designed biometric validation system secures the ID document by extracting the template from an enrolled frontal face image and encoding it into an MRC, which is further printed on the document in the specified area (Figure 1). The choice of MRC and encoding strategy is provided by the requirement that the bona fide MRC cannot be generated by an unauthorized issuer, which complicates producing fraudulent identification documents for deceitful impersonating.

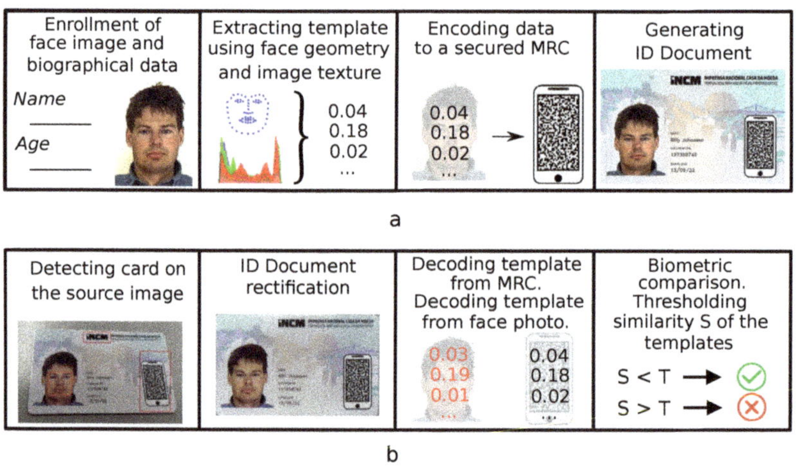

Figure 1. Pipeline of the proposed system: (**a**) ID card generating; and (**b**) ID card validating.

The validation of such document is performed by extracting two biometric templates from the frontal face photo and the MRC which are printed on its exterior. The templates are compared to determine if they belong to the same identity. The pipeline of the process in such a match-on-document scenario does not require network access, and the validation may be performed totally offline. However, the proposed approach in practical applications indeed may be expanded straightforwardly by including biographic data (name, date of birth, etc.) into the MRC.

It should be pointed out that the formulation of the face recognition problem in our method is specific. The main task being solved is to protect the particular instance of the issued document at the moment of its personalization with variable data. With this formulation, we do not focus on matching several photos of a single identity, although the designed method may be expanded for deployment in such a scenario. That is stated due to the fact that any newly issued document will contain the secured encoded biometric template, which corresponds to the particular face image to be printed on this document. However, in the scope of this work, the mentioned limitation is provided only by the choice of the training and test datasets. That is why we clarify the formulation of a problem as the protection of the particular face image of the ID document from various biometric impostor attacks.

Furthermore, one must notice that the matching between the two templates is not expected to be a perfect one, since the minutiae extracted from the frontal face are always different each time the validation occurs, due to image color and radiometric distortion, lighting conditions and the camera's intrinsic and extrinsic parameters, thus complicating the problem of matching different identities.

Summing up, the research and development of security solutions for match-on document scenario is important and for instance have been performed in our initial study [5]. In this work, which is the extended version of the paper presented at the BIOSIG2020 conference, we continue this investigation towards a compact offline mobile solution to secure ID and travel documents. By employing the secured MRC to embed the ID document with a carefully designed facial descriptor, we perform the document validation without storing biometric samples and templates in the remote database.

In comparison with the previous work, we modify the facial biometric template by including texture components. Our implementation of the facial descriptor is compact and optimized for usage in mobile devices. We also estimate the effect of template compression following the concept of match-on-document verification. Finally, we present several collected print/capture datasets which may be useful for analyzing the performance.

The paper is organized in the following way. In Section 2, we review some recently published works related to our research. Section 3 represents the discussion of the improved face descriptor implementation approach of differential validating and compressing the designed descriptors. Section 4 is devoted to the choice and application of machine-readable code (MRC) to our work. In Section 5, we present the details of the acquired in-house datasets. The discussion on the experimental results is performed in Section 6.

2. Related Work

The conventional pipeline of processing the input digital image (which may be acquired in different ways) in face recognition systems usually includes face detection [6] with optional alignment [7], followed by the face representation [8]. The last stage may be generally formulated as transforming the preprocessed face image into the low-dimensional feature space where various recognition tasks can be easily performed. This recognition scenario is usually defined by the practical purposes of the system. Intense efforts have been expended for the search of a better feature domain that possesses high face discriminative power and enough separability for distinguishing images corresponding to disjoint identities.

2.1. AAM in Face Recognition

Significant improvements in engineered methods for face recognition were achieved with the development of various techniques for automatic detection of special landmarks that allow localizing semantic regions on the face image. The list of selected face features is usually included in the standardized active appearance model (AAM) [9]. Such an approach gives huge opportunities for analyzing the face structure and processing the raw face images which have become very useful for face recognition applications.

For example, Abdulameer et al. [10] used facial features extracted with the use of AAM for the purposes of face verification that was performed with the trained classifier. Ouarda et al. [11] analyzed geometric face distances and characteristics of the semantic face shapes for face recognition purposes. The face recognition method reported by Juhong et al. [12] is based on face geometric invariances.

Another approach to face recognition with the use of an active appearance model is based on detecting semantic regions and extracting local texture features for further analysis. For instance, Ahonen et al. [13] considered both shape and texture information to discriminate face images. The face descriptor in this method is based on Local Binary Pattern (LBP) histograms extracted from the partitioned image. The dissimilarity metric between the descriptors is estimated by the nearest neighbor classifier.

Many improvements were introduced to this technique. For instance, Shen et al. [14] adopted discriminative LBP features for different color channels to enhance the performance for images with severe changes in illumination and face resolutions.

Another popular technique for face recognition that deals with image texture is the histogram of oriented gradients (HOG) method, which usually implies sub-sampling images to small blocks and counting proportions of gradient orientation in these localized segments of an image. The extracted coefficients may further serve as discriminative features of the image.

The example of such an approach was reported by Shu et al. [15] who analyzed different factors that affect the HOG-based face descriptor and the performance and computational efficiency in comparison with other texture-based techniques.

Deniz et al. [16] considered the HOG descriptor with sub-sampling based on the facial landmarks. Various methods for increasing the robustness of extracted HOG features were considered by analyzing the impact of facial feature location error and replacing the detected features with the regular grid.

Other texture analysis techniques are generally less popular but also attract the interest of face recognition research. The Gabor feature method is widely used in computer vision for pattern recognition tasks. Applying special Gabor filters, it is possible to extract the directivity and frequency of the content within the vicinity of some point or region. For example, Yunfeng et al. [17] used concatenated Gabor wavelet coefficients, which are extracted in the vicinity area of each detected facial landmark. Further, the descriptors are distinguished with a support vector machine classifier.

In [5], the face recognition solution for the match-on-document scenario is introduced. It employs the process of encoding the biometric template into the secured MRC to be printed on the document. By comparison with this work, we revisit and improve the implementation of the facial biometric template, considering the methods of its compressing, which are important for the target platform (mobile devices).

2.2. Face Recognition with Machine Learning

Modern face recognition intensively uses recent achievements in machine learning. These tools are usually served to learn the discrimination of face descriptors, for example by solving a classification task to estimate the similarity between images. The face representation in these approaches is usually engineered and based on low-level face image information [10,13].

Another approach to face recognition came with convolutional deep networks which give the ability to efficiently learn discriminative face features themselves even from unconstrained images. These learned features generally do not include local image characteristics and realize the face description in a high-level manner [8]. Significant popularity in face recognition systems was gained by metric learning approaches which are focused on straightforwardly optimizing the face representation. For example, Schroff et al. [18] introduced a triplet loss for face recognition which minimizes the distance in the feature domain between samples of the same identity and maximizes it for the disjoint identities. However, such methods usually require unreasonably large datasets accompanied by a carefully designed sample mining strategy for successful convergence.

The best performance in unconstrained face recognition is achieved by methods that consider the problem as a classification task separating images by their identities. These approaches usually utilize softmax-based classification loss [8] which allows learning the discriminative face features, which may be further used for distinguishing tasks with trivial similarity metrics. Nowadays, investigation is focused on modifying softmax loss by different means. The main purpose of most of the published improvements is enlarging the inter-class variance and reducing the inter-class compactness [19]. For example, particular attention was paid to constraining the target feature distribution with a margin of a different kind [20,21].

Some deep learning methods also find their application in face recognition systems for document security. As an example, Shi et al. [22] proposed a method for 1-1 authentication for the differential automatic border control scenario. In their system, the face photo on the ID document is validated with the help of life face capture. The two images are processed by separate networks to estimate the similarity of their deep representations.

However, the intricacy and the lack of a clear sense of extracted deep representations may be an obstacle in practice while embedding these approaches in portable devices. The computational complexity requirements of deep learning approaches are still high for most of the smartphones on the market.

At the same time, the differential manner of match-on-document scenario implies 1-1 verification of templates extracted from an original digital image and its copy, which is printed on the ID document. With such a scenario, the engineered features, which may catch characteristic peculiarities of the particular image, have better usage perspectives than learned high-level features. However, machine learning tools indeed may be applied for distinguishing such engineered face feature representations.

2.3. Industry Solutions

Industry solutions also follow the advances in facial recognition for protecting ID and travel documents with modern techniques. Two noticeable products were developed by Jura (Digital IPI) [23] and IDEMIA (Lasink and DocSeal) [24]. These approaches attempt to embed face recognition systems into ubiquitous portable devices (e.g., smartphones). Such validation solutions may broadcast the convenience of authenticity verification of documents and products, while at the same time allowing to reduce the requirements for the sophisticated equipment. Their main idea is to augment the document with printed elements that store encoded personalized data to be further extracted and decoded with the use of a portable device with a digital camera.

In our work, we adopt a similar approach for protecting ID and travel documents. However, the above methods are included in proprietary commercial systems and are not publicly available, which does not allow completing benchmarks and comparison. That is why existing benchmarks (e.g., the NIST FRVT challenge [25]) have some submission restrictions and usually accept the solutions in the compiled form without the requirements of submitting the source code.

3. Materials and Methods

In the scope of this work, we followed the motivation of developing a simple and compact method for offline document validation that includes an in-house solution for face description and their verification. We also followed the trends of modern validation applications that rely on biometric recognition and emphasized portability and ubiquity.

3.1. Facial Biometric Template

In our investigation, we focus on the search for a method for facial description which is optimized for encoding into MRC and embedding into mobile devices. We combine our facial biometric template from several types of discriminative features which include information about face geometry and texture.

The process of extracting features from the frontal face image starts with the applying active appearance model and detecting facial landmarks (Figure 2). We employ the standard appearance model that includes 68 facial landmarks and entirely specifies face semantic regions to be further processing by the algorithm.

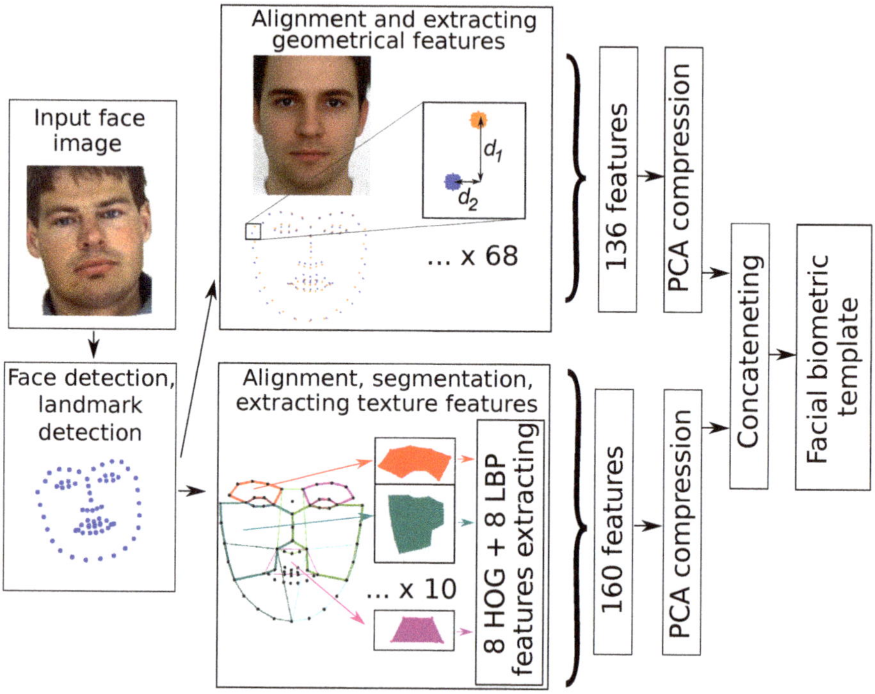

Figure 2. The process of extracting facial biometric template from the input face image.

The coordinates of the detected set of landmarks carry all discriminative geometric information given by the chosen model. However, due to the uncertainty of face parameters (e.g., size or pose) on the source image, these raw data cannot be directly used for making biometric comparisons and require some normalization procedure.

To achieve that, we define some fixed set of landmarks that serves as a reference for alignment. In that case, if two sets of landmarks from different arbitrary face images are aligned to the defined supporting set, they also become aligned to each other. In our work, we choose this supporting set by extracting facial landmarks from averaged face image in Figure 2.

The alignment of the input set of points $\{x_i, y_i\}$ with the supporting set, is implemented as a coordinate transformation to $\{x_i', y_i'\}$ by rotating, scaling and shifting operations, defined by Equation (1). The α rotation is determined to achieve the horizontal face pose. Scaling is performed by the relation of values of face contour perimeters (P_{sup} corresponds to the supporting set and P to the input set), which is the selection of points with indices in the range [0–26] (depicted with blue color in Figure 2). The shift $S(s_x, s_y)$ is defined as the difference between the centers of supporting and scaled input sets of points.

$$\begin{bmatrix} x_i' \\ y_i' \end{bmatrix} = \frac{P_{sup}}{P} * \begin{bmatrix} cos(\alpha) & -sin(\alpha) \\ sin(\alpha) & cos(\alpha) \end{bmatrix} * \begin{bmatrix} x_i \\ y_i \end{bmatrix} + \begin{bmatrix} s_x \\ s_y \end{bmatrix} \tag{1}$$

A facial biometric template for two aligned sets of landmarks is composed as a result of element-wise subtraction of their coordinates. However, the result values signify some pixel distance between coordinates and thus depend on the image characteristics. To avoid that, we normalize the template elements to the constant perimeter P_{sup}. Due to the parameters of the employed active appearance model, the resultant descriptor includes 136 values.

Many other geometry-based descriptor implementations usually consider subsets on the landmarks or rely on hand-crafted features which are manually designed by selecting special relations within the active appearance model. Our implementation retains the geometric data in its entirety and gives a profound geometric description. However, the impact of discriminative power for different template elements is not equal and requires proper analysis or weighting which we achieve by learning methods.

To increase the template robustness against specific biometric distortion attacks (e.g., when the fake face image is warped to fit the geometry of the original image), texture features are also included in it. The texture descriptor in our method is based on the combined usage of HOG and LBP techniques. In order to define the image regions from which the features are extracted, we perform the segmentation of aligned input face image by ten characteristic semantic sections. The contours of the sections are depicted in Figure 2. The particular choice of the region's borders is based on intensive experiments of searching for a better selection. For each region, we extract eight HOG and eight LBP texture features with conventional computer vision tools. As a result of extracting sixteen features for each of ten regions, we get 160 texture components that are combined with geometric ones in a complete biometric template which includes $D\_size = 296$ features $\{d_i\}$.

3.2. PCA Template Compression

The designed template provides a holistic description of the face, which may include some redundancy. To eliminate it and make the system more compact, we employ compression techniques and evaluate the effect of template compression on the performance of validation. As an automatic approach for compressing the designed descriptor, we use the well-known principal components analysis (PCA) algorithm.

PCA is used for reducing the dimensionality of a template by projecting its elements onto a lower number of principal components, while at the same time maximizing the variance of the data.

3.3. Differential Template Verification

The process of document validation follows the differential scenario when the comparison is performed for two facial biometric templates extracted from this document. The first one $\{d\_test_i\}$ is extracted from the printed face photo, which is potentially counterfeited. Another template $\{d\_orig_i\}$, which is securely encoded with MRC, acts as a trusted reference. The match comparison decision signifies the genuineness of the tested document sample.

The superficial comparison indeed may be performed by applying the Euclidean distance metric (Equation (2)). This simple similarity score can be used to make the validation decision by comparison with the fixed threshold. However, different template elements can have different impact weight on the verification decision, which is not accounted for in this trivial linear estimation.

$$E = \sum_{i=1}^{D\_size} |(d\_test_i - d\_orig_i)| = \sum_{i=1}^{D\_size} |d\_sub_i| \tag{2}$$

Instead of tuning the similarity metric parameters manually, at this stage, we rely on the learning approaches. For such robust verification, we train a binary match/non-match classifier which is designed as a fully connected artificial neural network with a sigma activation function along with the network architecture.

As input, the classifier takes the absolute values of Its first layer receives the absolute result of element-wise subtraction of biometric templates d_sub_i.

The final layer of the architecture includes a single node that outputs a scalar S in the range [0,1], where 0 corresponds to the ideal match decision. We train the classifier in the regression manner with the use of standard sequential back-propagation [26].

In order to normalize the values of network input and fit it better with the first layer activation function at the initialization stage, we introduce the coefficient N (see Equation (3)). In our experiments, the best optimization is achieved by setting $N = 0.015$ for geometrical template components and $N = 0.1$ and $N = 0.05$ for HOG and LBP texture components.

$$d\_inp_i = max(1, \frac{|d\_sub_i|}{N}) \tag{3}$$

4. MRC Application

The match-on-document scenario implies embedding the document sample with additional machine-readable data, which can be implemented in various ways. In our work, we follow the trend to employ computer vision tools to make the document readable with a digital camera. Such an approach may be conveniently implemented with a machine-readable code printed on the surface of the document (see Figure 3). For this work, we utilize the Graphic Code [27] that can be customized for security purposes.

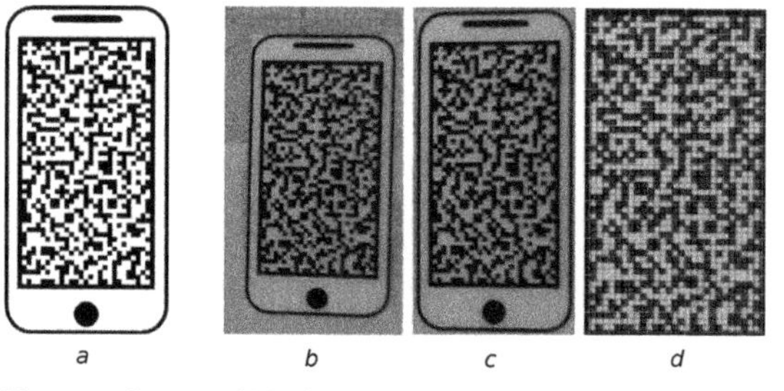

Figure 3. Stages of MRC processing: (**a**) Generated MRC sample to be printed on the ID document; (**b**) detected MRC on the ID document printed sample; (**c**) frontalized MRC image; and (**d**) MRC message reconstruction [5].

We assume that the MRC sample for a particular biometric template is secured to store the trusted reference data for differential comparison, which is crucial for our approach. However, we refer to the detailed consideration of security issues to corresponding work [27]. We can summarize this discussion as follows.

During the process of creating the MRC sample, several layers of security and attack robustness can be introduced. These methods usually follow the symmetrical security approach, where various parameters of the system remain obscured and act as a key for both encryption and decryption. While the algorithm of assembling MRC remains open, the Graphic Code ensures the security performance by specifying its private internal parameters (e.g., unit cell characteristics and dictionary). To increase the security level, various cryptography methods over the data themselves may also be employed. For instance, to magnify the computation complexity of cryptanalysis, the message to be encoded may be encrypted to ciphered text. However, the usage of only these symmetrical methods leads to the overall risks, when compromising the application on a single portable device poses a threat to the full system.

This may be mitigated by adapting any asymmetric encryption approach. Indeed, to prove the originality of the printed MRC during the decoding process, it is required just to validate the document issuer's authority. To achieve that, the message with a biometric template can be protected with the digital signature which is a well-tested method for similar applications. This method requires the issuer to generate its private and public keys. The hash, which is extracted from the template data, is signed with a private key and added to the result message, which is encoded into the MRC. In that scenario, having the public key, the authority of the issuer of the document may be verified. The offline mode of this verification can be maintained by uploading the public key of the issuer once during the initial installation.

4.1. Encoding

The Graphic Code allows an arbitrary choice of the outline image to ease the coordination with standards, which are usually applied to the appearance of the ID document. As an example in the scope of this work, we use one depicted in Figure 3a. The required alphabet defines the correspondence between N (120 in this work) characters and various unit cells composed of 3×3 pixels. To encode the biometric template to the MRC, the result message is transferred into the alphabet space by quantization process. To compose the MRC instance, each character in the message is replaced by the corresponding pattern basing on the defined dictionary. The total amount of information I that is carried by the encoded template results to $\sim$260 bytes, which is estimated by Equation (4).

$$I = K \cdot \log_2 N \tag{4}$$

Practical application may also require the encoding of some biographical data (ID card number and name) in addition to the biometric template to ease automatic document processing. To verify the correct decoding, a set of check digits is concatenated with the message. If any empty cells are left, they may be replaced with random non-dictionary unit cells.

4.2. Decoding

During the process of decoding, the captured image of the detected MRC is processed with conventional computer vision algorithms to achieve the properly aligned binary form suitable for decoding (Figure 3b,c). Next, the rectified image (Figure 3d) is aligned and examined to find unit cells defined in the dictionary for combining the result message.

The print/capture process introduces various distortions to the image of the MRC. Errors that occur due to various illumination conditions, reflection and MRC surface attrition may be detected with the use of check digits. We performed various tests with the various acquisition parameters and MRC deformations, to prove the overall robustness of the decoding algorithm. However, in practical applications, most of the inaccuracies can be compensated by processing the stream of frames captured by a digital camera.

5. Datasets

It is important to note that the deployment of a mobile face recognition system for document security purposes implies dealing with images that are captured by the portable digital camera from the physically printed documents.

As an example, some works directed on document scanning scenario (which is constrained with the absence of perspective transformations) introduce collected print-scan datasets to deal with that problem [28] or methods for generating such print/scan datasets automatically [29]. However, the last option usually can barely cover all aspects of illumination and acquisition distortions with various capture devices.

The document acquisition with a portable digital camera introduces even more variable noise to the captured images, which applies to the perspective inaccuracies and more severe lighting distortions.

Following the original purpose of protecting the ID document at the moment of its personalization, for all images from the chosen original dataset, we collected their print/capture digital copies trying to cover possible noise and distortion variations during the process. Acquired images were then automatically frontalized (see examples in Figures 4 and 5).

The differential verification with such an approach implies estimating the similarity between the original and the captured images in a single verification iteration. In order to retain such differential manner of the processing scenario, we labeled the sets of collected images similarly for original and captured ones.

A similar print/capture dataset (Print/Capture Aberdeen [30]) was obtained by Medvedev et al. [5]. We extend that approach to several other datasets that contain frontal face images complied with travel document standards [4] (Utrecht [30] and AR [31]), including various level of acquisition parameter variations. As a result, we obtain several print/capture datasets (https://github.com/visteam-isr-uc/trustfaces-template-verification (accessed on 30 June 2021)) which we call as follows :

- Print/Capture Aberdeen v2 (89 identities, 15 k captures);
- Print/Capture Utrecht (67 identities, 16 k captures); and
- Print/Capture AR (135 identities, 29 k captures).

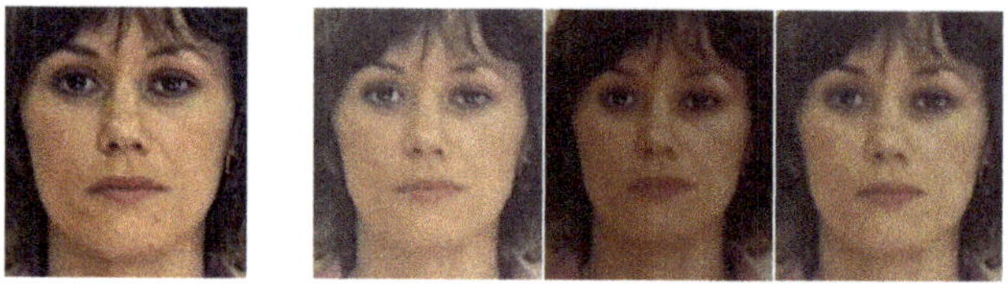

a b

Figure 4. Example images from the Print-Capture Aberdeen dataset: (**a**) original digital image; and (**b**) captured photo image.

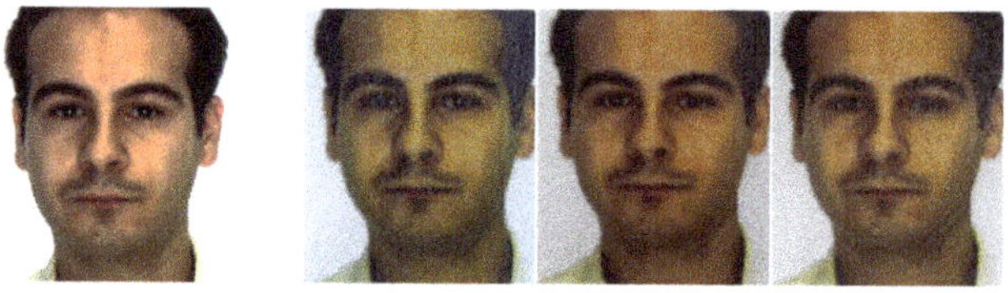

a b

Figure 5. Example images from the Print-Capture AR dataset: (**a**) original digital image; and (**b**) captured photo image.

In the process of training on the templates extracted from such datasets, the network learns the proper weighting for the particular template components. At the same time, such an approach optimizes the further face verification process by learning existing irregularities related to printing, digital capture, misalignment and rectification process.

At the same time, practical deployment in the mobile application assumes handling the stream of frames from which only a few are selected for processing when the bad quality ones (over illuminated or occluded) may be skipped. Such practical details are usually important to be accounted in the early research stage. That is why one has to be careful choosing the strategy for selecting images to be included in the dataset, reducing ones that will likely be skipped during the deployment.

We achieve that by carefully designing a rectification process that is primarily directed on the eliminating perspective deformation of the document on the captured image.

5.1. Document Rectification

In a comparison with the automatic border control (ABC) scenario, the document validation with a mobile device deals with variations of the document alignment. That is why document rectification is usually a mandatory step in the validation pipeline. This operation allows obtaining frontalized and standardized images of the document for further processing and extracting data embedded in it.

This operation indeed may be performed in various ways. We assume that the ID document is flat and perform rectification by the trivial perspective transformation (Figure 6). The parameters of the transformation matrix are estimated by matching the detected features of the specific regions of the document [32]. Despite the fact that during that process we require a specific document appearance, it is used only as a reference to achieve the main goal—frontalize the face images. This automatic process indeed also introduces additional warping noise to the result images.

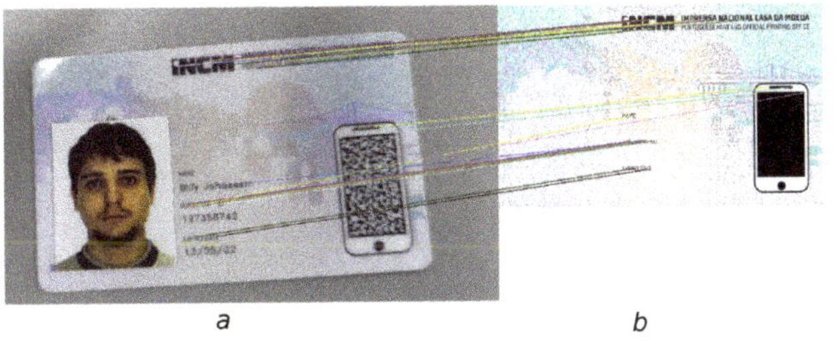

Figure 6. Rectification process with the featured detected by the document appearance: (a) captured document image; and (b) original card layout.

5.2. Train and Test Protocols

As stated above, the face recognition task in the target scenario is to verify the particular face image sample. That is why we define the training and testing protocols as follows. From the labeled sets of original and printed/captured images, we select pairs for extracting templates and computing their element-wise subtracts $\{d\_sub_i\}$. This set is also labeled in binary form, depending on their cross identity label. If images in the pair belong to a single identity, this pair is labeled as bona fide. Pairs with images from different identities are labeled as a biometric impostor attack.

In order to make the extracted data balanced in terms of the presence of match and non-match pairs, we significantly reduce the number of non-match ones to be included in the resulting protocol. This choice is randomized based on the overall dataset statistics. In our experiments, the result data were divided into train and test protocol parts with split proportions of 80% and 20%, respectively. The identities lists are disjoint in these two parts.

6. Results

In a recent work [5], we focused on the search for a better classifier architecture to optimize it in terms of the balance between efficiency and complexity. In this study, we mostly focused on estimating the impact of template compression on the overall system performance. We employed an ANN-based classifier with the following combination of hidden layer numbers: 300-400-200-100. The template was compressed by selecting a number of the first PCA components (compressed features) (Figures 7–10). To demonstrate the compression effect for each template part, we separately applied PCA to the geometric and texture sections. For geometric template, we first took 120 and 100 components. For texture template, we took 140 and 120 components (features).

With these settings, we performed intensive experiments and trained the set of classifiers for compressed collections of templates according to defined protocols. In each iteration, we defined the number of epochs equal to 20 and chose the best result at the end of the training.

As a metric for evaluating the performance, we estimated ROC (receiver operating characteristic) curves and computed their corresponding AUCs (area under curve) (see Figures 7 and 8 and Table 1).

We also estimated the performance with false acceptance rate and false rejection rate metrics (see Figures 9 and 10).

From the obtained results, one can see that the compression of the designed geometrical features does not significantly affect the performance. Indeed, this is related to the redundancy of defined descriptors as the locations of neighbor landmarks are highly correlated. At the same time, the drop in the performance with the compression of the texture features is more significant as they have much less redundancy. These features are extracted holistically from the semantic regions which do not intersect.

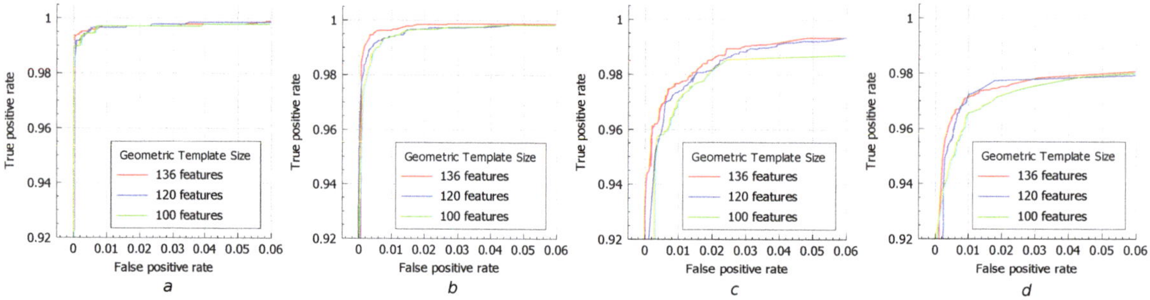

Figure 7. ROC curves of ANN classifier for various various compression of geometric features (PCA components): (**a**) Print/Capture Aberdeen; (**b**) Print/Capture Aberdeen v2; and (**c**) Print/Capture Utrecht; and (**d**) Print/Capture AR.

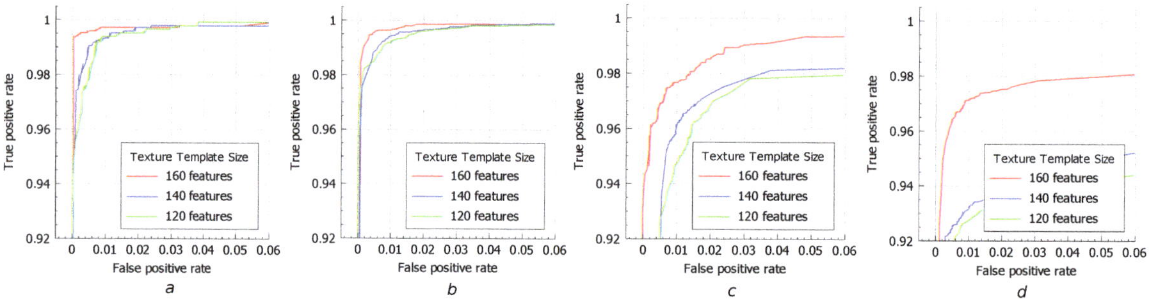

Figure 8. ROC curves of ANN classifier for various compression of texture features (PCA components): (**a**) Print/Capture Aberdeen; (**b**) Print/Capture Aberdeen v2; (**c**) Print/Capture Utrecht; and (**d**) Print/Capture AR.

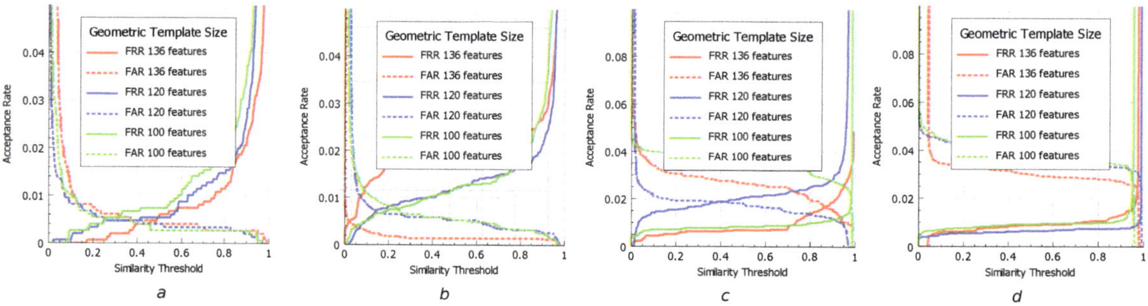

Figure 9. FAR/FRR of ANN classifier for various various compression of geometric features (PCA components): (**a**) Print/Capture Aberdeen; (**b**) Print/Capture Aberdeen v2; (**c**) Print/Capture Utrecht; and (**d**) Print/Capture AR.

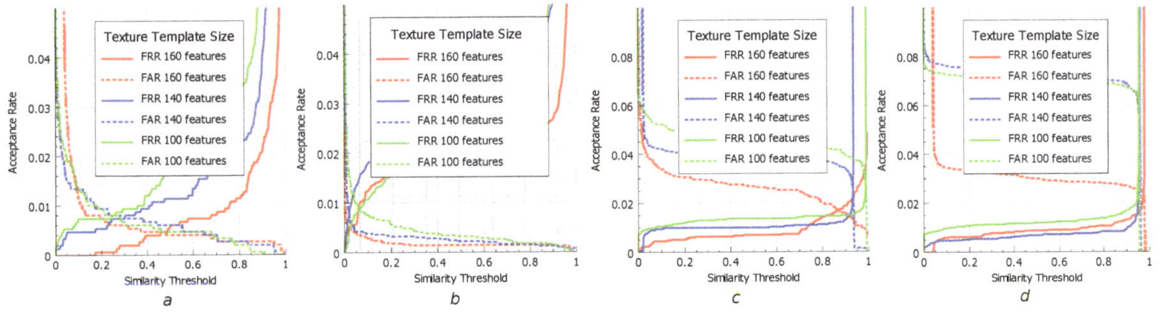

Figure 10. FAR/FRR of ANN classifier for various compression of texture features (PCA components): (**a**) Print/Capture Aberdeen; (**b**) Print/Capture Aberdeen v2; (**c**) Print/Capture Utrecht; and (**d**) Print/Capture AR.

Table 1. Performance characteristics (AUC/FNMR@FMR = 0.01/Equal Error Rate(EER)) of classifiers for various sizes of compressed templates and datasets.

Template Size Geometric + Texture	Print/Capture Aberdeen	Print/Capture Aberdeen v2	Print/Capture Utrecht	Print/Capture AR
136 G + 160 T.	0.999438/0.0027 0.0047	0.999179/0.0036 0.0048	0.99592/0.023 0.016	0.994624/0.029 0.023
120 G + 160 T.	0.999465/0.0034 0.0047	0.999394/0.0059 0.0073	0.998035/0.027 0.018	0.992643/0.073 0.018
100 G + 160 T.	0.999452/0.0027 0.0054	0.999277/0.0063 0.0077	0.995276/0.033 0.020	0.994577/0.078 0.024
136 G + 140 T.	0.999262/0.0068 0.0074	0.998991/0.0061 0.0077	0.992228/0.044 0.024	0.987724/0.030 0.043
136 G + 120 T.	0.999297/0.0075 0.0081	0.99912/0.0081 0.0087	0.991232/0.059 0.026	0.982557/0.038 0.047

The expected feature of the results is that the experiments on the different datasets demonstrate slightly different performance, which occurs due to the different level of illumination condition variations (exposure, relative position of light source and camera and applied shadows) during the process of their harvesting.

Another observation is that, depending on the template size, the parameters of the architecture (sizes of the hidden layer), which may be optimized to achieve the best accuracy, indeed depend on the template size (the size of the input layer). However, here we do not follow that suggestion, limiting ourselves to only estimating the compression effect with the fixed parameters of the setup.

7. Conclusions

This paper is devoted to the development of an efficient method for protecting ID and travel documents by augmenting them with a secured facial biometric template to be encoded in the machine-readable code. The approach is optimized for portable devices (e.g., smartphones) in terms of the CPU usage and solves the frontal face verification problem in the offline match-on-document scenario. Our demo application on an iPhone 7 is able to perform the complete card validation process (including detection, rectification, extracting templates and verification) in 0.2 s.

We introduce the improved facial biometric descriptor and estimate the effect of its compression on the performance of the system in various experiments.

As an additional contribution of this work, we introduce several print/capture datasets that may be useful for the research related to face recognition for mobile document security applications. They can be employed to analyze the robustness of face recognition algorithms to various distortions caused by the combined impact of printer and digital camera.

The overall results show the high performance of the developed method against biometric impostor attacks. At the same time, it may be customized with the use of biographical data or adapted for other biometric characteristics (such as fingerprints and iris). The method may be applied without sophisticated equipment, in a very cheap and convenient way. Our future work will be directed towards increasing the robustness of the developed facial template, more detailed analysis of the performance with a multi-fold approach and adapting deep learning techniques for the match-on-document scenario.

8. Patents

The results of our work on the project TrustFaces were published in the patent [33].

Author Contributions: Conceptualization and methodology, N.G. and L.C.; software, L.C.; investigation, data curation and writing—original draft preparation, I.M.; writing—review and editing, F.S.; and supervision, project administration and funding acquisition, N.G. All authors have read and agreed to the published version of the manuscript.

Funding: The authors would like to thank the Portuguese Mint and Official Printing Office (INCM) and the University of Coimbra for the support of the project TrustFaces.

Institutional Review Board Statement: Not applicable. The study involved the usage of the human face images, which were taken from the publicly available datasets.

Informed Consent Statement: Not applicable. The study involved the usage of the human face images, which were taken from the publicly available datasets.

Data Availability Statement: One of the results of presented work is a set of print/capture face datasets that are harvested with the use of publicly available images. (https://github.com/visteam-isr-uc/trustfaces-template-verification (accessed on 30 June 2021)).

Conflicts of Interest: The authors declare no conflict of interest. The funders had no role in the design of the study; in the collection, analyses, or interpretation of data; in the writing of the manuscript, or in the decision to publish the results.

Abbreviations

The following abbreviations are used in this manuscript:

MRC	Machine Readable Code
ANN	Artificial Neural Network
ABC	Automatic Border Control
ROC	Receiver Operating Characteristic
AUC	Area Under Curve

References

1. Thieme, M. *Biometrics Market and Industry Report 2009–2014*; Tech. Rep.; International Biometric Group: New-York, NY, USA, 2008.
2. Ramachandra, R.; Busch, C. Presentation Attack Detection Methods for Face Recognition Systems: A Comprehensive Survey. *ACM Comput. Surv.* **2017**, *50*. [CrossRef]
3. Kumar, S.; Singh, S.; Kumar, J. A comparative study on face spoofing attacks. In Proceedings of the 2017 International Conference on Computing, Communication and Automation (ICCCA), Greater Noida, India, 5–6 May 2017; pp. 1104–1108.
4. ISO/IEC 19794-5:2011. *Information Technology—Biometric Data Interchange Formats—Part 5: Face Image Data*; ISO/IEC JTC 1/SC 37 Biometrics; 2011. Available online: https://www.iso.org/standard/50867.html (accessed on 13 May 2021).
5. Medvedev, I.; Gonçalves, N.; Cruz, L. Biometric System for Mobile Validation of ID And Travel Documents. In Proceedings of the 2020 International Conference of the Biometrics Special Interest Group (BIOSIG), Darmstadt, Germany, 16–18 September 2020; pp. 1–5.
6. Ranjan, R.; Sankar, S.; Bansal, A.; Bodla, N.; Chen, J.C.; Patel, V.; Castillo, C.; Chellappa, R. Deep Learning for Understanding Faces: Machines May Be Just as Good, or Better, than Humans. *IEEE Signal Process. Mag.* **2018**, *35*, 66–83. [CrossRef]
7. Jin, X.; Tan, X. Face Alignment In-the-Wild: A Survey. *Comput. Vis. Image Underst.* **2016**. [CrossRef]
8. Taigman, Y.; Yang, M.; Ranzato, M.; Wolf, L. DeepFace: Closing the Gap to Human-Level Performance in Face Verification. In Proceedings of the 2014 IEEE Conference CVPR, Columbus, OH, USA, 24–27 June 2014; pp. 1701–1708.
9. Cootes, T.; Edwards, G.; Taylor, C. Active Appearance Models. *Pattern Anal. Mach. Intell. IEEE Trans.* **2001**, *23*, 681–685. [CrossRef]

10. Abdulameer, M.; Sheikh Abdullah, S.; Ali Othman, Z. Face recognition technique based on active appearance model. *Int. Rev. Comput. Softw.* **2013**, *8*, 2733–2739.
11. Ouarda, W.; Trichili, H.; Alimi, A.M.; Solaiman, B. Face recognition based on geometric features using Support Vector Machines. In Proceedings of the 2014 6th International Conference SoCPaR, Tunis, Tunisia, 11–14 August 2014; pp. 89–95.
12. Juhong, A.; Pintavirooj, C. Face recognition based on facial landmark detection. In Proceedings of the 2017 10th Biomedical Engineering International Conference (BMEiCON), Hokkaido, Japan, 31 August–2 September 2017; pp. 1–4. [CrossRef]
13. Ahonen, T.; Hadid, A.; Pietikäinen, M. Face Recognition with Local Binary Patterns. *Eur. Conf. Comput. Vis.* **2004**, *3021*, 469–481. [CrossRef]
14. Shen, Y.; Chiu, C. Local binary pattern orientation based face recognition. In Proceedings of the 2015 IEEE International Conference on Acoustics, Speech and Signal Processing (ICASSP), South Brisbane, Australia, 19–24 April 2015; pp. 1091–1095. [CrossRef]
15. Shu, C.; Ding, X.; Fang, C. Histogram of the Oriented Gradient for Face Recognition. *Tsinghua Sci. Technol.* **2011**, *16*, 216–224. [CrossRef]
16. Deniz, O.; Bueno, G.; Salido, J.; De la Torre, F. Face recognition using Histograms of Oriented Gradients. *Pattern Recognit. Lett.* **2011**, *32*, 1598–1603. [CrossRef]
17. Li, Y.; Ou, Z.; Wang, G. Face Recognition Using Gabor Features and Support Vector Machines. In *Advances in Natural Computation;* Wang, L., Chen, K., Ong, Y.S., Eds.; Springer: Berlin/Heidelberg, Germany, 2005; pp. 119–122.
18. Schroff, F.; Kalenichenko, D.; Philbin, J. FaceNet: A unified embedding for face recognition and clustering. In Proceedings of the 2015 IEEE Conference CVPR, Boston, MA, USA, 7–12 June 2015; pp. 815–823.
19. Sun, J.; Yang, W.; Xue, J.; Liao, Q. An Equalized Margin Loss for Face Recognition. *IEEE Trans. Multimed.* **2020**, *22*, 2833–2843. [CrossRef]
20. Deng, J.; Guo, J.; Xue, N.; Zafeiriou, S. ArcFace: Additive Angular Margin Loss for Deep Face Recognition. In Proceedings of the 2019 IEEE/CVF Conference on Computer Vision and Pattern Recognition (CVPR), Long Beach, CA, USA, 16–20 June 2019; pp. 4685–4694.
21. Liu, W.; Wen, Y.; Yu, Z.; Li, M.; Raj, B.; Song, L. SphereFace: Deep Hypersphere Embedding for Face Recognition. In Proceedings of the 2017 IEEE Conference on Computer Vision and Pattern Recognition (CVPR), Honolulu, HI, USA, 21–26 July 2017; pp. 6738–6746. [CrossRef]
22. Shi, Y.; Jain, A.K. DocFace: Matching ID Document Photos to Selfies. In Proceedings of the 2018 IEEE 9th International Conference on Biometrics Theory, Applications and Systems (BTAS), Redondo Beach, CA, USA, 22–25 October 2018; pp. 1–8. [CrossRef]
23. Koltai, F.; Adam, B. Enhanced optical security by using information carrier digital screening. In *Optical Security and Counterfeit Deterrence Techniques V*; van Renesse, R.L., Ed.; International Society for Optics and Photonics, SPIE: Bellingham, WA, USA, 2004; Volume 5310, pp. 160–169. [CrossRef]
24. Jones, R.L.; Eckel, R.A. Line Segment Code for Embedding Information in an Image. U.S. Patent Application No. 16/236,068, 4 July 2019.
25. Grother, P.; Ngan, M. Face Recognition Vendor Test (FRVT) Performance of Face Identification Algorithms NIST IR 8009. 2014. Available online: https://www.nist.gov/publications/face-recognition-vendor-test-frvt-performance-face-identification-algorithms-nist-ir (accessed on 13 May 2021).
26. LeCun, Y.; Bottou, G.O.; Muller, K.R. *Efficient Backprop, in Neural Networks—Tricks of the Trade*; Springer Lecture Notes in Computer Sciences; Springer: Berlin/Heidelberg, Germany, 1998.
27. Cruz, L.; Patrão, B.; Gonçalves, N. Graphic Code: A New Machine Readable Approach. In Proceedings of the 2018 IEEE International Conference AIVR, Taichung, Taiwan, 10–12 December 2018; pp. 169–172.
28. Ferrara, M.; Franco, A.; Maltoni, D. Face morphing detection in the presence of printing/scanning and heterogeneous image sources. *arXiv* **2019**, arXiv:1901.08811.
29. Mitkovski, A.; Merkle, J.; Rathgeb, C.; Tams, B.; Bernardo, K.; Haryanto, N.E.; Busch, C. Simulation of Print-Scan Transformations for Face Images based on Conditional Adversarial Networks. In Proceedings of the 2020 International Conference of the Biometrics Special Interest Group (BIOSIG), Darmstadt, Germany, 16–18 September 2020; pp. 1–5.
30. School of Natural Sciences University of Stirling. Psychological Image Collection of Stirling. 1998. Available online: http://pics.stir.ac.uk (accessed on 13 May 2021).
31. Martinez, A.; Benavente, R. The AR Face Database. Tech. Rep. 24 CVC Technical Report. 1998. Available online: http://www2.ece.ohio-state.edu/~aleix/ARdatabase.html (accessed on 13 May 2021).
32. Itseez. Open Source Computer Vision Library. 2015. Available online: https://github.com/itseez/opencv (accessed on 13 May 2021).
33. Gonçalves, N.M.; Cruz, L.; Medvedev, I. Method for Validation of Authenticity of an Image Present in an Object, Object with Increased Security Level and Method for Preparation Thereof, Computer Equipment, Computer Program and Appropriate Reading Means. Portugal Patent No. WO2020251380A1, 17 September 2020.

applied sciences

MDPI

Review

Re-Identification in Urban Scenarios: A Review of Tools and Methods

Hugo S. Oliveira [1], José J. M. Machado [2] and João Manuel R. S. Tavares [2,*]

1 Faculdade de Engenharia, Universidade do Porto, Rua Dr. Roberto Frias, s/n, 4200-465 Porto, Portugal; hugo.soares@fe.up.pt
2 Departamento de Engenharia Mecânica, Faculdade de Engenharia, Universidade do Porto, Rua Dr. Roberto Frias, s/n, 4200-465 Porto, Portugal; jjmm@fe.up.pt
* Correspondence: tavares@fe.up.pt; Tel.: +351-22-041-3472

Abstract: With the widespread use of surveillance image cameras and enhanced awareness of public security, objects, and persons Re-Identification (ReID), the task of recognizing objects in non-overlapping camera networks has attracted particular attention in computer vision and pattern recognition communities. Given an image or video of an object-of-interest (query), object identification aims to identify the object from images or video feed taken from different cameras. After many years of great effort, object ReID remains a notably challenging task. The main reason is that an object's appearance may dramatically change across camera views due to significant variations in illumination, poses or viewpoints, or even cluttered backgrounds. With the advent of Deep Neural Networks (DNN), there have been many proposals for different network architectures achieving high-performance levels. With the aim of identifying the most promising methods for ReID for future robust implementations, a review study is presented, mainly focusing on the person and multi-object ReID and auxiliary methods for image enhancement. Such methods are crucial for robust object ReID, while highlighting limitations of the identified methods. This is a very active field, evidenced by the dates of the publications found. However, most works use data from very different datasets and genres, which presents an obstacle to wide generalized DNN model training and usage. Although the model's performance has achieved satisfactory results on particular datasets, a particular trend was observed in the use of 3D Convolutional Neural Networks (CNN), attention mechanisms to capture object-relevant features, and generative adversarial training to overcome data limitations. However, there is still room for improvement, namely in using images from urban scenarios among anonymized images to comply with public privacy legislation. The main challenges that remain in the ReID field, and prospects for future research directions towards ReID in dense urban scenarios, are also discussed.

Keywords: person ReID; computer vision; deep neural networks; image enhancement

Citation: Oliveira, H.S.; Machado, J.J.M.; Tavares, J.M.R.S. Re-Identification in Urban Scenarios: A Review of Tools and Methods. *Appl. Sci.* **2021**, *11*, 10809. https://doi.org/10.3390/app112210809

Academic Editors: João M. F. Rodrigues, Pedro J. S. Cardoso and Marta Chinnici

Received: 7 September 2021
Accepted: 10 November 2021
Published: 16 November 2021

Publisher's Note: MDPI stays neutral with regard to jurisdictional claims in published maps and institutional affiliations.

1. Introduction

The task of object ReID on image cameras has been studied for several years by the computer vision and pattern recognition communities [1], with the primary goal to ReID a query object among different cameras.

Multi-object ReID, based on a wide range of surveillance cameras, is nowadays a vital aspect in modern cities, to better understand city movement patterns among the different infrastructures [2], with the primary intention of rapidly mitigate abnormal situations, such as tracking car thieves, wanted persons, or even lost children.

This is still a challenging task, since an object's appearance may dramatically change across camera views due to the significant variations in illumination, poses or viewpoints, or even cluttered backgrounds [2] (Figure 1). According to the state-of-the-art research studies, existing object ReID methodologies can be divided into two main categories: image-based and video-based object ReID.

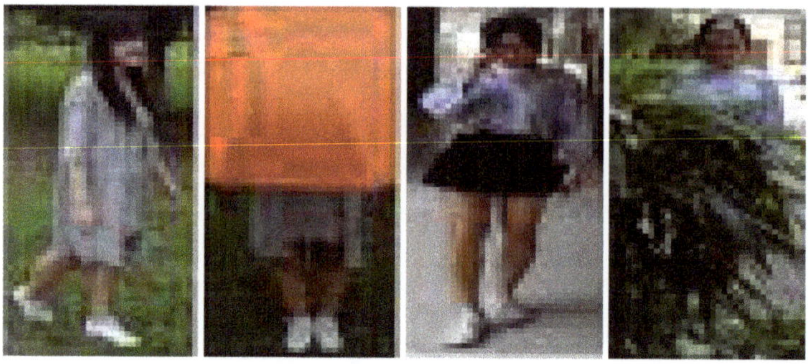

(a) ID 1 (b) ID 1 (occluded) (c) ID 2 (d) ID 2 (occluded)

Figure 1. Some common problems found in object ReID.

The former category focuses on matching a probe image of one object, with an image of the object with the same ID among gallery sets, which is mainly established based on image content analysis and matching. In contrast, the latter category focuses on matching two videos, exploiting different information, such as temporal, and motion-based information. A gallery corresponds to a collection of object images gathered from different perspectives over time. In both approaches, the pairs of objects to be matched are analogous. However, in real scenarios, object ReID needs to be conducted between the image and video. For example, given a picture of a criminal suspect, the police would like to quickly locate and track the suspect from hundreds of city surveillance videos. The ReID under this scenario is called image-to-video person ReID, where a probe image is searched in a gallery of videos acquired from different surveillance cameras.

Although videos contain more information, image-to-video ReID share the same challenges with image-based and video-based objects ReID, namely, similar appearance, low resolution, substantial variation in poses, occlusion, and different viewpoints. In addition, an extra difficulty resides on the match between two different datasets, one static and another dynamic, i.e., image and video, respectively.

Image and video are usually represented using different features. While only visual features can be obtained from a single image, both visual features and spatial–temporal features can be extracted from a video. Recently, CNN has shown potential for learning state-of-the-art image feature embedding [3,4] and Recurrent Neural Network (RNN) yields a promising performance in obtaining spatial–temporal features from videos [5,6].

In general, there are two major types of deep learning structures for object ReID; namely, verification models and identification models. Verification models take a pair of data as input and determine whether they belong to the same object or not, by leveraging weak ReID labels that can be regarded as a binary-class classification or similarity task [7]. In contrast, identification models aim at feature learning by treating object ReID as a multi-class classification task [4], but lack direct similarity measurement between input pairs. Due to their complementary advantages and limitations, the two models have been combined to improve the ReID performance [8]. However, in the image-to-video object ReID task, a cross-modal system, directly using DNN and the information provided by the target task still cannot perfectly bridge the "media gap", which means that representations of different datasets are inconsistent. Therefore, most of the current works directly rely on weights from pre-trained deep networks as the backbone to obtain initial values for the target model and initiate the pre-trained network structure to facilitate the training of the new deep model. Figure 2 depicts the common base architecture for person ReID.

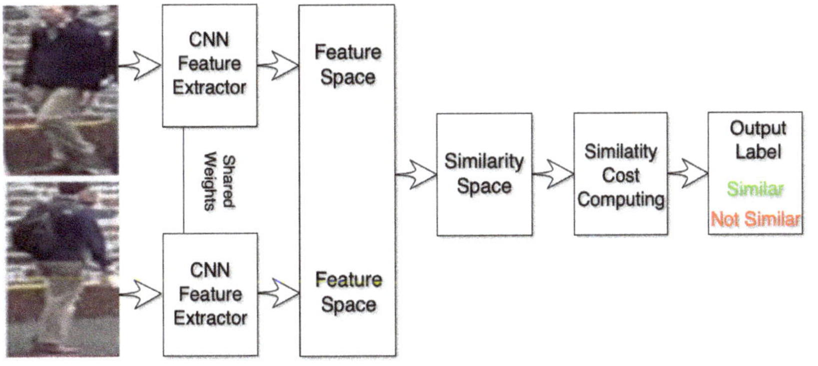

Figure 2. Common base architecture for person ReID.

With this problem in mind, we focus on identifying the promising techniques and methods for ReID that lead to a unified use in public urban scenarios, in adversarial and challenging conditions. This research article is organized toward identifying the main methodologies and key aspects. The following section presents an overview of the deep learning method for object ReID, and the most widely employed models proposed for ReID tasks. The main goal is to improve existing methods and derive new approaches to the ReID tasks. In the methods section, we detail the search used for this review article. The results section presents the main findings achieved with the selected works grouped by the person ReID and the multi-object ReID, with temporal constraints in consideration. The final section provides a critical discussion of the results and draws the conclusions.

2. Deep Learning for Object Re-Identification

A turning point in the history of machine learning and computer vision was reached by researchers of the University of Toronto, who proposed a new image classification approach and achieved excellent results in the ImageNet [9] competition [10]. The winning proposal, defined as AlexNet, consisted of a CNN composed of a set of stacked deep layers and dense layers that enabled the reduction of the error drastically. However, the first appearances of CNN dated from 1990, when Lecun et al. [11] proposed a CNN method addressing the task of hand-written digit recognition to alleviate the work of the postal office.

A CNN multi-layer contains at least one layer to perform convolution operations from image inputs, by using filters of kernels that are translated across and down the input matrix, to generate a feature representation map of the original image input. The characteristics of these filters can be widely different, and each one of them is composed of learnable parameters, updated through a gradient descent optimization scheme. The same layer can employ other filters. For the same part of the image, each filter produces a set of local responses, enabling correlation of specific pixel information with the content of the adjacent pixels. After the proposal by [10], many different network architectures were developed to address such a problem, each one with its inner characteristics, to name a few, the VGG [12], Inception [13], and ResNet [13] architectures. Many other network architectures were proposed, but in summary, they all share some building blocks.

In regard to the ReID task, the most used backbone model architecture is ResNet [14], due to its flexibility and ease of reusing and implementation to solve new problems. Most of the ReID tasks explore the use of pre-trained ResNet as backbones for feature extraction for the object ReID task.

ResNets can address the vanishing gradient problem when the networks are too deep, making the gradients quickly shrink to zero after several chain rule applications, leading toward "not updating" the weights and, therefore, harming of the learning process. With

ResNets, the gradients can flow directly through the skip connections backward from later layers to the initial filters.

ResNets can have different sizes, depending on how big each model layer is, and how many layers it contains. As an example, ResNet 34 [15] contains one convolution and pooling layer step, followed by four similar layers (Figure 3). Each layer follows the same pattern by performing 3×3 convolutions with a fixed feature map dimension, and by bypassing the input every two convolutions. A concern with the special characteristics of ResNet 34 is the fact that the width W and height H dimensions remain constant during the entire single layer. The size reduction is achieved by the stride size used in the convolution kernels, instead of the polling layers commonly used in other models.

Every layer of a ResNet is composed of several blocks, enabling it to go deeper. This deepness is achieved by increasing the number of operations within a block, while maintaining the total number of layers. Each operation comprises a convolution step, batch normalization, and a ReLU activation to a particular input; except for the last operation of the block, which does not contain a ReLU activation function.

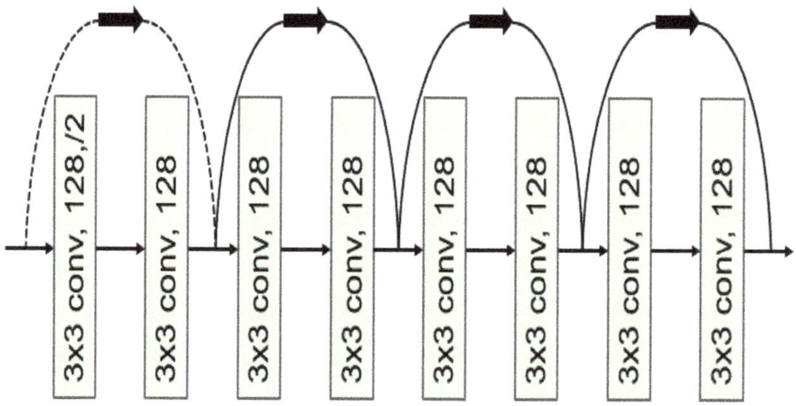

Figure 3. A common ResNet block architecture.

The aforementioned characteristics make ResNets particularly suitable for object ReID, since it enables shallow lower-level features to be reused at higher-level stages. This scheme allows exploring relevant information for ReID task from all layer feature maps, instead of relying only on more abstract summarized features provided by the higher layers.

3. Evaluation Metrics

Cumulative Matching Characteristic curve (CMC) is a common evaluation metric for person or object ReID methods. It can be considered a simple single-gallery-shot setting, where each gallery identity only has one instance. Given a probe image, an algorithm will rank the entire gallery sample according to the distances to the probe, with the CMC top-k accuracy given as:

$$Acc_k = \begin{cases} 1 & \text{if top-}k\text{ ranked gallery samples contain the query identity,} \\ 0 & \text{otherwise,} \end{cases} \quad (1)$$

which is a shifted step function. The final Cumulative Matching Characteristics (CMC) curve is built by averaging the shifted step functions over all the queries.

Another commonly used metric is the mean Average Precision (mAP), which is very often employed on each image query, and defined as:

$$MAP = \frac{\sum_{q=1}^{Q} AveP(q)}{Q}, \quad (2)$$

where Q is the number of queries

4. Methodology

The current bibliographic analysis involved two steps: (a) collecting related work and (b) a detailed review and analysis of the gathered work.

The research methodology consisted of a keyword-based search of conference papers or journal articles from scientific databases; namely, IEEE Xplore and Science Direct, and from web scientific indexing services: Web of Science, Google Scholar, and arXiv. As search keywords, one performed the following query: ["deep learning"] AND ["reid" OR "re-identification" OR "person reid" OR "object reid"] The gathered information excluded filtered out articles referring to DNN, but did not apply to the ReID domain. Articles were initially identified from this process.

Restricting the search to articles in combination with connected papers, a web tool, the initial number of articles was lessened to 47. In the second step, the 21 articles selected from the previous step were analyzed one-by-one for the task of a person ReID, considering the following research questions:

1. What was the ReID- or multi-object ReID problem addressed?
2. What was the general approach and type of DNN-based models employed?
3. What were the datasets and models proposed by the authors? Were there any variations observed by the authors?
4. Was any pre-processing of data or data augmentation technique used?
5. What was the overall performance in (depending on the adopted metric)?
6. Did the authors test their model performances on different datasets?
7. Did the authors compare their approaches with other techniques? If yes, what was the difference in performance?

5. Person Re-Identification

Person ReID is the problem of matching the same individuals across multiple image cameras or across time within a single image camera. The computer vision and pattern recognition research communities have paid particular attention to it due to its relevance in many applications, such as video surveillance, human–computer interactions, robotics, and content-based video retrieval. However, despite years of effort, person ReID remains a challenging task for several reasons [16], such as variations in visual appearance and the ambient environment caused by different viewpoints from different cameras.

Significant changes in humans pose—across time and space—background clutter and occlusions; different individuals with similar appearances present difficulties to the ReID tasks. Moreover, with little or no visible image faces due to low image resolution, the exploitation of biometric and soft-biometric features for person ReID is limited. For the person ReID task, databases and different approaches have been proposed by several authors, which are summarized in the following sections.

5.1. Person Re-Identification Databases

The recognition of human attributes, such as gender and clothing types, has excellent prospects in real applications. However, the development of suitable benchmark datasets for attribute recognition remains lagged. Existing human attribute datasets are collected from various sources or from integrating pedestrian ReID datasets. Such heterogeneous collections pose a significant challenge in developing high-quality fine-grained attribute recognition algorithms.

Among the public databases that have been proposed for person ReID, some examples can be found in the open domain, such as the Richly Annotated Pedestrian (RAP) [17], which contains images gathered from real multi-camera surveillance scenarios with long-term collections, where data samples are annotated, not only with fine-grained human attributes, but also with environmental and contextual factors. RAP contains a total of

41,585 pedestrian image samples, each with 72 annotated attributes, as well as viewpoints, occlusions, and body part information.

Another example is the VIPeR [18] database, which contains 632 identities acquired from 2 cameras, forming a total of 1264 images. All images were manually annotated, with each image having a resolution of 128 × 48 pixels.

A summary of the public available databases for person ReID, with main characteristics, is presented in Table 1.

Table 1. Global overview of the found public available databases for person ReID.

Dataset	# Identities	# Cameras	# Images	Label Method	Size	Tracking Sequences
VIPeR	632	2	1264	Hand	128 × 48	NO
ETH1,2,3	853,528	1	8580	Hand	Vary	YES
QMUL iLIDS	119	2	476	Hand	Vary	NO
GRID	1025	8	1275	Hand	Vary	NO
CAVIAR4reid	72	2	1220	Hand	Vary	NO
3DPeS	192	8	1011	Hand	Vary	NO
PRID2011	934	2	24,541	Hand	128 × 64	YES
WARD	70	3	4786	Hand	128 × 48	YES
SAIVT-Softbio	152	8	64,472	Hand	Vary	YES
CUHK01	971	2	3884	Hand	160 × 60	NO
CUHK02	1816	10 (5 pairs)	7264	Hand	160 × 60	NO
CUHK03	1467	10 (5 pairs)	13,164	Hand	Hand/DPM	NO
RAiD	43	4	6920	Hand	128 × 64	NO
iLIDS-VID	300	2	42,495	Hand	Vary	YES
MPR Drone	84	1	-	ACF	Vary	NO
HDA Person Dataset	53	13	2976	Hand/ACF	Vary	YES
Shinpuhkan Dataset	24	16	-	Hand/ACF	128 × 48	YES
CASIA Gait Database B	124	11	-	Background subtraction	Vary	YES
Market-1501	1501	6	32,217	Hand/DPM	128 × 64	NO
PKU-reid	114	2	1824	Hand	128 × 64	NO
PRW	932	6	34,304	Hand	Vary	NO
Large scale person search	11,934	-	34,574	Hand	Vary	NO
MARS	1261	6	1,191,003	DPM+GMMCP	256 × 128	YES
DukeMTMC-reid	1812	8	36,441	Hand	Vary	NO
DukeMTMC4reid	1852	8	346,261	Doppia	Vary	NO
Airport	9651	6	39,902	ACF	128 × 64	NO
MSMT17	4101	15	126,441	Faster RCNN	Vary	NO
RPIfield	112	12	1,601,581	ACF	Vary	NO

DPM—deformable part models, ACF—pyramid features, GMMCP—generalized maximum multi clique.

Although many other datasets suitable for object ReID can be found, the ones listed in Table 1 are widely used by most of the authors as benchmarks for performance evaluation and comparison of the proposed works.

5.2. Person Re-Identification Methods

In this section, deep learning-based person ReID methods are grouped into four main categories, as represented in Figure 4, including methods for feature learning, sequence learning, generative learning, and deep learning metrics. These categories encompassed several methods, and they are discussed in the following in terms of their main aspects and experimental results.

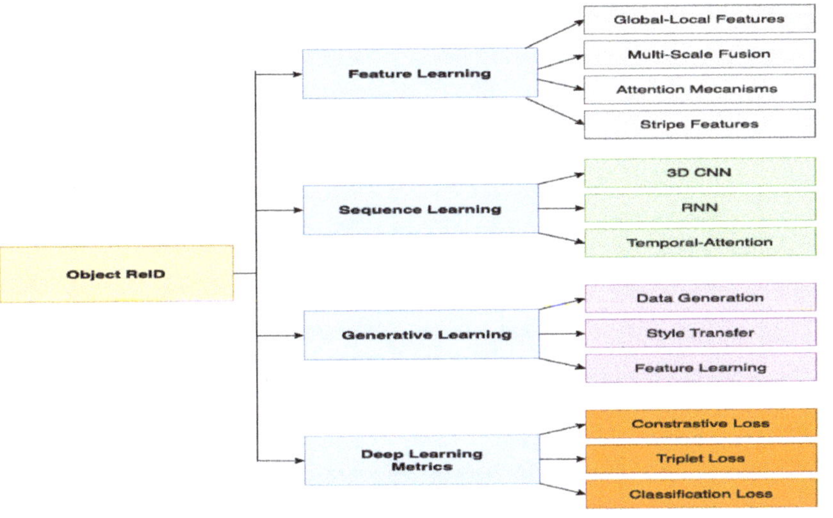

Figure 4. Deep learning-based person re-identification methods.

5.2.1. Feature Learning

Considering the features extracted from images, person ReID methods can explore the extracted feature from a global and local perspective, to better represent the object of interest. Global feature learning usually provides a single feature from the target image, making it difficult to capture detailed information of the person image. To overcome this problem, distinguishable local features are also used to capture subtle invariant features that are often combined in a fusion scheme.

An example of the use of a fusion scheme to combine global and local features for person ReID is presented by [19]. It consists of the formulation of an image-to-video person ReID as a classification-based information retrieval problem, where a model of "person appearance" is learned from the gallery images, and the identity of the interested person is determined by the probability that the corresponding probe image belongs to one of the gallery images.

To learn a model of person appearance, two kinds of features, Kernel Descriminator (KDES) [20] and CNN, are extracted from each person's image. Then, a Support Vector Machine (SVM) model is employed to learn the model. For ReID, three fusion schemes, early fusion, product rule, and query-adaptive late fusions, are proposed to aggregate the features. A ranking scheme in descending order of similarity is employed between the query image and the learned model to determine the most likely image pair. The work was evaluated in two benchmark datasets, CAVIAR4reid [21] and RAID [22], and by using CMC [23] as evaluation measures for person ReID. The model achieved a CMC of 96.11% when using the CAVIAR4reid dataset, and contained balanced cases combined with data augmentation during training, while employing late adaptive feature fusion. In contrast, when using the same dataset containing imbalanced cases and the same training and fusion methods, the model obtained an CMC of 93.33%. As for the RAID database, the model presented a CMC of 94.29% when using data augmentation during training and late feature fusion.

The same author refined in [24] its previous image-to-video person ReID framework [19] by adding two extra features: the Gaussian of Gaussian (GOG) [25] and learned features from Residual Neural Network (ResNet) [14]. The same feature fusion methodology employed in its previous work [19] was used, and the newly added features were evaluated using the same CAVIAR4reid and RAID databases. The model on CAVIAR4reid, containing balanced cases, and using late adaptive fusion combined with data augmentation during training attained an CMC of 86.39%. In contrast, on CAVIAR4reid and with imbalanced cases, as well as the same fusion scheme and augmentation, the model achieved a CMC of 91.94%. Conducted experiments on the RAID database using the same fusion scheme and augmentation techniques achieved a CMC of 92.80%, proving the effectiveness of the newly added features. When compared to the previous work, the performance fell by approximately 2% on both datasets, mainly due to DNN co-adaptation that led to some degree of overfitting. While these works are somehow complementary, the suggested approach can overcome the difficulties in learning cross-scale features by learning multi-scale complementary features.

Jointly learning of local and global features using a CNN is proposed by [26], exploring advantages of jointly learning local and global features in a CNN, in order to obtain correlated local and global features in different contexts scenarios for person ReID. A deep two-branch CNN architecture was proposed, with one branch being responsible for learning localized features (local branch) and the second directed to learning global feature (global branch); the two components were not independent, but synergistically correlated and jointly learned, concurrently. The proposed joint learning multi-loss (JLML) CNN model consists of two branch CNN networks.

The local branch aims to learn the most discriminating local visual features of the surroundings of a people bounding box. In contrast, the second branch is responsible for learning the most discriminating global level features from the entire person's image. The joint learning scheme is employed for concurrently optimizing per-branch discriminative feature representations, and discovering correlated complementary information between local and global features by subjecting both local and global branches to the same identity label supervision. For sharing low-level features, a multi learning methodology is explored, as in the work proposed by [27]. An inter-branch common learning inter-permutation is to be shared on the first convolution layer, with the intuition that lower convolution layers capture low levels features, such as edged and corners that are common patterns to all images, while the complementary discriminative features from local and global representations are learned independently, and related to a given identity label. Moreover, a structure sparsity-induced regularization [28] is introduced to discourage the use of irrelevant features while encouraging discriminative features, to learn concurrently on both local and global contexts, and to maximize a shared identity matching objective. The final global feature representation corresponds to a sparsity measure with LASSO [29]. Cross-entropy is then used as the loss function for both global and local branches, in order to optimize person identity classification and a pairwise person ReID. Concerning the distance metrics for person ReID, a 1024-D deep feature representation is employed using only a generic distance metric without camera-pair specific distance learnable metrics. The models were evaluated on the CUHK01, VIPeR [30], CUHK03, Market-1501, and GRID datasets. On CHUK03, with the proposed method achieving a CMC of 83.2% Rank-1 using labeled objects and 80.6% with automatically detected objects. On Market-1501, the method achieved an CMC Rank-1 of 85.1% on single query, and 89.7% for multi-query. On CHUK01, the method achieved a CMC for Rank-1 of 91.2% when applying an 871/100 dataset split and 76.7% using a 486/485 split. For a GRID dataset, the method obtained a CMC Rank-1 of 37.5%. Finally, as to the VIPeR dataset, the method achieved a CMC Rank-1 of 50.2%. In most datasets, the proposed method surpassed the compared state-of-the-art approaches except for the VIPeR dataset. The combination of local and global features potentiates model generalization, avoiding overfitting to image-specific features.

In [31], a novel deep ReID CNN is proposed for omni-scale feature learning (OSNet). The model is based on residual blocks composed of multiple convolutional streams, with each detecting feature at a certain scale. A novel unified aggregation gate is then introduced to dynamically fuse multi-scale features with output-dependent channel-wise weights. To efficiently learn spatial-channel correlations and to avoid over-fitting, pointwise and depth-wise convolutions are used. Depth-wise separable convolutions are also adopted to reduce the number of parameters. Person matching is based on the $\zeta 2$ distance from 512-D feature vectors extracted from the last layer. Two OSNet models were trained for comparison proposes, with the first model being trained from scratch during 350 epochs using Stochastic Gradient Descent (SGD) as an optimizer. In contrast, a second model was fine-tuned using ImageNet [9] weights and AMSGrad optimizer [32]. The images were resized to 256 × 128, and common data augmentation techniques, such as random flip, random crop, and random image patch, and random erase [33], were applied. Model experiments were conducted on six widely used person ReID datasets: Market-1501, CUHK03, DukeMTMC-reid, MSMT17, VIPeR, and GRID datasets. Collected results showed that the model achieved overall supremacy when compared with most of the state-of-art methods, attaining a CMC Rank-1 of 94.8% on Market-1501, 72.3% on CUHK03, 88.6% on Duke, and 78.7% on MSMT17. The proposed fusion scheme shows the effectiveness of the omni-scale features in different scales to comply with a large range of possible viewpoints from image pairs.

To improve the capabilities of attention mechanisms and obtain fine detailed features for person ReID, ref [34] proposes a feature refinement process in combination with filter network, by weakening the high response features and eliminating the interference raised by the background information. The model includes a network formed by the weaken feature convolution blocks based on ResNet, in combination with a multi-branch scheme. An attention mechanism is also set in place to act as an attention feature map in the convolution module, with higher values corresponding to regions where the model has paid more attention.

Extensive experiments were conducted in the Market-1501, DukeMTMC-reID, CUHK03 and MSMT17 person ReID benchmarks datasets, with the proposed model achieving a mAP of 94.2% in Market-1501.

An automatic search for a CNN architecture, specifically suited for the ReID problematic was proposed by [35], and denoted as Auto-ID. The method is based on the neural architecture search (NAS) [36] to automate the process of architecture design without human effort, directed to the task of ReID, by using a retrieval search-based algorithm. This design enables to obtain more optimal architectures that make the best use of human body structure information for person ReID, eliminating human expert efforts in the manual design of CNN models for the task. The model starts by integrating structural body cues into the input tensors, and then by vertically splitting the input feature tensor into four body part features, averaging each tensor part into a vector, and transforming each of the tensors into a new part feature vector using a linear layer. The obtained part vectors interact between them via a self-attention mechanism, enabling each part of the vectors to incorporate more specific body part information. Each obtained part vector is repeated and concatenated to recover the original spatial shape as the input tensor. Finally, the formed global feature tensor is fused with the original input tensor, using a one-by-one convolutional layer. A class-balance data sampler to equal the sample batch data for the triplet loss is used to overcome the original sensitivity of triplet loss to the batch data. This sampler first samples some identities, uniformly, and then, for each identity, it randomly samples the same number of images. To better explore the benefits from the cross-entropy and triplet losses, a mixture retrieval loss between sample loss and triplet loss is considered. Experiments were conducted on the Market-1501, CUHK03, and MSMT17 public databases, with the proposed Auto-ID model achieving a CMC rank-1 of 95.4% on Market-1501, when using the re-ranking technique, 77.9% on CUHK03 with labeled examples, and 73.3% on detected ones. On MSMT17, the model achieved a CMC Rank-1 of 78.2%. The proposal work enables increasing the ReID performance by taking into consideration attention mechanisms combined with triplet loss methods.

A dropout technique was proposed by [4] for learning deep feature representations from multiple domains with CNN. Multi-domain learning is frequently directed toward solving the problem of using datasets across different domains simultaneously, by using all data they provide for the task of multi-domain learning, while robustly handling data domain discrepancies. The central problem in multiple learning relies on the fact that samples from the same domain follow the same underlying data distribution, which degrades the model performance since some neurons may gain focus on some domain representation, while discarding the remainder domains, and leading to a bad model generalization. To overcome this problem, a domain guided dropout algorithm was proposed to avoid the model to learn only from domains where samples follow the same underlying data distribution. The CNN models are trained from scratch using all data domains using a single Softmax loss, creating a solid baseline. For each domain, a forward pass was performed on all data domains, and the impact that each neuron had on the objective function was quantified. After a few epochs of training, the standard network dropout layer was replaced by the proposed domain guided dropout layer, and the training process continued for several epochs to guide the neurons to the effective domain, enabling the CNN to learn more discriminative features for all of them. Experiments were conducted using the CUHK03, CUHK01, CUHK03, and PRID datasets, and compared with state-of-the-art methods using the CMC metrics. The proposed method outperformed the studied state-of-the-art methods, achieving a CMC Rank-1 of 75.3% on CUHK03, 66.6% on CUHK01, and 64.0% on PRID. The introduced domain dropout layer acts as regularized mechanism, avoiding the neurons co-adaptation to specific domains, reducing overfitting that degraded the ReID performance.

Attention mechanism has gained relevance in recent years, showing good performance in many fields, and it is often used as a local feature learning mechanism, which is useful for the task of ReID.

One example of an attention mechanism for deep learning networks was presented by [37] to provide simultaneously masks-free and foreground-focused samples for the inference phase. The main objective was to generate synthetic data that are composed of interleaved segments gathered from the original learning image set, while using class information only from specific segments. The proposed augmentation technique was evaluated using a baseline method proposed by [38], based on a deep learning-based classification framework using the ResNet-50 as a feature extractor, with weights initialized on ImageNet [9], along with a bag of tricks, known to be particularly effective for person ReID tasks. Since the richly annotated pedestrian (RAP) [17] dataset does not provide human body segmentation annotations, human binary segmentation masks were extracted using Mask-RCNN [39] to obtain the human body segmentation binary masks. Afterwards, fake images were generated to enlarge the dataset. For a matter of performance evaluation of the augmentation method, the default parameter settings detailed on the official project and the same weights were reused without modifications. The models were evaluated on two different loss schemes: Softmax and Triplet, with the results being slightly better when using triple loss on the RAP dataset. Accurately, the model presented a mAP Rank-1 of 62.9% when considering the upper body part, and 65.7% for the full body. The proposed augmentation technique enabled an increase in the performance of the baseline method by almost 20%.

Another attention mechanism based in Deep Learning (DL) models was proposed by [40] to overcome the problem of learning fine-grained pedestrian features that are useful for pedestrians ReID. A self-denominated HydraPlus-Net (HP-net), which multi-directionally feeds the multi-level attention maps to different feature maps and to different feature layers, is used. The method enables the model to capture multiple attention from a low-level to a semantic level by exploring the multi-scale selectiveness of attentive features, to enrich the final feature representations for the pedestrian image. The proposed approach was evaluated on three publicly standard datasets: CUHK03, VIPeR, and Market-1501 datasets. The model achieved a CMC Top-1 of 91.8 % on CUHK03 , 56.6% on ViPer, and 76.9% on Market-1501. The proposed method achieved top performance results, proving the effectiveness of attention mechanisms to capture relevant features maps for ReID tasks.

Attention mechanisms are also explored by [41], by introducing a bilateral complementary network (BiCnet), formed by a two-branch scheme; the first operating in the original image resolution, and the second called context branch, operating in downsampled resolution to capture long-range context. A specific attention mechanism, called diverse attention operation, was added to enforce consecutive frames to focus on different body characteristics regarding each identity. The mining of spatial clues was carried by a temporal kernel selection to jointly combine the short- and long-term temporal relations. Exhaustive experiments were conducted in the MARS, DukeMTMC-VideoReID, and LS-VID datasets, with the model obtaining a mAP of 0.860 on MARS dataset.

Attention mechanisms used to obtain more robust salient features from images are proposed by [42]. Since complex backgrounds can generate salient features that can degrade the performance of the ReID task, a joint weak saliency, in combination with an aware attention mechanism, is set in place to obtain refined global features, while weakening some of the saliency features. Similar to [34], this model employs a ResNet scheme from the weekend saliency block, where an attention mechanism is set in place, and final results of both processes are fused together to form the final feature. The performance of the method is evaluated using the Market-1501 and DukeMTMC-ReID datasets, with the method achieving a mAP of 89.2% in the Market-1501 dataset.

A performance evaluation of the reviewed methods for person ReID that explore Feature learning is presented in Table 2.

Table 2. Performance evaluation of the reviewed person ReID using feature learning methods.

Cat	Ref.	Main Technique(s)	# Data Success	Pros/Cons
Fusion	[19]	KDE and CNN features, late fusion, SVM model	CAVIAR4reid CMC 0.933	Robust, simple
	[24]	GOG and ResNet features, Data augmentation	CAVIAR4reid CMC 0.919	Simple, lack train data
	[26]	Joint learning multi-loss, two-branch CNN	CHUK03, CMC 0.832	Simple, efficient
	[31]	Residual blocks, multi-scale feature	Market-1501 CMC 0.948	Hard to train, can over fit
	[34]	Weaken feature convolution, ResNet	Market-1501 mAP 0.942	Robust, reusable
Strip	[35]	Neural architecture, search (NAS)	Market-1501 CMC 0.954	Hard to train, complex, not reusable
Drop	[4]	Modified dropout layer	CUHK03 CMC 0.666	Easy train, data domain, problematic
Attention	[37]	ResNet-50 as feature extractor, attention mechanism	RAP mAP 0.862	Simple to replicate, reusable
	[40]	ResNet-50 as feature extractor, multi-directional and level attention maps	CUHK03 CMC 0.918	Complex, not reusable state-of-the-art
	[41]	Two Branch, multi-scale and attention maps	MARS mAP 0.860	Complex, reusable state-of-the-art
	[42]	Attention, saliency maps ResNet	Market-1501 mAP 0.892	Complex, generalizes well, state-of-the-art

CMC—cumulative matching characteristic (higher the better), mAP—mean average precision (higher the better), all measures range: $[0.0, 1.0]$.

A performance comparison among the described works is depicted in Figure 5.

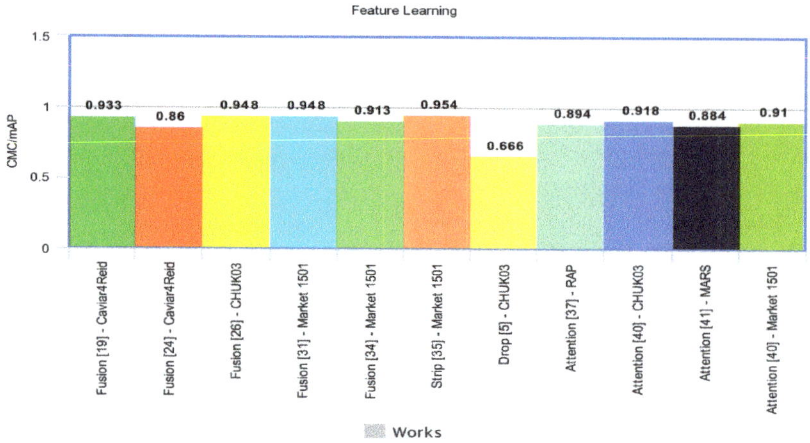

Figure 5. Performance summary of the evaluated feature learning methods.

Although the Fusion scheme of local and global features is a reasonable approach to explore, the use of attention mechanisms enables leveraging the performance of person ReID, by capturing important aspects that are relevant to objectives set-in place.

5.2.2. Deep Learning Metrics

Deep learning metrics is one commonly used strategy that aims to learn the dissimilarity or similarity between two given objects. The main objective is to learn a projection mapping from the original image into the embedding feature space, enabling to determine the degree of similarity between two-person images. This helps with learning the discriminate features by the design of loss-specific functions for the DNN model.

One standard metric corresponds to the contrast loss, which enables quantifying the similarity or dissimilarity between pairs of data, commonly used on the training of Siamese Networks, with the function expressed as:

$$L_c = yd(x_a - x_b)^2 + (1-y)\left[m - d(x_1 - x_b)^2\right]_+, \tag{3}$$

where $[\cdot]_+ = max(0, x)$, with x_a and x_b corresponds to the two image pairs of the Siamese Network, and a distance metric $d(x_a, x_b)$, usually the Euclidean distance, quantifies the degree of similarity among the pairs, m corresponds to a training parameter, and y is the corresponding matching label. When $y = 1$ the two input mages belong to the same ID (positive sample pair); on the other hand, when $y = 0$, it reflects the opposite case (negative sample pair).

One example of a Siamese Network for person ReID is explored by [43], by applying different distance metrics to corresponding feature maps. Defined as MSP-CNN, the approach starts by using image pairs as network input, with all images going through the same share-weighted deep CNN network, formed by small convolution filter layers followed by a simple inception module [44]. To attain the distinct characteristics from the diverse feature maps, similarity constraints are applied to both low-level and high-level feature maps during the training stage to effectively learn discriminative feature representations at different levels. The objective function was designed to emphasize low-level features that are frequently related to schoolbags, T-shirts, and higher-level special textures, which are shared among persons from the same personality to propagate those relevant features to the upper layers. At the higher-level feature maps, a Euclidean distance after L2normalization [45] is used to represent the abstract global similarities. The approach enables the CNN to extract robust feature representations without any complicated distance metric to be learned in the process, contrary to those found in more traditional hand-crafted systems. This enables easily incorporating constraints, forming a unified multi-task network with similar constraints. Model evaluations were conducted using the CUHK03 [46] and Market-1501 [47] datasets, and the small CUHK01 dataset [48], being evaluated by the use of the CMC metric. Results on the CUHK03 dataset show that the proposed model achieved an CMC Rank-1 accuracy of 85.7% when using manual hand-labeled object boxes, and 83.6% when using a deformable parts model detector for object extraction. While the proposed model obtains competitive results, it still requires some degree of human annotation to achieve good performance.

A novel filter pairing neural network (FPNN) was proposed by [46], which is composed of six layers to jointly handle misalignment, photometric and geometric transforms, occlusions, and background clutter in person ReID tasks. The network was initially composed by a convolution and max-pooling layer that operated on two pair of RGB or Lab Color space (LAB) images from different cameras, generating the responses from local patches as local features. Each feature map is partitioned into $H_1 \times W_1$ stripe sub-regions, and only the maximum response in each sub-region is taken into account, with the max-pooling layer outputting a $H_1 \times W_1 \times K_1$ feature map. The computed feature maps are processed by a patch matching layer to match the filter responses from local patches across the different views. Considering that each input image contains M horizontal patches, these image patches are only compared with the corresponding stripe from the other pair images, forming displacement matrices to encode the spatial patterns of each matching patch under the different features representations. The patch matching is then further refined by dividing the patch displacement matrices into T groups, and within each group, a max out-grouping layer is used. Only prominent feature activations are passed to the next layer, allowing each feature to be represented by multiple redundant channels, enabling the modeling of a mixture of photometric transforms. For body parts, convolution and a max-polling layer are added to the patch displacement matrices to obtain the displacement matrices of body parts on a larger scale. For the final identity recognition, a Softmax function is used to measure the degree of similarity between two input person images, given the global geometric transforms detected on the previous layer. During the model training, several conventional techniques, such as dropout [49], data augmentation, and bootstrap were employed. The model was evaluated using the constructed CUHK03 [46], and the results were collected and compared with other state-of-the-art methods using the CMC metric, with the proposed method achieving a 20.65% when considering Rank-1 rate. The partial region patch makes the method suitable for partial pairs matching, enabling refining similarity metrics, according to the context of the image.

In contrast, an unsupervised learning approach is explored by [50], where an unsupervised incremental learning algorithm, denominated TFusion, aided by the transfer learning of pedestrian spatial–temporal patterns from an unlabeled target domain, is used for person ReID. The algorithm transfers the visual classifier trained on a small labeled source dataset to the unlabeled target dataset and learns pedestrian spatial–temporal patterns. A Bayesian fusion model is then used to combine the learned spatial–temporal patterns with extracted visual features to create an improved classifier. A learning-to-rank based on the mutual promotion procedure is used to optimize the classifiers based on the unlabeled data domain incrementally. The proposed framework explores a Siamese Network scheme [8] based on two ResNet50 [14] CNN pre-trained networks to extract visual features from different pair object images. The outputs of the Siamese Network are flattened into two one-dimensional vectors, with the model predicting the identities of each of the input pair images and their similarity score by using cosine similarity. A spatial–temporal pattern learning is formulated considering pedestrian patterns among different cameras and corresponding time intervals of objects that were previously considered similar by the model. In the last stage, the Bayesian fusion model combines the visual features with the spatial–temporal features to achieve a composite similarity score of the given pair of images. Exploring the fact that the Bayesian fusion model is based on the Bayes theorem, it is possible to access the likelihood of the scores of each image that belong to the same object. Model experiments were conducted using the GRID [51], Market-1501 [47], CUHK01 [48], and

VIPeR [30] datasets using a cross-validation strategy, where one of the datasets is selected as the source and another one as the target dataset to test, enabling performing cross-dataset person ReID evaluation. The results show that the proposed TFusion model achieved an CMC in Rank-1 of 64.10%, when using the GRID dataset, and 73.13% when using the Market-1501 dataset. The results, when compared to the work of [24] using fusion schemes, are much lower, mostly since the author relies only on ResNet50 as backbones, and uses simple cosine similarity metrics that may not capture other image similarity domains.

A deep convolutional network with layers, specially designed to address the problem of ReID, was proposed by [52], by outputting a similarity value, indicating whether the two input images are from the same person. The network encompasses two layers: the neighborhood difference layer for comparing convolutional image features from each patch and a subsequent layer where features are summarized. For the extracted features to be comparable across the two images in later layers, the first two layers are set to perform a tied convolution, with weights shared across the two views, ensuring that the same filters are used in both image pairs to compute the corresponding features. Two tied convolution layers enable providing a set of feature maps for each input image, from where relations between the two views are learned and supplied to a cross-input neighborhood difference layer, to compute the differences between the two views around a neighborhood of each feature location, generating a set of 25 neighborhood difference maps. Subsequently, a patch summary layer summarizes these neighborhood difference maps by producing a holistic representation of the differences in each view 5×5. The learning of the spatial relationships across neighborhood differences is achieved by employing a convolution layer using 25 filters of size 3×3 with stride 1, and the resulting features are passed through a max-pooling kernel to reduce the height and width. Finally, a fully connected layer captures the relations by combining information from patches that are far from each other and with a Softmax layer to output the similarity of both images. The proposed methods were evaluated using the CUHK03, CUHK01, and VIPeR datasets and the CMC curve. The model achieved a CMC Rank-1 accuracy of 54.74% on CUHK03-labeled and 44.96% on CUHK03-detected; in VIPeR, the method obtained a 34.81% Rank-1 accuracy, while in CUHK01, it achieved a Rank-1 recognition rate of 65%. The use of shared weights ensures fair feature selection; however, some specific image domains can be neglected, harming the ReID process.

Triplet loss is one of the most widely used deep learning metrics used in person ReID problems, aiming to minimize the intra-class distance while maximizing the intra- to intra-class distance of the given samples. The triplet loss can be expressed as:

$$L_{trip} = \left[m + d(x_a, x_p) - d(x_a, x_n)\right]_+. \tag{4}$$

When compared to contrast loss, the input of the triplet Loss consists of three images, with each triplet set containing a pair of a positive sample x_p, a negative sample x_n, with a corresponding anchor image x_a. x_a and x_p correspond to images with the same ID, while the pair x_a and x_n to images with different IDs. During model training, the distance between the same ID pairs x_a and x_p is minimized, while the distance between different ID pairs x_a and x_n is set apart. To increase the performance during training, a combination of classification loss and triplet loss [53] is used, enabling learning discriminatory features.

One example of the use of triplet loss is in the work by [53], which also explores a modification of the triplet loss, defined as TriNet, to perform end-to-end deep metric learning to tackle the person ReID problematic. Triplet loss has been proposed previously by [54], and vastly explored on FaceNet [45], where a CNN is used to learn an embedding for faces. Two approaches were explored in the proposed work, with the first being based on a ResNet-50 [14] architecture and the weights provided from the ImageNet pre-training procedure, with the last layer being discarded and replaced by two Fully Connected (FC) layers. The first contains 1024 units, followed by batch normalization [55] and Rectified Linear Unit (ReLU) [56], and the second goes down to 128 units, forming the final embedding dimension. The second approach consists of a network trained from scratch, denoted as LuNet, which follows the style of [57], but uses leaky ReLU [58], nonlinearities, multiple 3 × 3 max-polling's with stride 2, and omits the final average pooling of feature-maps in favor of a channel-reducing final res-block. Distinct training parameters were used to train both networks, being the TriNet trained with the modified batch triplet loss, by setting the batch size to 72 to circumvent memory issues, due to the number of parameters (25.74 M); while in the second network, LuNet contains 5.00 Million parameters, and was trained using a large batch size (128). Model experiments were conducted on the CUHK03, Market-1501, and MARS [59] datasets, and several triplet variations and model comparisons were evaluated. One advantage of the use of triplet loss is that it allows performing end-to-end learning between the input image and the target embedding space, directly optimizing the network for the final task. Person comparison is performed by computing the Euclidean distance of their embeddings. The proposed pre-trained TriNet achieved a CMC of 89.63% when using CUHK03 and labeled box sets, and of 87.58% when automatically detecting box sets. Additionally, the proposed LuNet achieved competitive performance. One of the main disadvantages of using pre-trained networks is the flexibility to try out new advances in deep learning or to make task-specific changes in a network.

Once traditional triplet loss randomly selects three images from the training set during training, on many occasions, the sample combinations may evidence the lack of complex sample combinations that correspond to the more difficult cases, degrading the generalization capabilities of the model. To overcome this, many researchers improved the triplet loss to mine hard samples.

A multi-channel parts-based on CNN under a modified triplet framework for person ReID was proposed by [60]. The CNN network consists of multiple channels to jointly learn both the global full-body and local body-part features of the input image. The person ReID modeled network is trained using a modified triplet loss function to pull the feature instances of the same person together, while setting those instances further, corresponding to different persons in the learned feature space. Three CNN with the same sets of weights and biases are used, with the triplets from image I_1 space being mapped into a learning feature space from I_i. The multi-channel CNN model is composed of the following distinct layers: one global convolution layer, one full-body convolution layer, four body-part convolution layers, five channel-wise full connection layers, and one network-wise full connection layer. A global convolution layer acts as the first layer of the CNN network. It is split into four equal parts, with each part forming the first layer of the independent body-part channel, responsible for learning features of the corresponding body parts. Moreover, a full-body channel that considers the entire global convolution layer as its first layer is added to learn the global full-body features of the persons. The four body-part channels, together with the full-body channel, constitute five independent channels that are trained separately from each other. The final outputs of the channel-wise full connection layers, from the five separate channels, are concatenated into one vector and fed into the final fully connected layer. For model training, a novel data augmentation technique is performed by cropping the center of each image region of 80 × 230 pixels and introducing a small random perturbation to augment the training data, a technique close to [61]. Experiments were conducted in the VIPeR, i-LIDS [62] and PRID2011 [63] datasets. The models were assessed using the CMC metric for quantitative evaluation on each of the referred datasets, and several model variations were also evaluated, with the best-proposed model variation formed by the full version of the proposed multi-channel CNN model trained with the modified triplet loss function achieving a CMC Rank-1 accuracy of 60.4% on i-LIDS, a 22.0% CMC Rank-1 on PRID2011, 47.8% on VIPeR, and 53.7% on CUHK01. The proposed method showed promising performances in competitive scenarios and positive ReID person in partially occluded environments.

A performance evaluation of the reviewed methods for person ReID using deep learning metrics is given in Table 3.

Table 3. Performance evaluation of the reviewed person ReID using deep metric methods.

Cat	Ref.	Main Technique(s)	# Data Success	Pros/Cons
Contrast Loss	[43]	Unsupervised, ResNet50 features, Siamese networks, Bayesian fusion	CUHK03, CMC 0.857	Good dataset generalization, Cross domains, Complex
	[46]	Filter pairing neural network	CHUK03, CMC 0.206	Bad performance Complex, Not robust
	[50]	Unsupervised, ResNet50 features, Siamese networks, Bayesian fusion, spatial–temporal model	CUHK03, CMC 0.857	Good dataset generalization, Cross domains, Complex
	[52]	Siamese networks, Tied convolution	CUHK03 CMC 0.547	Simple, reusable
Triplet Loss	[53]	Triplet loss, pre-trained ResNet	CUHK03 CMC 0.876	Simple to replicate, architecture poses restraints
	[60]	Three CNN with shared weights, modified triplet loss	i-LIDS CMC 0.604	Simple train Scalable Efficient

CMC—Cumulative matching characteristic (higher, the better), mAP—mean average precision (higher, the better), all measures range: [0.0, 1.0].

A performance comparison among the described works is depicted in Figure 6.

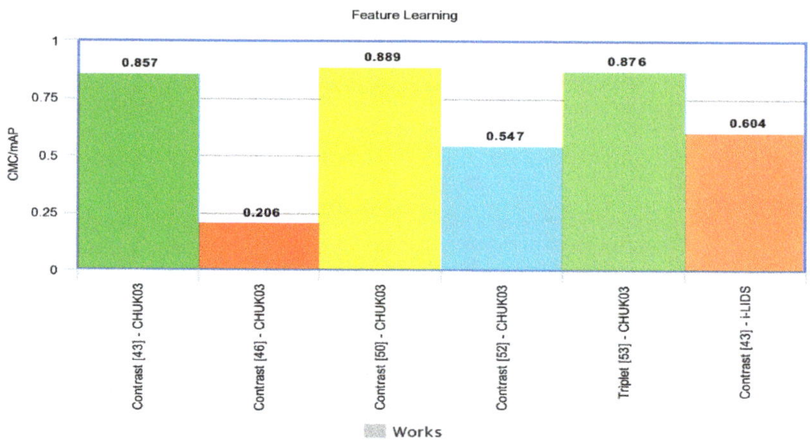

Figure 6. Performance summary of the evaluated deep learning metric methods.

Concerning the use of deep learning metrics, the use of contrast loss and triplet loss are the most common and usual methods employed in the person ReID task. This preference is mainly related to the simplicity of the methods, without major modifications to the existing pre-trained backbone, with the Siamese scheme being extremely suitable for pair image comparison.

5.2.3. Sequence Learning for ReID

One common approach to capture the spatial–temporal cues for the task of ReID is to explore a sequence of videos or a small set of images to train RNN models that can be employed in the person ReID task. Many approaches explore 3D CNN to capture temporal and spatial features simultaneously [64].

The 3D CNN, in combination with a non-local attention mechanism, was proposed by [64] for person ReID, inspired by video action recognition models that involve the identification of different actions from video tracks. For the task, 3D convolutions on video volume, instead of using 2D convolutions across frames, are used to extract spatial and temporal features simultaneously. To handle misalignments, a non-local block is employed to capture spatial–temporal long-range dependencies, resulting in a network being able to learn useful spatial–temporal information as a weighted sum of the features in all space and temporal positions from the input feature map. Triplet loss function with hard mining proposed by [53] and a Softmax cross-entropy loss function with label smoothing regularization are employed to train the network. As for the network, 3D convolutions are replaced with two consecutive convolution layers, one one-dimensional (1D) convolution layer acting purely on the temporal axis, followed by a two-dimensional (2D) convolution layer to learn spatial features on the residual block. The modified 3D ResNet-50 is pre-trained on kinetics [65] to enhance the generalization performance of the model, and the final classification layer is replaced to output person identity. Experiments were performed using three datasets, namely, the iLIDS-VID, PRID-2011, and MARS datasets, and the results were compared with the ones of several state-of-the-art methods and of an established baseline model, which corresponds to a ResNet50 trained with Softmax cross-entropy loss and triplet with hard mining on an image-based person ReID. The proposed framework showed competitive results, outperforming several state-of-the-art approaches by a large margin on multiple metrics, attaining a mAP of 84.3% on the MARS dataset.

In [66] is proposed a two-stream convolution network to extract spatial and temporal cues for video-based person ReID. A temporal stream network was built by inserting several multi-scale 3D (M3D) convolution layers into a 2D CNN network. The M3D convolution network introduces a fraction of parameters into the 2D CNN to gain the ability of multi-scale temporal feature learning. In addition, a temporal stream was included using residual attention layers to refine the temporal features further. The jointly learning of spatial–temporal attention masks in a residual manner enables the identification of the discriminative spatial regions and temporal cues. Model evaluations were performed on three widely used benchmarks datasets: the MARS, PRID2011, and iLIDS-VID datasets, with the proposed model obtaining a mAP on 0.740 on the MARS dataset.

RNN in the form of Long Short-Term Memory (LSTM) of Gated Recurrent Unit (GRU) are commonly employed to capture temporal or spatial features. In ReID tasks, often the use of RNN is applied into sequences of images or video frames to capture spatial features extracted from CNN.

One example of use of LSTM is presented by [6], to progressive aggregate frame-wise human region representation at each frame extracted from the Local Binary Patterns (LBPs) detector, yielding a sequence feature representation. LSTM enables remembering and propagating previously accumulated representative features while forgetting irrelevant ones. The proposed RNN acts as a feature aggregation, generating highly discriminating sequence-level object representations. The evaluations of the models were conducted using the iLIDS-VID and PRID 2011 datasets, obtaining a Rank 1 of 49.3 on iLIDS-VID.

A RNN to jointly use spatial and temporal features is presented by [67], enabling exploring all relevant information useful for the person ReID task. The method encompasses a temporal attention mechanism to automatically pick the most discriminating features in a specific frame obtained from a CNN, while integrating surrounding information. Experiments were carried out using the iLIDS-VID, PRID 2011, and MARS datasets, with the proposed model achieving a Rank-1 of 70.6 on MARS.

A performance evaluation of the reviewed methods for person ReID using sequence learning is given in Table 4.

Table 4. Performance evaluation of the reviewed person ReID Sequence methods.

Cat	Ref.	Main Technique(s)	# Data Success	Pros/Cons
3D CNN	[64]	3D CNN, Attention, triple loss	MARS mAP 0.834	Simple to replicate, reusable state-of-the-art
	[66]	3D- Two stream CNN, Residual attention	MARS mAP 0.740	Replicable, SOTA
RNN	[6]	LSTM, LBP features	iLIDS-VID Acc1 0.493	Not robust, Old
	[67]	LSTM, CNN features	MARS Rank1 0.706	Simple, replicable

CMC—cumulative matching characteristic (higher, the better), mAP—mean average precision (higher, the better), all measures range: [0.0, 1.0].

A performance comparison among the described works is presented, as depicted in Figure 7.

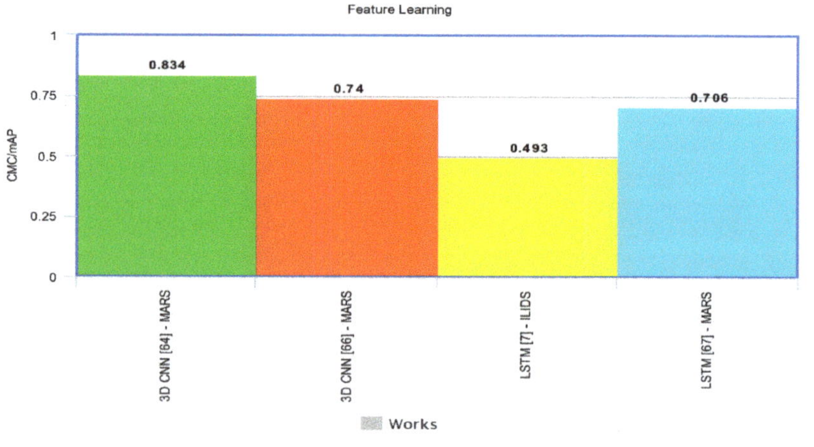

Figure 7. Performance summary of the evaluated sequence model methods.

While the use of RNN networks seems to e a natural choice to capture relevant features for the person ReID task, recently, authors have relied on 3D CNN to automatically capture these dependencies, obtaining models that are less complex to train and achieve superior performance when compared with traditional RNN, such as LSTM or GRU.

5.2.4. Generative Learning for ReID

One of the main difficulties in ReID tasks is the small diversity of images from the same object with different surroundings and conditions, which poses difficulties for models to generalize well to unseen image contexts. Among the identified datasets, one of their main limitations concerns the uniform illumination, and similar image poses. The use of generative learning, mainly by Generative Adversarial Network (GAN) to increase the amount of training data while presenting the model with more complicated cases, is one common practice in the field of computer vision, and was recently used in ReID tasks. GAN commonly employ the use of two neural networks that compete against each other to become more accurate and output more precise predictions. They are composed of the generator responsible for generating artificial data that can be mapped into the unknown training data distribution. In contrast, a discriminator tries to identify with the generator outputs that correspond to the real examples or the generated ones. The GAN training is performed in an adversarial way, improving the capabilities of the generator, simultaneously representing the natural data distribution, and the discriminator to identify the artificially generated images, overcoming the training data limitations. One of the main problems in person ReID concerns the reduced number of images from different poses.

To overcome this limitation, a feature distilling generative adversarial network (FD-GAN) is proposed by [68] in combination with a Siamese CNN structure to learn identity-related and pose-unrelated representations. In addition, a novel same-pose loss was also formulated and integrated, requiring the appearance of the same person's generated images to be similar. The proposed FD-GAN explores the Siamese scheme, where an image encoder, an image generator, an identity verification classifier, and two adversarial discriminators are included. The corresponding branch of the network takes a person image and a target pose landmark map as inputs. The image encoder at each branch initially transforms the input person image into feature representations. Then, the identity verification classifier is used to supervise the feature learning for person ReID. The image generator starts by taking the encoded person features and target pose map as inputs, and outputs another image of the same person in a different target pose. The target pose map is represented by an 18-channel map, with each channel representing the location of one pose landmark's location, and with the one-dot landmark location being converted to a Gaussian-like heatmap. The encoding is performed by using a 5-block convolution-Batch Normalization (BN)-ReLU subnetwork, generating a 128-dimensional pose feature vector. The visual features, target pose features, and an additional 256-dimensional noise vector, sampled from standard Gaussian distribution, are then concatenated and input into a series of 5 convolution-BN-dropout-ReLU upsampling blocks to output the generated person images. Concerning training, it was performed in three main stages. Initially, the Siamese network baseline built on ResNet-50, using the weights from ImageNet [9], was established. The network was firstly optimized with SGD and trained during 80 epochs. In the second training stage, the encoder and validation classifier were fixed, and the generator was integrated. Adam optimizer [69] was employed to optimize the generator, while the identity discriminator and posed discriminator were optimized with SGD. Lastly, a global fine-tuning was done on the model through all blocks in an end-to-end fashion. For performance evaluation, Market-1501, CUHK03, and DukeMTMC-ReID datasets were used, with mAP and CMC Rank-1 accuracy metrics being adopted for performance evaluation on all the three datasets, and the proposed FD-GAN obtained a CMC of 90.5% on Market-1501, 92.6% on CUHK03, and 80.0% on DukeMTMC-ReID. The inclusion of the FD-GAN and pose encoder enables a substantial increase in model performance.

Another standard limitation concerns the lack of diversity in the image domain, namely, different images gathered and subjected to other illumination conditions. A style transfer GAN is proposed by [70] to serve as a data argumentation approach to smooth camera style disparities. The method employs CycleGAN [71] to style transfer trained labeled mages from different cameras aiming an increase of the diversity of training examples, avoiding model overfitting. Focusing on a better handling of noise, a smooth label regularization is introduced. The style transfer method is evaluated on the ReID task using the Market-1501 and DukeMTMC-ReID datasets, with the proposed model achieving a mAP of 71.55% in the Market-1501.

The vast domain range of images poses difficulties to the ReID tasks. To reduce the impact of image domain diversity, ref [72] proposes a joint learning scheme to improve domain adaptation, to disentangle ID-related/unrelated features, which enforces adaptation to focus on the ID-related features space only. The disentangle module is responsible for encoding cross-domain images into a shared appearance and two separated structure spaces, with the adversarial alignment being performed by the adaptation module. Extensive experiments using the Market-1501 and DukeMTMC-ReID datasets were performed, with the model achieving a Rank-1 of 83.1 in the Market-1501.

A careful evaluation of methodologies for person ReID was performed by [38]. The evaluation starts by setting a baseline backbone architecture based on ResNet-50 [14] with weights initialized on ImageNet [9], and changing the dimension of the fully connected layer to the number of identities in the training set. Similar assumptions were made for all experiments for training as described in the article [38]. Several training tricks were evaluated, such as warm-up learning rate [73], to bootstrap the network for better performance; random erasing augmentation [33], where an image I in a mini-batch has the probability of undergoing random erasing of p_e; label smoothing [44] to prevent the model from overfitting the training ID, where a small constant ϵ is introduced to avoid the overfit of the training set; last stride [74] to obtain a higher spatial resolution, enriching the granularity of features. To embedded different features distances to accommodate different class distributions in different sub-spaces on the ID loss during inferring stage several strategies are employed, such as BNNeck, which adds only BN layer after features and before classifier FC layers; and, finally, a center loss [75], to simultaneously learning deep features of each class, while penalizing the distances between the deep features and their corresponding class centers, avoiding the drawbacks of the triplet loss. The performance and contribution of each of the tricks were evaluated, with the best model using all the described tricks used, achieving a CMC Rank-1 of 94.5% on Market-1501 and CMC Rank-1 of 86.9% on DukeMTMC-ReID. While the study shows the effectiveness of the DNN tricks to the ReID task, in [76] is proposed a k-reciprocal encoding method to re-rank the ReID results, to increase the accuracy. The main underline consideration concerns the fact that if a gallery image is on par with the probe in the k-reciprocal nearest neighbors, it is more likely to be a true match. In detail, given an image, a k-reciprocal feature is calculated by encoding its k-reciprocal nearest neighbors into a single vector used for re-ranking under the Jaccard distance.

A performance evaluation of the reviewed methods for person ReID using Generative learning and other complementary methods is given in Table 5.

Table 5. Performance evaluation of the reviewed person ReID using generative methods.

Cat	Ref.	Main Technique(s)	# Data Success	Pros/Cons
GAN	[68]	GAN, ResNet-50, pre-trained, pose generator	Market-1501 CMC 0.905	Uses GAN, hard to train state-of-the-art
	[70]	GAN, style transfer, smooth regularization	Market-1501 mAP 0.715	Uses GAN, simple replicable
	[72]	GAN, joint learning, domain adaptation	Market-1501 Rank 1 0.831	Uses GAN, complex replicable
Others	[38]	Evaluation of techniques, Pre-trained, Modified Triple loss	Market-1501 CMC 0.945	Simple to reuse, reusable explanatory

CMC—cumulative matching characteristic (higher, the better), mAP—mean average precision (higher, the better), all measures range: $[0.0, 1.0]$.

A performance comparison among the described works is depicted in Figure 8.

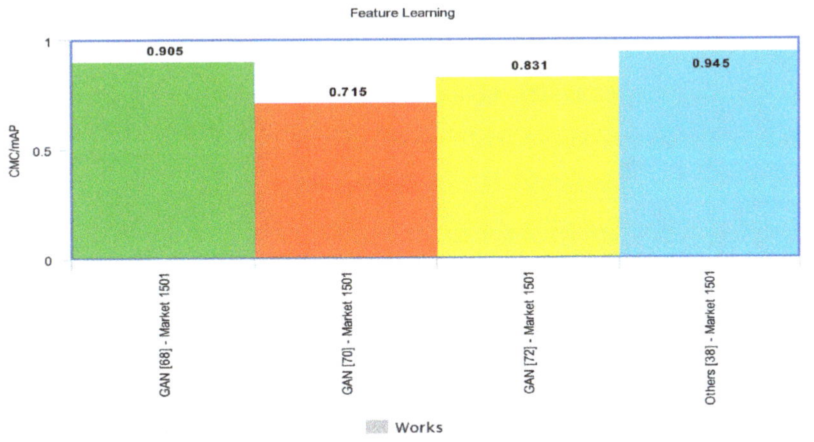

Figure 8. Performance summary of the evaluated generative model methods.

5.2.5. Summary of Person ReID

The different techniques reviewed in this section focus on solving the person ReID problematic. The study about the different methods, which researchers among the literature have proposed, reveals that very few methods can achieve accurate results on a wide range of datasets that contain different varying position, occlusions genres, shapes, and the illumination of the person in the scene. The performance of the enlisted methods is useful for comparison purposes, giving insight on how to devise a robust, yet simple, person ReID method that can achieve high accuracy. The best performing models are mostly based on pre-trained deep neural network models for feature extraction, combined with schemes and modified triplet loss for person ReID. New approaches are focusing on 3D CNN networks by transfer learning their embedding and reusing them into person ReID. Other methods explore baseline models and complement then using data augmentation and other tricks to improve their performance for ReID tasks. Moreover, a clear trend in the ReID research community relates to the use of GAN methods, in combination with ResNets, to increase model robustness against different object poses, overcoming the number of pair examples in the training dataset, and achieving superior results. A recent trend in person ReID is the use of attention mechanisms to capture relevant features, leading to a significant improvement in model performance.

An important issue regarding the person ReID is the use of biometric characteristics, such a human faces or people's skin. In [77], an important study was conducted using obfuscated and non-obfuscated person faces, and most of the case models performed in the majority of benchmark datasets were better when trained and used people's faces as expected. However, this can lead to biased models (discriminating ones). The public usage of these models can collide with current policies in law practices in several countries; it is helpful to always have a side-by-side comparison of both modalities.

6. ReID and Spatial–Temporal Multi Object ReID Methods

One of the main difficulties of object ReID is to operate in distributed scenarios and account for spatial–temporal constraints for multi-object ReID. Main techniques explore the use of RNN models to construct tracklets to assign IDs to objects, enabling to robustly handle occlusions. In contrast, others rely on 3D CNN to attain temporal dependencies of the object in track.

6.1. *Multi Object ReID Datasets with Trajectories*

Multi-Object ReID considers attributes, such as shape and category combined with trajectories, and is one of the objectives that go towards the objective to perform multi-object

ReID in a real urban scenario. However, the development of suitable benchmark datasets for attribute recognition remains sparse. Some object ReID datasets contain trajectories collected from various sources, and such heterogeneous collection poses a significant challenge in developing high-quality fine-grained multi-object recognition algorithms.

Among the publicly available datasets for ReID, one example is the NGSIM dataset [78], a publicly available data set with hand-coded Ground Truth (GT) that enables evaluating multi-camera, multi-vehicle tracking algorithms on real data, quantitatively. This dataset includes multiple views of a dense traffic scene with stop-and-go driving patterns, numerous partial and complete occlusions, and several intersections.

Another example is the KITTI Vision Benchmark Suite [79], which is composed of several datasets for a wide range of tasks, such as stereo, optical flow, visual odometry, 3D object detection, and 3D tracking, complemented with accurate ground truth provided by Velodyne laser 3D scanner and real GPS localization system, Figure 9. The datasets were captured by driving around the mid-size city of Karlsruhe, in Germany, in rural areas, and on highways, and on average, there are up to 15 cars and 30 pedestrians per image. A detailed evaluation metric and evaluation site are also provided.

Figure 9. Example of the multi-object tracking and segmentation system (MOTS).

A summary of the public available databases for multi-object car ReID with trajectories is presented on Table 6.

Table 6. Global overview of the public available databases for multi-object car ReID with trajectories.

Dataset	# Identities	# Cameras	# Images	Label Method	Size	Tracking Sequences
NGSIM						
KITTI	-	-	-	R-CNN	1392 × 512	Yes
UA-DETRAC	825	24	1.21 M	Manual	960 × 540	Yes
VehicleID	26,267	-	221 K	Manual	Vary	Yes
VeRi-776	776	18	50 k	Manual	Vary	Yes
CompCar	1687	-	18 k	Manual	Vary	Yes
PKU-Vehicle	-	-	18 M	Manual	Cropped	Yes
MOT20-03	735	-	356 k	R-CNN	1173 × 880	Yes
MOT16	-	-	476 k	R-CNN	1920 × 1080	Yes
TRANCOS	46,796	-	58 M	HOG	Vary	Yes
WebCamT	-	212	60 k	-	-	No

R-CNN—region proposals with CNN, HOG—histogram oriented gradients.

In addition, the PASCAL VOC project [80], provides standardized image datasets for object class recognition, with annotations that enable evaluation and comparison of different methods. Another useful dataset is ImageNet [10]. This image database is organized according to the WordNet hierarchy (currently only the nouns), where each node is represented by hundreds and thousands of images. Currently, each node contains an average of over 500 images. The dataset encompasses a total of 21,841 non-empty sets, forming a total number of images of around 14 million, with several images with bounding box annotations of 1 million. In addition, it contains 1.2 million images with pre-computed SIFT features [81]. Pre-trained networks on these two datasets are commonly reused as backbones for feature extraction to perform several tasks, such as object ReID.

6.2. Spatial–Temporal Constrained and Multi Object ReID Methods

In this section, deep learning-based vehicle ReID methods are grouped into four main categories, as represented in Figure 10; which includes methods for feature learning, sequence learning, deep learning metrics, and tracking, with these categories encompassing several methods. The main aspects of each category method and their experimental results are discussed in the following.

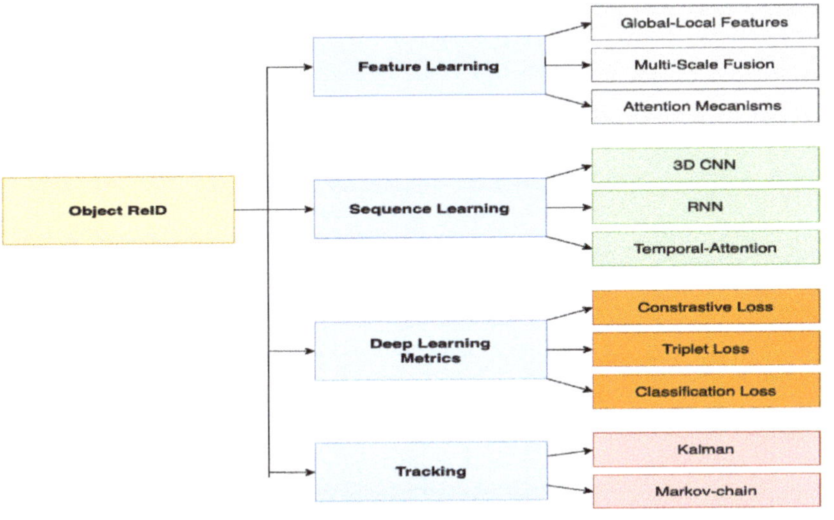

Figure 10. Deep learning-based vehicle re-identification methods.

6.2.1. Deep Learning Metrics for Vehicle ReID

Similar to person ReID, triplet loss are a excellent tool for vehicle ReID tasks. A variation of the triplet loss commonly used in deep learning methods, defined as a group-sensitive-triplet embedding (GS-TRE), was proposed in [82] to recognize and retrieve vehicles, where the intraclass variance is elegantly modeled by incorporating an intermediate representation "group" between samples and each vehicle in the triplet network learning. The main objective is to address the car ReID problematic, namely the fact that common deep metric learning with a triplet network common configuration ignores the impact of intra-class variance-incorporated embedding on the performance of vehicle ReID, where robust fine-grained features for large-scale vehicle ReID have not been fully studied. In addition, a clustering strategy to derive group labels, and in particular, an online clustering method, is employed, and a mean-valued triplet loss [83] is also proposed to enhance the learning of discriminative features. The performance of the proposed group-sensitive-triplet embedding (GS-TRE) was evaluated on the VeRi-776 and VehicleID datasets, with the model obtaining a mAP of 0.743 on the VehicleID dataset, with the modified triplet loss well suitable to help in the ReID task.

An improvement of triplet loss is presented by [84], focusing on two aspects: first, a stronger constraint, namely classification-oriented loss augmented with the original triplet loss; second, a new triplet sampling method based on pairwise images is proposed in combination with a classification-oriented loss to implicitly impose a constraint for the embedded features of the images of the same vehicle to be similar, and also by ensuring negative samples in one triplet act as positive samples in another triplet. The system architecture consists of three parts: a shared deep CNN to learn a mapping from raw images to Euclidean space, with the distance reflecting the relevance between the images, a triplet stream for calculating the distances and providing the constraint of the triplet loss, and a classification stream for ID level supervision provided by the classification-oriented loss, with the image triplets being generated by the proposed triplet sampling method. The deep CNN for feature extraction was fine-tuned from VGG CNN using the pre-trained weights from the ILSVRC-2012 dataset [10]. Stochastic gradient descent was employed during the training process. Results were gathered based on the VeRi dataset using the CMC and mAP, with model obtaining a mAP of 0.5740 on this dataset. While the results show a lower level of performance, making evident that simple triplet usage is not sufficient for robust ReID systems.

Contrast loss is a useful tool for pair-wise marching, enabling the model to focus on discriminate features. A novel deep learning-based approach to progressive vehicle ReID, called PROVID, was proposed by [85]. The approach starts by addressing the ReID as two distinct search processes: coarse-to-fine search that operates in the feature space and near-to-distant search to address real-world scenarios. The first searching process employs the appearance attributes of the vehicle for coarse filtering, while simultaneously exploring Siamese Neural Network architecture for license plate verification for vehicle identification. The near-to-distant search process enables to retrieve vehicles′ identity by searching from near to faraway cameras and from close to a distant time. To account for the spatial–temporal domain, an assumption is that two images have a higher probability of being the same object, if they have a small space or time distance among frames, and a lower probability of being the same vehicle if they have large space or time distance. With this in mind, for each query image i and test image j, a spatial–temporal similarity $ST(i,j)$ is defined, and with a re-ranking strategy in combination to model the spatial–temporal information with the appearance and plate features. The model was evaluated using the VeRi-776 dataset, showing that the proposed method achieved a 0.277 mAP. The results are much lower when compared with related works. In addition, the use of discriminating features, such as the vehicle identification plate, may cause problems in the usage on public surveillance systems and comply with legislation in place.

A unified multi-object tracking (MOT) framework is presented in [86], enabling exploring the full potential of the long-term and short-term cues for handling complex cases in MOT scenes. For better association, a switcher-aware classification (SAC) is proposed, exploring the potential of the identity-switch causer (switcher). Specifically, the proposed method incorporates a single object tracking (SOT) subnet to capture short-term cues, a ReID subnet to extract long-term cues and a switcher-aware classifier to make matching object decisions, using extracted features from the main target and the switcher. The main objective of short-term cues is to help find false negatives, while long-term cues avoid critical mistakes when occlusion occurs, with the SAC learning used to combine multiple cues in an effective way to improve robustness. The SOT subnet and the ReID subnet are trained independently. For the SOT subnet, image pairs of targets are generated according to the GT of the videos, and the pairs are extended to include part of the background according to the training schema of Siamese-RPN. On the other hand, for the ReID subnet, each target is regarded as one class, with the network trained to predict the class of the input target. Extensive model evaluations were performed using the challenging MOT16 benchmarks [87], achieving a CLEAR MOT of 71.2%, proving the effectiveness of the switcher to robustly assign long-term occluded objects to corresponding tracklets.

Performance evaluation of vehicle ReID using deep learning metrics is resumed in Table 7.

Since it is similar to person ReID, several authors explore the Siamese and triplet loss for the ReID task. Because the scheme is similar in what concerns the job itself, the leaned features are very distinct, hampering the use of Siamese and triplet loss, both for person and vehicle ReID in a single unified framework.

Table 7. Performance evaluation of the reviewed vehicle ReID using deep leaning metric methods.

Cat	Ref.	Main Technique(s)	# Data Success	Pros/Cons
Triplet	[82]	CNN features, group-sensitive-triplet emb.	VehicleID mAP 0.743	Reproducible, ranking problems
	[84]	VGG features, triplet sampling method	VeRi-776 mAP 0.574	reproducible, robust
Contrast loss	[85]	Siamese Neural Net, spatial–temporal similarity	VeRi-776 mAP 27.77	Simple, not robust
	[86]	Siamese-RPN, switcher -aware classification (SAC)	MOT16 CLEAR 0.712	Complex, trajectory ID handled

mAP—mean average precision (higher, the better; range: $[0.0, 100.0]$).

6.2.2. Sequence Models for Vehicle ReID

To capture temporal features, RNN are commonly employed.

In [88], two end-to-end deep architectures, defined as the spatially concatenated CNN and CNN LSTM bi-directional loop, were proposed to address the problematic of vehicle viewpoint uncertainty. The models exploit the great advantages of CNN and LSTM to learn transformations across different viewpoints of vehicles, enabling to attain multi-view vehicle representation containing all viewpoints; information that can be inferred from the only one input view, and then used for learning to measure distance. The evaluation of the model was performed using a new proposed toy car ReID dataset with images from multiple viewpoints of 200 vehicles and the public multi-view car, VehicleID, and VeRi datasets. Conducted experiments showed that the proposed model could achieve a mAP of 18.13 on the VeRi dataset.

In [89], a deep spatial–temporal neural network is proposed to solve the task of sequentially counting vehicles from low-quality videos acquired by city cameras (citycams). Citycam videos are characterized by low resolution, low frame rate, high occlusion, and broad perspective, making most existing methods lose efficacy. To overcome the limitations of the current methods and incorporate the temporal information of traffic video, a novel FCN-rLSTM network was proposed to jointly estimate vehicle density and vehicle count by connecting Fully Convolutional Network (FCN) with LSTM in a residual learning approach. The design enables leveraging the strengths of FCN for pixel-level prediction and the strengths of LSTM to learn complex temporal dynamics. The residual learning connection reformulates the vehicle count regression as a learning residual function concerning the sum of densities in each frame, leading to a significant reduction in network training. To preserve feature map resolution, a hyper-atrous combination was proposed to integrate atrous convolution on the FCN, and combine feature maps of different convolution layers. FCN-rLSTM enables refined feature representation and a new end-to-end trainable mapping from pixels to vehicle count. The proposed method was extensively evaluated on different counting tasks using three datasets, with experimental results demonstrating their effectiveness and robustness. In particular, the proposed FCN-rLSTM reduced the Mean Absolute Error (MAE) from 5.31 to 4.21 on the TRANCOS dataset, showing the abilities of LSTM in combination with FCN to track objects in low-resolution videos.

A practical vehicle tracking framework and trajectory-based weighted ranking method, which significantly improves the performance of cars ReID, were proposed in [90]. The proposed approach makes use of a ResNet50 [14] as the backbone for feature extraction, trained using the set of AI City Challenge [91] and VisDrone2018 [92] datasets, and only considering only the vehicle category. In the inference phase, the image is resized into 1440×800, to capture small vehicles in the video. By using the unified multi-object tracking framework proposed by [86], long-term and short-term cues are fully used as a detector with high recall, considering only boxes with higher confidence as input for the multiple target tracking algorithm. The similarity between the two features space is calculated through cosine similarity with the overall loss function containing the cluster loss, trajectory consistency loss, and classification loss. During the inference phase, a re-ranking with spatial–temporal cue is used, with all trajectories encoded in a feature vector of 2048 dimensions. A density clustering DBSCAN is used to gather similar vehicles with the different ID that belong to the same class. Finally, a ranking with weighted features and trajectory information is set in place to identify individual trajectories from the class group. Conducted experiments using the AI city challenge dataset achieved a mAP of 0.730, showing competitive tracking results in real urban scenarios.

In [93] is proposed an extension to the prevalent task of multi-object tracking and segmentation (MOTS). It explores widely annotated dense pixel-level annotations of two existing tracking datasets using a semi-automatic annotation procedure, containing the masks for 977 distinct objects (cars and pedestrians) in 10,870 video frames. To tackle detection, tracking, and segmentation, i.e., the MOTS task, a neural network is employed jointly with a baseline method built upon the famous Mask R-CNN [39] architecture, which extends the faster R-CNN [94] detector with a mask head.

The TrackR-CNN model provides mask-based detection and association features. Both are used as input to a tracking algorithm that decides which detection to select, and how to integrate temporal context information. The temporal context of the input video is explored by the integration of 3D convolutions (to account for time), into Mask R-CNN on top of a ResNet-101 [14] backbone. The 3D convolutions layers are used to extract the backbone features and to obtain a temporal augmentation context. These new augmented features are then used by the Region Proposal Network (RPN) for the ReID task. For evaluation purposes, the proposed method was set as a baseline that jointly addresses detection, tracking, and segmentation with a single CNN. Conducted experiments demonstrated the relevance of the constructed datasets, enabling them to achieve considerable improvements in performance when trained on MOTS annotations. The datasets, proposed metrics, and baselines, such as MOTSA and sMOTSA, were considered with the baseline model achieving a sMOTSA of 52.7% and MOTSA of 66.9% on the KITTI MOTS dataset, enabling to account for temporal multi-object ReID.

Table 8 summarizes the discussed works and establish comparisons among the performance and used datasets.

Table 8. Performance evaluation of the reviewed vehicle ReID using sequence learning methods.

Cat	Ref.	Main Technique(s)	# Data Success	Pros/Cons
LSTM	[88]	Spatially Concatenated CNN, CNN-LSTM bi-directional loop	VeRi-776 mAP 18.13	Simple, applicable
	[89]	CNN features, gFCN-rLSTM network + Atrous	TRANCOS MAE 4.21	Reproducible, ranking problems
	[90]	ResNet50, LSTM + clustering DBSCAN	AI City MAE 0.730	Complex, trajectory problems
3D	[93]	Mask R-CNN, 3D convolutional layers	KITTI MOTS 0.669	Robust, Short term ID handled

mAP—mean average precision (higher, the better; range: [0.0, 100.0]).

6.2.3. Feature Learning for Vehicle ReID

Feature learning are sometimes implicit when using pre-trained backbones, such as ResNets. However, many works explore the use of specialized schemes to explore the potentialities of global and local features, often by using fusion schemes.

In [95], the authors proposed an approach on vehicle ReID without any knowledge about localization or movements of the cars. This method obtains real-time traffic information based on linear regression with SVM, according to feature vectors, which consist of a color histogram and oriented gradients. First, the vehicles are detected in the video by an object classifier model that creates 3D bounding boxes around the cars. Only the side and front (or back) faces vehicle images are extracted. The extracted image is then fitted into a grid and color histograms to be found in another vehicle image set by simulating a different camera view to be used for the first-round regression. Vehicles with positive first-round regression results are then tested on the second-round regression, where the average Histogram of Oriented Gradients (HOG) vector is used. Cars with both regression results positive are added to another set and are considered as highly potential positive ReID candidates. Experiments were performed on pre-selected semi-automatically 1232 image pairs likely to be matching vehicles, and using a web interface and crowd-sourced people's opinion of vehicles that "are likely to be the same vehicle". The findings showed that 60% of matches could be retrieved (TPR), with only about 10% of False Positive (FP) being included. The proposed method lacks robustness and relies on great percentages on non-robust features that are not optimal to be used in a variety of urban scenarios.

A two-branch CNN scheme was presented in [96] to learn deep features and the distance metric simultaneously. The proposed model uses the late fusion scheme to combine attributes and color features (FACT). It is the late fusion scheme that starts by ranking scores of all test images with the semantic feature learned by GoogLeNet [97] separately. Conducted experiments were performed against other methods, with the rank scores being calculated by the Euclidean distance. The model evaluation was performed on the VeRi dataset, with the proposed model achieving a mAP of 19.92.

In [98], the authors proposed a new spatially constrained similarity measure (SCSM) to handle object rotation, scaling, viewpoint change, and appearance deformation for object ReID in combination with a robust re-ranking method with the k-nearest neighbors of a given query for automatically refining the initial search results. The retrieval system is implemented with SIFT descriptors [81] and fast approximate k-means clustering [99] creating a bag of words (BoW) classification scheme. Extensive performance evaluations on INRIA dataset achieves a mAP of 0.762.

In [100], the authors presented a forecasting mechanism to forecast pedestrian destinations in a large area with a limited number of observations. To address the challenges posed by a limited number of observations (e.g., sparse cameras), and change in pedestrian appearance cues across different cameras, a new descriptor is defined as social affinity maps (SAMs) to link broken, or unobserved trajectories of individuals in the crowd. To continuously track the pedestrians, a Markov-chain model is used to connect every intermediate track x_t^i in trajectory T, to subsequent track x_t^{i+1} with a given probability encoded as priors over Origin and Destination (OD) preferences. In addition, the proposed work also introduces a dataset of 42 million trajectories collected in train stations. The conducted experiments were performed using SAM features, and results showed that the performance of OD forecasting with a different number of in-between cameras increased, and more accurate trajectories were predicted, obtaining an overall OD error rate of 0.672. The SAM enables the extraction of relevant features to maintain continuous track in occluded pedestrians over time.

The fine-grained recognition of vehicles, mainly in traffic surveillance applications, is addressed in [101]. The approach is based on recent advancements in fine-grained recognition: automatic part discovery and bilinear pooling. In contrast to other methods that focus on fine-grained recognition of vehicles, viewpoints are not limited only to a frontal/rear viewpoint, but it allows the vehicles to be seen from any viewpoint. The approach is based on 3D bounding boxes built around the vehicles that are automatically constructed from traffic surveillance data. An CNN based on ResNet50 is used for the estimation of the directions towards the identified vanishing points by feeding the vehicle image into a ResNet50 with three separate outputs regarding the probabilities for directions of vanishing points in quantized angle space. A new annotated dataset BoxCars116k is proposed, focusing on images gathered from surveillance cameras. Several experiments were conducted, with the proposed method significantly improving the CNN classification accuracy, achieving a 12% increase (80.8%) on bounding box determination.

A summary regarding the methods for vehicle ReID is presented in Table 9.

Table 9. Performance evaluation of the reviewed vehicle ReID using feature learning methods.

Cat	Ref.	Main Technique(s)	# Data Success	Pros/Cons
Fusion	[95]	SVM, HOG	Own -	Not replicable,
	[96]	GoogLeNet, Feature fusion	VeRi-776 mAP 19.92	Simple, Baseline
	[98]	SIFT + BOW, re-ranking	INRIA mAP 0.762	OLD fashion, Not SOTA
	[100]	Social Affinity Maps (SAM), Markov-chain model	Own –	Complex, Only indoors
	[101]	3D box prediction, ResNet	BoxCars116k ACC 0.808	Not useful, simple

mAP—mean average precision (higher, the better; range: [0.0, 100.0]).

6.2.4. Tracking for Vehicle ReID

Continuously tracking of objects of interest is a common requirement for any ReID system that has been addressed using different strategies ranging form Kalman filtering based, Optical flow, and many combinations of those with other mechanism.

In [102], the authors proposed an object tracking and 3D reconstruction method to perform 3D object motion estimation. Object tracking and 3D reconstruction are often performed together, with tracking used as input for the 3D reconstruction. To improve tracking performance, a novel method is proposed to close this gap, by first tracking and reconstructing to track. The proposed multi-object tracking, segmentation, and dynamic object fusion (MOTSFusion) approach exploits the 3D motion information extracted from dynamic object reconstructions to track objects through long periods with complete occlusion and recover missing detections. The method first builds up short tracklets using 2D optical flow and then fuses them into dynamic 3D object reconstructions. The precise 3D object motion of these reconstructions is used to merge tracklets through occlusion into long-term tracks and to locate objects in the absence of detection. Conducted experiments were performed on the KITTI platform [93,103], with the reconstruction-based tracking reducing the number of ID switches of the initial tracklets by more than 50%. CLEARMOT [104] was adopted as evaluation metric for bounding box tracking to rank it in terms of MOTA [103], which incorporates FP, False Negative (FN), ID switches (IDS) and sMOTSA to account the segmentation Intersection over Union (IoU) accuracy, with the method achieving a MOTA of 84.83%. The method enables to robustly incorporate long-term occluded objects in an optimized manner.

A recent and widely used tracking-by-detection algorithm is the DeepSort [105]. Simple online and real-time tracking (SORT) enables tracking multiple objects for more extended periods. The method relies on object detector backbones, such as Yolo [106], where an appearance integration is integrated into the Kalman filter to effectively track uniquely similar objects in strong occluded environments, while reducing the number of tracking or ID switch. When an object in track by the Kalman filter, the update of the filter is performed considering the dynamics, the association mechanism based on the Mahalanobis distance, with an extra feature concerning the object appearance, for each object bounding box appearance, a gallery of associated appearance descriptors is kept for each track. Cosine similarity on new observation is used to compute the distance of the current object in track and the new observation, enabling to correctly associate the observation even for an object with high dynamics that cannot be handled solely by the Kalman filter association metrics. The experimental evaluation shows that the inclusion of deep features into the Kalman filter reduces the number of identity switches by 45%.

Table 10 summarizes the discussed works and establish comparisons among the performance and used datasets.

Table 10. Performance evaluation of the reviewed vehicle ReID Tracking methods.

Cat	Ref.	Main Technique(s)	# Data Success	Pros/Cons
	[102]	3D reconstruction, 2D optical flow	KITTI MOTA 0.848	Robust, Short term ID handled
	[105]	CNN features, Kalman + Association	- -	Robust, Short term ID handled

mAP—mean average precision (higher, the better; range: [0.0, 100.0]).

6.2.5. Summary of Vehicle ReID Methods

Several research works concerning object ReID with spatial–temporal constraints can be identified in the literature. However, the ReID in non-overlapping cameras with tracking is commonly accepted to be difficult task, and the range of works that address this problem is not vast. However, there is a clear trend in the use of deep learning features or 3D CNN commonly used in action recognition, explored in the ReID task to capture spatial–temporal invariant features to improve ReID generalization performance in unseen objects over time; however, ReID with spatial–temporal constraints is a difficult task to accomplish soon, mainly in urban scenarios due to the infinite number of partial occlusions, uneven, and dynamic illumination conditions.

7. Methods for Image Enhancement

Severe weather conditions, such as rain and snow, adversely may affect the visual quality of the images acquired under such conditions; thus, rendering them useless for further usage and sharing. In addition, such degraded images usually drastically affect the performance of vision systems. Mainly, it is essential to address the problem of single image de-raining. However, the inherent ill-posed nature of the situation presents several challenges.

In [107], it is proposed an image de-raining conditional generative adversarial network (ID-CGAN) that account for quantitative, visual, and also discriminative performance into the objective function. The proposal method explores the capabilities of conditional generative adversarial networks (CGAN), in combination with additional constraint to enforce the de-rained image to be indistinguishable from its corresponding GT clean image. A refined loss function and other architectural novelties in the generator–discriminator pair were also introduced, with the loss function aimed towards the reduction of artifacts introduced by GAN, ensuring better visual quality. The generator sub-network is constructed using densely connected networks, whereas the discriminator is designed to leverage global and local information and between real/fake images. Exhaustive experiments were conducted against several State-of-the-Art (SOTA) methods using synthetic datasets derived from the UCID [108] and BSD-500 [109] datasets, and with external noise artifacts added. The experiments were evaluated on synthetic and real images using several evaluation metrics such as peak signal to noise ratio (PSNR), structural similarity index (SSIM) [110], universal quality index (UQI) [111], and visual information fidelity (VIF) [112], with the proposed model achieving an PSNR (DB) of 24.34. Moreover, experimental results evaluated on object detection methods, such as FasterRCNN [94], demonstrated the effectiveness of the proposed method in improving the detection performance on images degraded by rain.

A single-image-based rain removal framework was proposed in [113] by properly formulating the rain removal problem as an image decomposition problem based on the morphological decomposition analysis. The alternative to applying a conventional image decomposition technique, the proposed method first decomposes an image into the low and high-frequency (HF) components by employing a bilateral filter. The HF part is then decomposed into a "rain component" and a "non-rain component" using sparse coding. The model experiments were conducted on synthetic rain images built using an image software, with the model achieving a VIF of 0.60. While the method has some degree of performance with common rain conditions, it has difficulties to handling more complex rain dynamic scenarios.

In [114], an effective method based on simple patch-based priors for both the background and rain layers is proposed, which is based on the Gaussian mixture model (GMM) to accommodate multiple orientations and scales of the rain streaks. The two GMMs for the background and rain layers, defined as GB and GR, are based on a pre-trained GMM model with 200 mixture components. The method was evaluated using synthetic and real images, and the results compared to SOTA methods, with the proposed method achieving an SSIM of 0.88.

In [115], it is proposed a DNN architecture called DerainNet for removing rain streaks from an image. The architecture is based on a CNN, enabling the direct map of the relationship between rainy and clean image detail layers from the data. For effective enhancement, each image is decomposed into a low-frequency base layer and a high-frequency detail layer. The detail layer corresponds to the input to the CNN for rain removal to be combined at a final stage with the low-frequency component. The CNN model was trained using synthesized images with rain, with the model achieving an SSIM of 0.900, increasing by 2% the performance in comparison to [114] using GMM.

A performance evaluation of the reviewed image de-raining methods is given in Table 11.

Table 11. Performance evaluation of the reviewed image de-raining methods.

Reference	Main Techniques	# Data Success	Pros/Cons
[107]	GANS, conditional GAN	UCID PSNR 24.34	Robust, SOTA
[113]	Bilateral filter, image decomposition	Systeticg VIF 0.60	Simple, Parameter dependent
[114]	GMM, image decomposition	Systetic SSIM 0.880	Simple, Pre-trained dependent
[115]	CNN, HF component layer	Systetic SSIM 0.900	Simple, Robust

PSNR—peak signal to noise ratio (higher, the better; range: [0.0, −]), VIF—Information Fidelity (higher the better, range: [0.0, 1.0]), SSIM—structural similarity index (higher, the better; range: [0.0, 1.0]).

While many other methods can be found, the aforementioned ones highlight the most common approaches to the image enhancement problematic when operating in urban scenarios, where illumination conditions are not constant, due to rain, fog, and illumination, which potentially hamper the performance of the ReID methods.

8. Conclusions

A detailed overview of SOTA methods to date were presented in this paper, including comparisons to identify the main advantages and problems the methods present. In addition, the most commonly used image datasets and their main characteristic were identified.

Image enhancement is a vital component of any computer vision system. It can improve the performance of the initial object detectors and classification, leading to improved ReID systems. Most of the works explore the use of pre-trained DNN, acting as a backbone for feature extraction, with most of them exploring a residual network, enabling to easily reuse the extracted feature maps for ReID model variations. However, the person and vehicle ReID are addressed separately, and fewer research studies have proposed long-term tracking with ReID simultaneously.

The published articles for person ReID concerned DNN using Siamese networks are the most prominent, exhibiting good performance results. Most of the identified works explore novel augmentation and dropout techniques during training, framed with different triplet loss variations. Most of the research studies have obtained competitive results in the common ReID datasets. However, without proper generalization evaluation in real scenarios where light conditions are more challenging.

In this review, the focus was only on object ReID methods; however, a reliable system comprehends two stages for the task, the detection process and the ReID mechanism by itself. However, a fully end-to-end object ReID requires high precision of the object detection, as well as unlabeled ones, and difficulties on the effective combinations of object detection and ReID in a fully integrated ReID system are directions that require attention in the near future.

From this review, it is possible to conclude that there is a lot of room for improvement regarding the multi-object ReID and long-term tracking that is still not explored in the scientific community, combined with the object detection stage, is still an open problem, and commonly not addressed in the identified ReID works.

Author Contributions: Conceptualization, funding acquisition, and supervision by J.M.R.S.T.; investigation, data collection, formal analysis, and writing original draft preparation by H.S.O.; writing review and editing by H.S.O., J.J.M.M., and J.M.R.S.T. All authors have read and agreed to the published version of the manuscript.

Funding: This article is a result of the project Safe Cities—"Inovação para Construir Cidades Seguras", with reference POCI-01-0247-FEDER-041435, co-funded by the European Regional Development Fund (ERDF), through the Operational Programme for Competitiveness and Internationalization (COMPETE 2020), under the PORTUGAL 2020 Partnership Agreement.

Institutional Review Board Statement: Not applicable.

Informed Consent Statement: Not applicable.

Data Availability Statement: Not applicable.

Conflicts of Interest: The authors declare no conflict of interest.

References

1. Wu, L.; Wang, Y.; Gao, J.; Li, X. Deep adaptive feature embedding with local sample distributions for person re-identification. *Pattern Recognit.* **2018**, *73*, 275–288. [CrossRef]
2. Zhang, W.; Ma, B.; Liu, K.; Huang, R. Video-based pedestrian re-identification by adaptive spatio-temporal appearance model. *IEEE Trans. Image Process.* **2017**, *26*, 2042–2054. [CrossRef]
3. Varior, R.R.; Haloi, M.; Wang, G. Gated Siamese Convolutional Neural Network Architecture for Human Re-Identification. In Proceedings of the European Conference on Computer Vision, Amsterdam, The Netherlands, 8–16 October 2016; Springer: Berlin/Heidelberg, Germany, 2016; pp. 791–808.
4. Xiao, T.; Li, H.; Ouyang, W.; Wang, X. Learning deep feature representations with domain guided dropout for person re-identification. In Proceedings of the IEEE Conference on Computer Vision and Pattern Recognition, Vegas, NV, USA, 27–30 June 2016; pp. 1249–1258.
5. McLaughlin, N.; Martinez del Rincon, J.; Miller, P. Recurrent convolutional network for video-based person re-identification. In Proceedings of the IEEE Conference on Computer Vision and Pattern Recognition, Vegas, NV, USA, 27–30 June 2016; pp. 1325–1334.
6. Yan, Y.; Ni, B.; Song, Z.; Ma, C.; Yan, Y.; Yang, X. Person re-identification via recurrent feature aggregation. In Proceedings of the European Conference on Computer Vision, Amsterdam, The Netherlands, 8–16 October 2016; Springer: Berlin/Heidelberg, Germany, 2016; pp. 701–716.
7. Yi, D.; Lei, Z.; Liao, S.; Li, S.Z. Deep metric learning for person re-identification. In Proceedings of the 2014 22nd International Conference on Pattern Recognition, Washington, DC, USA, 24–28 August 2014; pp. 34–39.
8. Zheng, Z.; Zheng, L.; Yang, Y. A discriminatively learned cnn embedding for person reidentification. *ACM Trans. Multimed. Comput. Commun. Appl. (TOMM)* **2017**, *14*, 1–20. [CrossRef]
9. Deng, J.; Dong, W.; Socher, R.; Li, L.J.; Li, K.; Fei-Fei, L. Imagenet: A large-scale hierarchical image database. In Proceedings of the 2009 IEEE Conference on Computer Vision and Pattern Recognition, Miami, FL, USA, 20–25 June 2009; pp. 248–255.
10. Krizhevsky, A.; Sutskever, I.; Hinton, G.E. Imagenet classification with deep convolutional neural networks. *Adv. Neural Inf. Process. Syst.* **2012**, *25*, 1097–1105. [CrossRef]
11. LeCun, Y.; Boser, B.; Denker, J.; Henderson, D.; Howard, R.; Hubbard, W.; Jackel, L. Handwritten digit recognition with a back-propagation network. *Adv. Neural Inf. Process. Syst.* **1989**, *2*, 396–404.
12. Simonyan, K.; Zisserman, A. Very deep convolutional networks for large-scale image recognition. *arXiv* **2014**, arXiv:1409.1556.

13. Szegedy, C.; Ioffe, S.; Vanhoucke, V.; Alemi, A. Inception-v4, inception-resnet and the impact of residual connections on learning. *arXiv* **2016**, arXiv:1602.07261.
14. He, K.; Zhang, X.; Ren, S.; Sun, J. Deep residual learning for image recognition. In Proceedings of the IEEE Conference on Computer Vision and Pattern Recognition, Vegas, NV, USA, 27–30 June 2016; pp. 770–778.
15. Canziani, A.; Paszke, A.; Culurciello, E. An analysis of deep neural network models for practical applications. *arXiv* **2016**, arXiv:1605.07678.
16. Gong, S.; Cristani, M.; Yan, S.; Loy, C.C.; Re-Identification, P. Springer Publishing Company. *Incorporated* **2014**, *1447162951*, 9781447162957.
17. Li, D.; Zhang, Z.; Chen, X.; Ling, H.; Huang, K. A richly annotated dataset for pedestrian attribute recognition. *arXiv* **2016**, arXiv:1603.07054.
18. Gray, D.; Tao, H. Viewpoint Invariant Pedestrian Recognition with an Ensemble of Localized Features. In Proceedings of the European Conference on Computer Vision, Marseille, France, 12–18 October 2008; Springer: Berlin/Heidelberg, Germany, 2008; pp. 262–275.
19. Nguyen, T.B.; Le, T.L.; Nguyen, D.D.; Pham, D.T. A Reliable Image-to-Video Person Re-Identification Based on Feature Fusion. In Proceedings of the Asian Conference on Intelligent Information and Database Systems, Dong Hoi City, Vietnam, 19–21 March 2018; Springer: Berlin/Heidelberg, Germany, 2018; pp. 433–442.
20. Pham, T.T.T.; Le, T.L.; Vu, H.; Dao, T.K. Fully-automated person re-identification in multi-camera surveillance system with a robust kernel descriptor and effective shadow removal method. *Image Vis. Comput.* **2017**, *59*, 44–62. [CrossRef]
21. Cheng, D.S.; Cristani, M.; Stoppa, M.; Bazzani, L.; Murino, V. Custom pictorial structures for re-identification. *BMVC* **2011**, *1*, 6.
22. Das, A.; Chakraborty, A.; Roy-Chowdhury, A.K. Consistent Re-Identification in a Camera Network. In Proceedings of the European Conference on Computer Vision, Zurich, Switzerland, 6–12 September 2014; Springer: Berlin/Heidelberg, Germany, 2014; pp. 330–345.
23. Moon, H.; Phillips, P.J. Computational and performance aspects of PCA-based face-recognition algorithms. *Perception* **2001**, *30*, 303–321. [CrossRef] [PubMed]
24. Nguyen, T.B.; Le, T.L.; Ngoc, N.P. Fusion schemes for image-to-video person re-identification. *J. Inf. Telecommun.* **2019**, *3*, 74–94. [CrossRef]
25. Matsukawa, T.; Okabe, T.; Suzuki, E.; Sato, Y. Hierarchical gaussian descriptor for person re-identification. In Proceedings of the IEEE Conference on Computer Vision and Pattern Recognition, Vegas, NV, USA, 27–30 June 2016; pp. 1363–1372.
26. Li, W.; Zhu, X.; Gong, S. Person re-identification by deep joint learning of multi-loss classification. *arXiv* **2017**, arXiv:1705.04724.
27. Argyriou, A.; Evgeniou, T.; Pontil, M. Multi-task feature learning. *Adv. Neural Inf. Process. Syst.* **2007**, *19*, 41–48.
28. Kong, D.; Fujimaki, R.; Liu, J.; Nie, F.; Ding, C. Exclusive Feature Learning on Arbitrary Structures via $l_{1,2}$-norm. *Adv. Neural Inf. Process. Syst.* **2014**, *1*, 1655–1663.
29. Wang, H.; Nie, F.; Huang, H. Multi-view clustering and feature learning via structured sparsity. *Int. Conf. Mach. Learn.* **2013**, *28*, 352–360.
30. Gray, D.; Brennan, S.; Tao, H. Evaluating appearance models for recognition, reacquisition, and tracking. In Proceedings of the IEEE International Workshop on Performance Evaluation for Tracking and Surveillance (PETS), Rio de Janeiro, Brazil, 14 October 2007; Volume 3, pp. 1–7.
31. Zhou, K.; Yang, Y.; Cavallaro, A.; Xiang, T. Omni-scale feature learning for person re-identification. In Proceedings of the IEEE International Conference on Computer Vision, Seoul, Korea, 27 October–3 November 2019; pp. 3702–3712.
32. Reddi, S.J.; Kale, S.; Kumar, S. On the convergence of adam and beyond. *arXiv* **2019**, arXiv:1904.09237.
33. Zhong, Z.; Zheng, L.; Kang, G.; Li, S.; Yang, Y. Random erasing data augmentation. *arXiv* **2017**, arXiv:1708.04896.
34. Ning, X.; Gong, K.; Li, W.; Zhang, L.; Bai, X.; Tian, S. Feature refinement and filter network for person Re-identification. *IEEE Trans. Circ. Syst. Video Technol.* **2020**, *31*, 3391–3402. [CrossRef]
35. Quan, R.; Dong, X.; Wu, Y.; Zhu, L.; Yang, Y. Auto-ReID: Searching for a part-aware ConvNet for person re-identification. In Proceedings of the IEEE International Conference on Computer Vision, Seoul, Korea, 27 October–3 November 2019; pp. 3750–3759.
36. Liu, H.; Simonyan, K.; Yang, Y. Darts: Differentiable architecture search. *arXiv* **2018**, arXiv:1806.09055.
37. Yaghoubi, E.; Borza, D.; Alirezazadeh, P.; Kumar, A.; Proença, H. An Implicit Attention Mechanism for Deep Learning Pedestrian Re-identification Frameworks. *arXiv* **2020**, arXiv:2001.11267.
38. Luo, H.; Gu, Y.; Liao, X.; Lai, S.; Jiang, W. Bag of tricks and a strong baseline for deep person re-identification. In Proceedings of the IEEE Conference on Computer Vision and Pattern Recognition Workshops, Long Beach, CA, USA, 16–20 June 2019; pp. 4321–4329.
39. He, K.; Gkioxari, G.; Dollár, P.; Girshick, R. Mask r-cnn. In Proceedings of the IEEE International Conference on Computer Vision, Venice, Italy, 22–29 October 2017; pp. 2961–2969.
40. Liu, X.; Zhao, H.; Tian, M.; Sheng, L.; Shao, J.; Yi, S.; Yan, J.; Wang, X. Hydraplus-net: Attentive deep features for pedestrian analysis. In Proceedings of the IEEE International Conference on Computer Vision, Venice, Italy, 22–29 October 2017; pp. 350–359.
41. Hou, R.; Chang, H.; Ma, B.; Huang, R.; Shan, S. BiCnet-TKS: Learning Efficient Spatial–Temporal Representation for Video Person Re-Identification. In Proceedings of the IEEE/CVF Conference on Computer Vision and Pattern Recognition, Nashville, TN, USA, 21–24 June 2021; pp. 2014–2023.

42. Ning, X.; Gong, K.; Li, W.; Zhang, L. JWSAA: Joint weak saliency and attention aware for person re-identification. *Neurocomputing* **2021**, *453*, 801–811. [CrossRef]
43. Shen, C.; Jin, Z.; Zhao, Y.; Fu, Z.; Jiang, R.; Chen, Y.; Hua, X.S. Deep siamese network with multi-level similarity perception for person re-identification. In Proceedings of the 25th ACM International Conference on Multimedia, Mountain View, CA, USA, 23–27 October 2017; pp. 1942–1950.
44. Szegedy, C.; Vanhoucke, V.; Ioffe, S.; Shlens, J.; Wojna, Z. Rethinking the Inception Architecture for Computer Vision. In Proceedings of the IEEE Conference on Computer Vision and Pattern Recognition, Las Vegas, NV, USA, 27–30 June 2016; pp. 2818–2826.
45. Schroff, F.; Kalenichenko, D.; Philbin, J. Facenet: A unified embedding for face recognition and clustering. In Proceedings of the IEEE Conference on Computer Vision and Pattern Recognition, Boston, MA, USA, 7–11 June 2015; pp. 815–823.
46. Li, W.; Zhao, R.; Xiao, T.; Wang, X. Deepreid: Deep filter pairing neural network for person re-identification. In Proceedings of the IEEE Conference on Computer Vision and Pattern Recognition, Columbus, OH, USA, 23–28 June 2014; pp. 152–159.
47. Zheng, L.; Shen, L.; Tian, L.; Wang, S.; Wang, J.; Tian, Q. Scalable person re-identification: A benchmark. In Proceedings of the IEEE Conference on Computer Vision and Pattern Recognition, Boston, MA, USA, 7–11 June 2015; pp. 1116–1124.
48. Li, W.; Wang, X. Locally aligned feature transforms across views. In Proceedings of the IEEE Conference on Computer Vision and Pattern Recognition, Portland, OR, USA, 23–28 June 2013; pp. 3594–3601.
49. Srivastava, N.; Hinton, G.; Krizhevsky, A.; Sutskever, I.; Salakhutdinov, R. Dropout: A simple way to prevent neural networks from overfitting. *J. Mach. Learn. Res.* **2014**, *15*, 1929–1958.
50. Lv, J.; Chen, W.; Li, Q.; Yang, C. Unsupervised cross-dataset person re-identification by transfer learning of spatial-temporal patterns. In Proceedings of the IEEE Conference on Computer Vision and Pattern Recognition, Salt Lake City, UT, USA, 18–23 June 2018; pp. 7948–7956.
51. Loy, C.C.; Xiang, T.; Gong, S. Multi-camera activity correlation analysis. In Proceedings of the 2009 IEEE Conference on Computer Vision and Pattern Recognition, Miami, FL, USA, 20–25 June 2009; pp. 1988–1995.
52. Ahmed, E.; Jones, M.; Marks, T.K. An improved deep learning architecture for person re-identification. In Proceedings of the IEEE Conference on Computer Vision and Pattern Recognition, Boston, MA, USA, 7–11 June 2015; pp. 3908–3916.
53. Hermans, A.; Beyer, L.; Leibe, B. In defense of the triplet loss for person re-identification. *arXiv* **2017**, arXiv:1703.07737.
54. Weinberger, K.Q.; Saul, L.K. Distance metric learning for large margin nearest neighbor classification. *J. Mach. Learn. Res.* **2009**, *10*, 207–244.
55. Ioffe, S.; Szegedy, C. Batch normalization: Accelerating deep network training by reducing internal covariate shift. *arXiv* **2015**, arXiv:1502.03167.
56. Glorot, X.; Bordes, A.; Bengio, Y. Deep sparse rectifier neural networks. In Proceedings of the Fourteenth International Conference on Artificial Intelligence and Statistics, Ft. Lauderdale, FL, USA, 11–13 April 2011; pp. 315–323.
57. Baldassarre, F.; Morín, D.G.; Rodés-Guirao, L. Deep koalarization: Image colorization using cnns and inception-resnet-v2. *arXiv* **2017**, arXiv:1712.03400.
58. Xu, B.; Wang, N.; Chen, T.; Li, M. Empirical evaluation of rectified activations in convolutional network. *arXiv* **2015**, arXiv:1505.00853.
59. Zheng, L.; Bie, Z.; Sun, Y.; Wang, J.; Su, C.; Wang, S.; Tian, Q. Mars: A Video Benchmark for Large-Scale Person Re-Identification. In Proceedings of the European Conference on Computer Vision, Amsterdam, The Netherlands, 11–14 October 2016; Springer: Berlin/Heidelberg, Germany, 2016; pp. 868–884.
60. Cheng, D.; Gong, Y.; Zhou, S.; Wang, J.; Zheng, N. Person re-identification by multi-channel parts-based cnn with improved triplet loss function. In Proceedings of the iEEE Conference on Computer Vision and Pattern Recognition, Las Vegas, NV, USA, 27–30 June 2016; pp. 1335–1344.
61. Yuan, X.; He, P.; Zhu, Q.; Li, X. Adversarial examples: Attacks and defenses for deep learning. *IEEE Trans. Neural Netw. Learn. Syst.* **2019**, *30*, 2805–2824. [CrossRef] [PubMed]
62. Hu, Y.; Yi, D.; Liao, S.; Lei, Z.; Li, S.Z. Cross Dataset Person Re-Identification. In Proceedings of the Asian Conference on Computer Vision, Singapore, 1–5 November 2014; Springer: Berlin/Heidelberg, Germany, 2014; pp. 650–664.
63. Hirzer, M.; Beleznai, C.; Roth, P.M.; Bischof, H. Person Re-Identification by Descriptive and Discriminative Classification. In Proceedings of the 17th Scandinavian Conference on Image Analysis, Ystad, Sweden, 23–25 May 2011; Springer: Berlin/Heidelberg, Germany, 2011; pp. 91–102.
64. Liao, X.; He, L.; Yang, Z.; Zhang, C. Video-Based Person Re-Identification Via 3D Convolutional Networks and Non-Local Attention. In Proceedings of the Asian Conference on Computer Vision, Perth, Australia, 2–6 December 2018; Springer: Berlin/Heidelberg, Germany, 2018; pp. 620–634.
65. Kay, W.; Carreira, J.; Simonyan, K.; Zhang, B.; Hillier, C.; Vijayanarasimhan, S.; Viola, F.; Green, T.; Back, T.; Natsev, P.; et al. The kinetics human action video dataset. *arXiv* **2017**, arXiv:1705.06950.
66. Li, J.; Zhang, S.; Huang, T. Multi-scale 3d convolution network for video based person re-identification. In Proceedings of the AAAI Conference on Artificial Intelligence, Honolulu, HI, USA, 27 January–1 February 2019; Volume 33, pp. 8618–8625.
67. Zhou, Z.; Huang, Y.; Wang, W.; Wang, L.; Tan, T. See the forest for the trees: Joint spatial and temporal recurrent neural networks for video-based person re-identification. In Proceedings of the IEEE Conference on Computer Vision and Pattern Recognition, Honolulu, HI, USA, 21–26 June 2017; pp. 4747–4756.

68. Ge, Y.; Li, Z.; Zhao, H.; Yin, G.; Yi, S.; Wang, X. Fd-gan: Pose-guided feature distilling gan for robust person re-identification. In Proceedings of the 32nd Conference on Neural Information Processing Systems (NeurIPS 2018), Montreal, QC, Canada, 3–8 December 2018; pp. 1222–1233.
69. Kingma, D.P.; Ba, J. Adam: A method for stochastic optimization. *arXiv* **2014**, arXiv:1412.6980.
70. Zhong, Z.; Zheng, L.; Zheng, Z.; Li, S.; Yang, Y. Camera style adaptation for person re-identification. In Proceedings of the IEEE Conference on Computer Vision and Pattern Recognition, Salt Lake City, UT, USA, 18–23 June 2018; pp. 5157–5166.
71. Zhu, J.Y.; Park, T.; Isola, P.; Efros, A.A. Unpaired image-to-image translation using cycle-consistent adversarial networks. In Proceedings of the IEEE International Conference on Computer Vision, Venice, Italy, 22–29 October 2017; pp. 2223–2232.
72. Zou, Y.; Yang, X.; Yu, Z.; Kumar, B.V.; Kautz, J. Joint disentangling and adaptation for cross-domain person re-identification. In Proceedings of the Computer Vision–ECCV 2020: 16th European Conference, Glasgow, UK, 23–28 August 2020; Proceedings, Part II 16; Springer: Berlin/Heidelberg, Germany, 2020; pp. 87–104.
73. Fan, X.; Jiang, W.; Luo, H.; Fei, M. Sphereeid: Deep hypersphere manifold embedding for person re-identification. *J. Vis. Commun. Image Represent.* **2019**, *60*, 51–58. [CrossRef]
74. Sun, Y.; Zheng, L.; Yang, Y.; Tian, Q.; Wang, S. Beyond part models: Person retrieval with refined part pooling (and a strong convolutional baseline). In Proceedings of the European Conference on Computer Vision (ECCV), Munich, Germany, 8–14 September 2018; pp. 480–496.
75. Wen, Y.; Zhang, K.; Li, Z.; Qiao, Y. A Discriminative Feature Learning Approach for Deep Face Recognition. In Proceedings of the European Conference on Computer Vision, Amsterdam, The Netherlands, 11–14 October 2016; Springer: Berlin/Heidelberg, Germany, 2016; pp. 499–515.
76. Zhong, Z.; Zheng, L.; Cao, D.; Li, S. Re-ranking person re-identification with k-reciprocal encoding. In Proceedings of the IEEE Conference on Computer Vision and Pattern Recognition, Honolulu, HI, USA, 21–26 June 2017; pp. 1318–1327.
77. Dietlmeier, J.; Antony, J.; McGuinness, K.; O'Connor, N.E. How important are faces for person re-identification? In Proceedings of the 2020 25th International Conference on Pattern Recognition (ICPR), Milan, Italy, 10–15 January 2021; pp. 6912–6919.
78. Lu, X.Y.; Skabardonis, A. Freeway traffic shockwave analysis: Exploring the NGSIM trajectory data. In Proceedings of the 86th Annual Meeting of the Transportation Research Board, Washington, DC, USA, 21–25 January 2007.
79. Geiger, A.; Lenz, P.; Stiller, C.; Urtasun, R. Vision meets robotics: The kitti dataset. *Int. J. Robot. Res.* **2013**, *32*, 1231–1237. [CrossRef]
80. Everingham, M.; Van Gool, L.; Williams, C.K.; Winn, J.; Zisserman, A. The pascal visual object classes (voc) challenge. *Int. J. Comput. Vis.* **2010**, *88*, 303–338. [CrossRef]
81. Lowe, D.G. Distinctive image features from scale-invariant keypoints. *Int. J. Comput. Vis.* **2004**, *60*, 91–110. [CrossRef]
82. Bai, Y.; Lou, Y.; Gao, F.; Wang, S.; Wu, Y.; Duan, L.Y. Group-sensitive triplet embedding for vehicle reidentification. *IEEE Trans. Multimed.* **2018**, *20*, 2385–2399. [CrossRef]
83. Em, Y.; Gag, F.; Lou, Y.; Wang, S.; Huang, T.; Duan, L.Y. Incorporating intra-class variance to fine-grained visual recognition. In Proceedings of the 2017 IEEE International Conference on Multimedia and Expo (ICME), Hong Kong, China, 10–14 July 2017; pp. 1452–1457.
84. Zhang, Y.; Liu, D.; Zha, Z.J. Improving triplet-wise training of convolutional neural network for vehicle re-identification. In Proceedings of the 2017 IEEE International Conference on Multimedia and Expo (ICME), Hong Kong, China, 10–14 July 2017; pp. 1386–1391.
85. Liu, X.; Liu, W.; Mei, T.; Ma, H. Provid: Progressive and multimodal vehicle reidentification for large-scale urban surveillance. *IEEE Trans. Multimed.* **2017**, *20*, 645–658. [CrossRef]
86. Feng, W.; Hu, Z.; Wu, W.; Yan, J.; Ouyang, W. Multi-object tracking with multiple cues and switcher-aware classification. *arXiv* **2019**, arXiv:1901.06129.
87. Milan, A.; Leal-Taixé, L.; Reid, I.; Roth, S.; Schindler, K. MOT16: A benchmark for multi-object tracking. *arXiv* **2016**, arXiv:1603.00831.
88. Zhou, Y.; Liu, L.; Shao, L. Vehicle re-identification by deep hidden multi-view inference. *IEEE Trans. Image Process.* **2018**, *27*, 3275–3287. [CrossRef]
89. Zhang, S.; Wu, G.; Costeira, J.P.; Moura, J.M. Fcn-rlstm: Deep spatio-temporal neural networks for vehicle counting in city cameras. In Proceedings of the IEEE International Conference on Computer Vision, Venice, Italy, 22–29 October 2017; pp. 3667–3676.
90. He, Z.; Lei, Y.; Bai, S.; Wu, W. Multi-Camera vehicle tracking with powerful visual features and spatial-temporal cue. In Proceedings of the CVPR Workshops, Long Beach, CA, USA, 16–20 June 2019; pp. 203–212.
91. Naphade, M.; Anastasiu, D.C.; Sharma, A.; Jagrlamudi, V.; Jeon, H.; Liu, K.; Chang, M.C.; Lyu, S.; Gao, Z. The nvidia ai city challenge. In Proceedings of the 2017 IEEE SmartWorld, Ubiquitous Intelligence & Computing, Advanced & Trusted Computed, Scalable Computing & Communications, Cloud & Big Data Computing, Internet of People and Smart City Innovation (SmartWorld/SCALCOM/UIC/ATC/CBDCom/IOP/SCI), San Francisco, CA, USA, 4–8 August 2017; pp. 1–6.
92. Zhu, P.; Wen, L.; Bian, X.; Ling, H.; Hu, Q. Vision meets drones: A challenge. *arXiv* **2018**, arXiv:1804.07437.
93. Voigtlaender, P.; Krause, M.; Osep, A.; Luiten, J.; Sekar, B.B.G.; Geiger, A.; Leibe, B. MOTS: Multi-object tracking and segmentation. In Proceedings of the IEEE Conference on Computer Vision and Pattern Recognition, Long Beach, CA, USA, 15–20 June 2019; pp. 7942–7951.

94. Ren, S.; He, K.; Girshick, R.; Sun, J. Faster r-cnn: Towards real-time object detection with region proposal networks. In Proceedings of the 32nd Conference on Neural Information Processing Systems, Montreal, QC, Canada, 7–10 December 2015; pp. 91–99.
95. Zapletal, D.; Herout, A. Vehicle re-identification for automatic video traffic surveillance. In Proceedings of the IEEE Conference on Computer Vision and Pattern Recognition Workshops, Las Vegas, NV, USA, 26 June–1 July 2016; pp. 25–31.
96. Liu, X.; Liu, W.; Ma, H.; Fu, H. Large-scale vehicle re-identification in urban surveillance videos. In Proceedings of the 2016 IEEE International Conference on Multimedia and Expo (ICME), Seattle, WA, USA, 11–15 July 2016; pp. 1–6.
97. Szegedy, C.; Liu, W.; Jia, Y.; Sermanet, P.; Reed, S.; Anguelov, D.; Erhan, D.; Vanhoucke, V.; Rabinovich, A. Going deeper with convolutions. In Proceedings of the IEEE Conference on Computer Vision and Pattern Recognition, Boston, MA, USA, 7–12 June 2015; pp. 1–9.
98. Shen, X.; Lin, Z.; Brandt, J.; Avidan, S.; Wu, Y. Object retrieval and localization with spatially-constrained similarity measure and k-nn re-ranking. In Proceedings of the 2012 IEEE Conference on Computer Vision and Pattern Recognition, Washington, DC, USA, 16–21 June 2012; pp. 3013–3020.
99. Muja, M.; Lowe, D.G. Fast approximate nearest neighbors with automatic algorithm configuration. *VISAPP* **2009**, *2*, 2.
100. Alahi, A.; Ramanathan, V.; Fei-Fei, L. Socially-aware large-scale crowd forecasting. In Proceedings of the IEEE Conference on Computer Vision and Pattern Recognition, Columbus, OH, USA, 23–28 June 2014; pp. 2203–2210.
101. Sochor, J.; Špaňhel, J.; Herout, A. BoxCars: Improving Fine-Grained Recognition of Vehicles Using 3-D Bounding Boxes in Traffic Surveillance. *IEEE Trans. Intell. Transp. Syst.* **2018**, *20*, 97–108. [CrossRef]
102. Luiten, J.; Fischer, T.; Leibe, B. Track to reconstruct and reconstruct to track. *IEEE Robot. Autom. Lett.* **2020**, *5*, 1803–1810. [CrossRef]
103. Geiger, A.; Lenz, P.; Urtasun, R. Are we ready for autonomous driving? The kitti vision benchmark suite. In Proceedings of the 2012 IEEE Conference on Computer Vision and Pattern Recognition, Washington, DC, USA, 16–21 June 2012; pp. 3354–3361.
104. Bernardin, K.; Stiefelhagen, R. Evaluating multiple object tracking performance: The CLEAR MOT metrics. *EURASIP J. Image Video Process.* **2008**, *2008*, 1–10. [CrossRef]
105. Wojke, N.; Bewley, A.; Paulus, D. Simple online and realtime tracking with a deep association metric. In Proceedings of the 2017 IEEE International Conference on Image Processing (ICIP), Beijing, China, 17–20 September 2017; pp. 3645–3649.
106. Redmon, J.; Farhadi, A. Yolov3: An incremental improvement. *arXiv* **2018**, arXiv:1804.02767.
107. Zhang, H.; Sindagi, V.; Patel, V.M. Image de-raining using a conditional generative adversarial network. *IEEE Trans. Circ. Syst. Video Technol.* **2019**, *30*, 3943–3956. [CrossRef]
108. Schaefer, G.; Stich, M. UCID: An uncompressed color image database. In *Storage and Retrieval Methods and Applications for Multimedia 2004*; International Society for Optics and Photonics: Washington, DC, USA, 2003; Volume 5307, pp. 472–480.
109. Arbelaez, P.; Maire, M.; Fowlkes, C.; Malik, J. Contour detection and hierarchical image segmentation. *IEEE Trans. Pattern Anal. Mach. Intell.* **2010**, *33*, 898–916. [CrossRef]
110. Wang, Z.; Bovik, A.C.; Sheikh, H.R.; Simoncelli, E.P. Image quality assessment: From error visibility to structural similarity. *IEEE Trans. Image Process.* **2004**, *13*, 600–612. [CrossRef] [PubMed]
111. Wang, Z.; Bovik, A.C. A universal image quality index. *IEEE Signal Process. Lett.* **2002**, *9*, 81–84. [CrossRef]
112. Sheikh, H.R.; Bovik, A.C. Image information and visual quality. *IEEE Trans. Image Process.* **2006**, *15*, 430–444. [CrossRef]
113. Kang, L.W.; Lin, C.W.; Fu, Y.H. Automatic single-image-based rain streaks removal via image decomposition. *IEEE Trans. Image Process.* **2011**, *21*, 1742–1755. [CrossRef] [PubMed]
114. Li, Y.; Tan, R.T.; Guo, X.; Lu, J.; Brown, M.S. Rain streak removal using layer priors. In Proceedings of the IEEE Conference on Computer Vision and Pattern Recognition, Las Vegas, NV, USA, 27–30 June 2016; pp. 2736–2744.
115. Fu, X.; Huang, J.; Ding, X.; Liao, Y.; Paisley, J. Clearing the skies: A deep network architecture for single-image rain removal. *IEEE Trans. Image Process.* **2017**, *26*, 2944–2956. [CrossRef] [PubMed]

applied sciences

MDPI

Article

Modular Dynamic Neural Network: A Continual Learning Architecture

Daniel Turner *, Pedro J. S. Cardoso and João M. F. Rodrigues *

LARSYS & ISE, Universidade do Algarve, 8005-139 Faro, Portugal; pcardoso@ualg.pt
* Correspondence: danielluisturner@hotmail.com (D.T.); jrodrig@ualg.pt (J.M.F.R.);
Tel.: +351-289-800-100 (D.T. & J.M.F.R.)

Abstract: Learning to recognize a new object after having learned to recognize other objects may be a simple task for a human, but not for machines. The present go-to approaches for teaching a machine to recognize a set of objects are based on the use of deep neural networks (DNN). So, intuitively, the solution for teaching new objects on the fly to a machine should be DNN. The problem is that the trained DNN weights used to classify the initial set of objects are extremely fragile, meaning that any change to those weights can severely damage the capacity to perform the initial recognitions; this phenomenon is known as catastrophic forgetting (CF). This paper presents a new (DNN) continual learning (CL) architecture that can deal with CF, the modular dynamic neural network (MDNN). The presented architecture consists of two main components: (a) the ResNet50-based feature extraction component as the backbone; and (b) the modular dynamic classification component, which consists of multiple sub-networks and progressively builds itself up in a tree-like structure that rearranges itself as it learns over time in such a way that each sub-network can function independently. The main contribution of the paper is a new architecture that is strongly based on its modular dynamic training feature. This modular structure allows for new classes to be added while only altering specific sub-networks in such a way that previously known classes are not forgotten. Tests on the CORe50 dataset showed results above the state of the art for CL architectures.

Keywords: continual learning; neural networks; catastrophic forgetting; object recognition

Citation: Turner, D.; Cardoso, P.J.S.; Rodrigues, J.M.F. Modular Dynamic Neural Network: A Continual Learning Architecture. *Appl. Sci.* **2021**, *11*, 12078. https://doi.org/10.3390/app112412078

Academic Editor: Fabio La Foresta

Received: 16 October 2021
Accepted: 13 December 2021
Published: 18 December 2021

Publisher's Note: MDPI stays neutral with regard to jurisdictional claims in published maps and institutional affiliations.

1. Introduction

Learning to recognize new objects may be a simple task for humans (even for small children), but successful implementations of this in machines prove to be very difficult. While the challenge of teaching machines how to distinguish (i.e., classify) between a set of pre-learned objects has already seen many solutions which present impressive results, e.g., ResNet50 [1], the challenge of teaching the same machine on the fly how to learn to classify new objects on top of those already known remains very much unsolved [2].

In this context, artificial neural networks (NN) have been extensively researched and proven to function with outstanding levels of precision over a variety of applications. The go-to approaches for the first part of the problem (teaching a machine to classify an initial set of objects) are typically based on the application of deep neural networks (DNN). So, intuitively, the solution for the next part of the problem (teaching a machine new objects on the fly) is to elaborate on the existing solutions, but the problem with this is that the trained weights in neural networks are extremely fragile, meaning that any change to accommodate new objects can severely damage their capacity to perform the functions they were trained for. This phenomenon is known as catastrophic forgetting (CF) [3].

The usual approach for avoiding the problems associated with the addition of new classes (objects) (i.e., avoiding catastrophic forgetting) is to simply retrain the entire network from scratch, using data from all the classes that are intended for the network to be able to classify. The issue with this is that the training process tends to be very computationally heavy and, therefore, time-consuming, even when performed on powerful machines. This may not be a problem in some situations, but it is a problem if the goal is to consistently add new classes regularly. Given that when using currently accessible hardware, the training process often takes between hours and days, it is, therefore, impossible to expect networks to learn multiple classes per day with this approach. This type of learning problem, learning a new class without retraining the entire network, is known as continual learning (CL) [4], where several other names are also commonly used, such as sequential learning [5], lifelong learning [3] and incremental learning [6]. For a comprehensive explanation about this subject, see [7].

This paper presents a new continual learning architecture for object classification, the modular dynamic neural network (MDNN), with the intention of creating a framework capable of learning new classes (on the fly) without forgetting the previously learned ones. As already mentioned, the current state of the art (see next section) shows that the methods most effectively used for object classification are DNN based. The presented framework is also DNN based and is divided into two main components: (a) the ResNet50-based [1] feature extraction component; and (b) the modular dynamic classification component, which consists of multiple sub-networks structured in such a way that they can function independently from one another. In more detail, it consists primarily of a static feature extraction component that passes extracted image features onto a modular dynamic classification component. The modular dynamic classification component is made up of multiple neural networks that serve as binary classifiers which are joined together to form a tree-like structure, where internal classifications dictate the path to follow down the tree of networks. The modules function independently of one another, which means the structure can be dynamic and that, after initializing with at least two classes, new modules can be added as more classes are learned to avoid affecting other modules. Another important advantage of the implemented structure is that classifications can be made using only a percentage of the modules, meaning that the addition of new modules will not have the same negative impact on scalability as it would if every new module had to be processed for every classification. While the scalability of the presented framework has not yet been proven on large datasets, the use of a percentage of the network is undoubtedly lighter in terms of computational requirements compared to the use of the entire network.

The main contribution of this paper is a framework that can learn new classes without having to retrain its entire network of classifiers, with features that help it to cope with the scalability problems that generally come with this type of expanding architecture. The presented framework is in itself original, as the methodology behind its architecture does not directly fit into any of the categories normally used to define CL methods, such as regularization, memory/replay, architecture or parameter-isolation-based methods [3,8], where it is, in fact, a kind of mixture of the mentioned categories. It is partially regularization-based because parts of the network are selectively not interfered with while training; partially memory/replay based because some data are stored and re-used; partially architecture-based because the network of modules does expand over time; and partially parameter isolation-based because specific parts of the network are assigned specific tasks. On top of being original in itself, the presented framework possesses a feature that is of particular interest in terms of innovation, which is the modular dynamic training component that is the main reason it is able to learn continuously, as modules can be added over time which are responsible for recognizing classes or groups of classes, and the assignment of which classes are attributed to which modules is done automatically as the network grows progressively.

The test results on the CORe50 dataset [9] show state-of-the-art results when compared with other CL architectures and, in some cases, the MDNN was even able to almost match results obtained on the same data without CL restrictions (i.e., the final accuracy after learning classes incrementally was similar to that of another network presented with all the data at once). The tests made primarily evaluate the framework's ability to learn classes incrementally and to re-evaluate its accuracy on old classes as new ones are added to make sure they were not forgotten.

The present section introduced the context and goals. Section 2 includes the related work on continual learning along with some background concepts. Section 3 addresses the presented architecture, Section 4 shows the tests and results of the implemented CL architecture and, finally, the conclusions and future work are presented in Section 5.

2. Related Work

Neural networks are currently receiving a large amount of interest and are being applied to a large variety of real-world problems. They are currently the go-to choice for image detection, image recognition and the classification of persons, objects or scenes. While they are widely used and considered to be very effective, if the goal is to add new classes to the previously learned ones, and therefore continue the learning process from where they left off, they fail. This is because after learning new classes on the fly, the network's ability to recognize the previously known classes would be severely reduced; this phenomenon is known as catastrophic forgetting [3] (as already mentioned in the Introduction).

In this context, continual learning [4] means being able to update the prediction model for new tasks while still being able to reuse and retain knowledge from previous tasks. CL challenges assume an incremental setting, where tasks are received one at a time and, in addition, most studies on the matter also consider the non-storage of data to be an essential characteristic of a continual learner. Usually, continual learning studies refer to humans as ideal examples of continual learners, examples can be seen in [2,3,5,8]. This is because, as is the case with NN, a large part of the ideas behind CL are inspired by how researchers presume human brains work, including how the process of consolidation and reconsolidation from short- to long-term memory can occur [10].

Lesort et al. [11] presented possible learning strategies, opportunities and challenges for CL, but one of the major problems when dealing with developing CL architectures is the lack of datasets available to test them with. There is a huge amount of datasets to test regular object classification (popular examples include ImageNet [12] and CIFAR [13]), but none of them has the fundamental characteristics to test CL, such as rules dictating how many classes should be learned at a time, if new data from known classes should be presented later, and how frequently they should be tested. Lomonaco and Maltoni [9] proposed CORe50, a dataset and benchmark that is more suitable for testing CL methods when compared with the "usually" used image datasets for object classification. CORe50 takes a variety of factors into account, such as intermediate testing, the order of image capture, different levels of occlusion and illumination, the number of classes to be learned at a time and the use or non-use of new data for previously known classes. The same authors presented a leaderboard (CORe50 benchmark: https://bit.ly/3klHhhK, accessed on 15 October 2021), where they keep track of both published and unpublished results achieved on the CORe50 benchmark so that they can be easily compared with other CL strategies; however, these results are not necessarily up to date. She et al. [2] tested a variety of continual learning methods on this dataset in order to evaluate their performance in real-world environments and concluded that the currently employed algorithms are still far from ready to face such complex problems.

Maltoni and Lomonaco [14] proposed AR1, a CL approach combining architectural and regularization strategies, where they were able to sequentially train complex models such as CaffeNet and GoogLeNet by limiting the detrimental effects of catastrophic forgetting. The reported results on CORe50 [9] and CIFAR-100 [13] showed that AR1 was able to outperform other models such as elastic weight consolidation (EWC) [15], learning without forgetting (LwF) [16] and synaptic intelligence (SI) [17].

Additionally, Delange et al. [8] studied a variety of CL methods and proposed a taxonomy, where they categorized 29 different methods. They came to the conclusion that iCaRL [18] had the best performance for replay-based methods, MAS [5] had the best performance for regularization-based methods, and PackNet [19] had the best performance for parameter isolation methods. However, each method had its own advantages and limitations, e.g., PackNet demonstrated the best accuracy, but even though it can learn a large number of tasks, there is a limit based on the size of the model. For more details about each method please see [8]. It is also important to mention the approach by Requeima et al. [20], which consists of a multi-task classification that uses conditional neural adaptive processes; while it might not be directly considered a CL method, it can be applied in CL settings.

van de Ven and Tolias [21] applied an array of CL methods to three different CL scenarios with increasing difficulties in order to compare their performance in different situations. In the first scenario, each model is aware of the identity of the task being performed, meaning that they can opt to use the best-suited components for the given task. In the second scenario, the task's identity is no longer known, but the models are only expected to solve the task and not necessarily identify its nature. In the third scenario, the model must be capable of not only solving a given task, but also identifying it.

Parisi et al. [3] explored three different continual learning approaches (they referred to it as lifelong learning). The first approach is based on the retraining of the entire network using regularization, meaning that catastrophic forgetting is dealt with by having constraints applied to the training of the neural network's weights. The second approach trains specific sections of the network and expands when necessary, operating as a kind of dynamic architecture, adding new neurons which are dedicated to newly learned information. The third approach is made up of methods that model complementary learning systems for memory consolidation, where they make a distinction between memorizing and learning.

Pellegrini et al. [22] presented latent replay for real-time CL. The authors make a separation between low-level feature extraction and high-level feature extraction. They also control the rate at which each level is trained so that, when training on new information, the low-level feature extraction is modified only slightly or not at all. This strategy enables them to store intermediate data to be re-used alongside new data when new classes are being learned, avoiding the need to re-perform the low-level feature extraction on the stored data. Very recently, from 2021, two CL surveys were made available [8,23] that present overviews of this subject.

To summarize, while there may be references to CL as far back as the 1990s, where ideas and concepts are discussed, almost all the research with practical tests and applications is very recent. The fact that a large number of the research papers employ different terms for CL, and the fact that standardized categories are absent for CL methods or consistent descriptions of CL are indications of just how "new" this area of study is when it comes to practical applications. The state of the art shows, especially in recent studies, that there are various CL approaches with a lot of differences between them. They vary so much that it is hard to find a scenario where they could all be tested equally. Nevertheless, some of the mentioned papers share similarities with the proposed architecture, and those are discussed throughout the paper.

3. Modular Dynamic Neural Network

Throughout the literature, the main goals for building CL architectures are very similar: to solve the catastrophic forgetting problem while at the same time reducing memory consumption and allowing for scalability. In the present framework, those goals are considered to be of great importance, but for a CL method to be functional and applicable to real-world problems, avoiding catastrophic forgetting is considered to be the most important objective to follow (the same happens in the majority of the papers that deal with CL). With the elimination of catastrophic forgetting being the main focus (and to consider memory and network size constraints as secondary goals), the presented approach prioritizes accuracy.

The modular dynamic neural network (MDNN) architecture is comprised of modular sub-networks that, as it learns continuously, continuously grows and re-arranges itself. It is structured in such a way that the modules function independently of one another. This configuration means that, as new classes are learned, only certain sub-networks are modified, making it so that old knowledge is retained. The network is divided into two main components: (a) feature extraction and (b) modular dynamic classification.

Figure 1 shows a global flowchart of the MDNN training and production processes, which are later explained in detail in the present section, where it can be seen that the feature extraction component is general to both training and "production" (inference) and how, based on decisions made by algorithms that are discussed in the paper, some recursive processes can occur.

For the first component of the network, (a) feature extraction, a pre-trained (ImageNet) ResNet50 [1] was used as a backbone, which is known to provide excellent results for image classification; see [24]. This component is only used for extracting generic "low-level" features, while smaller and more basic (modular) networks exist for extracting class-specific features in the dynamic classification component. The feature extraction part of the presented architecture is the only component that is never altered when new classes are being learnedbecause the modules in the next component rely on it to be consistent, so if the feature extraction part changed, they would be invalidated.

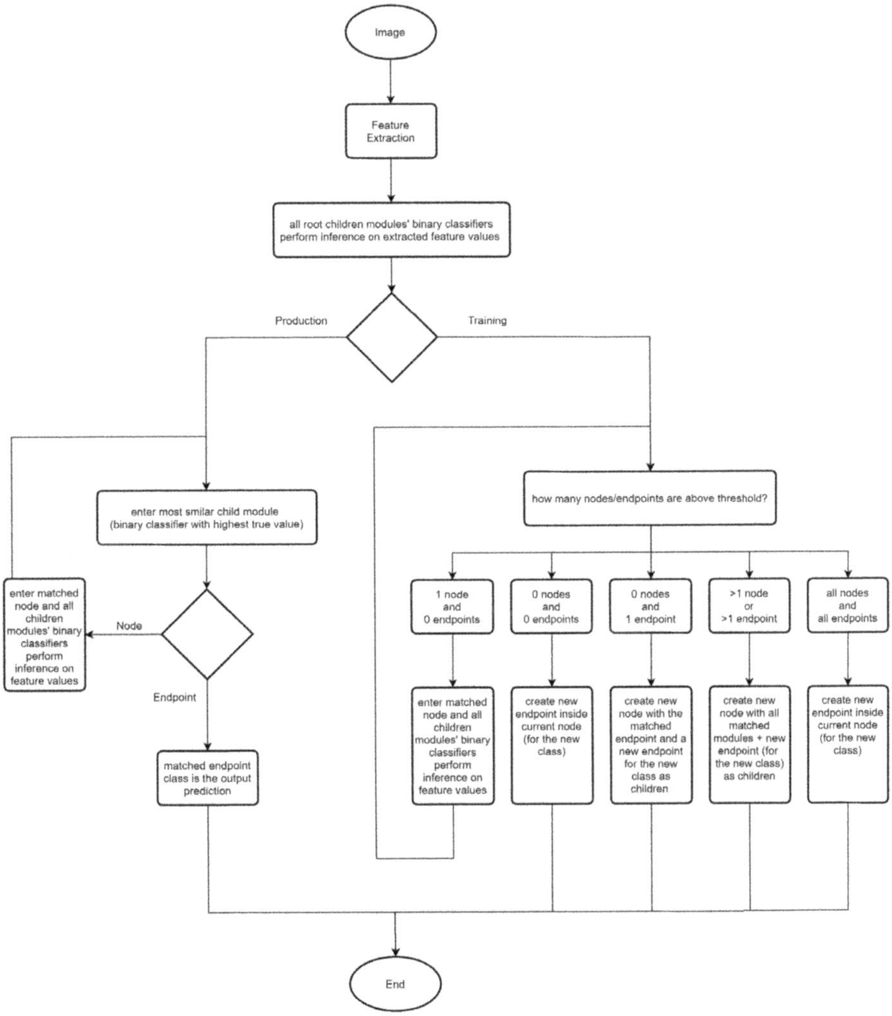

Figure 1. A global flowchart of the MDNN architecture (see the text for a detailed explanation).

The second component of the network, the (b) modular dynamic classification component, is made up of numerous small modules. These modules serve the purpose of classifying specific classes or groups of classes that, as new ones are learned, are automatically split into groups of modules and sub-modules based on their class's similarities. This grouping of classes can also be done manually if desired, but the standard operation of the network is to place them automatically based on the algorithms explained in Section 3.2. Figure 2 demonstrates how the modules fit into the proposed architecture: the modules with information inside brackets contain their own sub-modules (e.g., $[X_1, X_2, \ldots, X_{n_x}]$) or groups of sub-modules in the case of nested brackets (e.g., $[[X_1, X_2, \ldots, X_{n_x}], \ldots]$), where X_1 to X_{n_x} are modules with no children and n_x represents the number of sub-modules belonging to their parent. The modules containing their own sub-modules are designated as *node* modules and they have one binary classifier per direct child module. The modules with no children are named *endpoint* modules and contain nothing but feature data obtained during the training process.

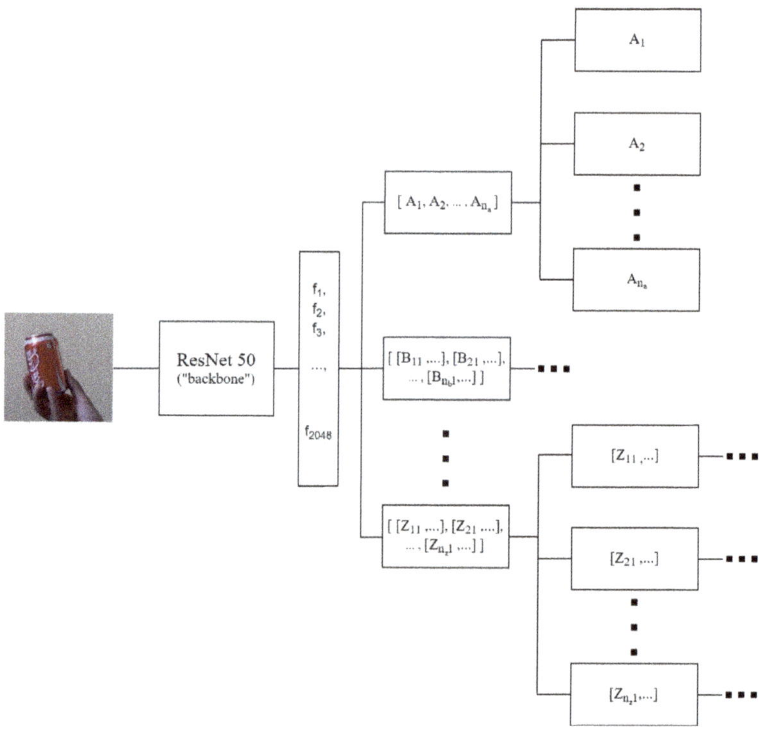

Figure 2. Generic representation of the modular dynamic neural network architecture.

As the network learns new classes, the extracted feature values are stored within the *endpoint* modules for future use, specifically for when some binary classifiers have to be retrained in order to avoid confusion with a new class. It is crucial to emphasize that the data preserved from each class are not the input data in their original format ("image"), but the resulting feature values obtained by the feature extraction component, which have smaller dimensions. In the case of the ResNet50, after being re-sized using nearest-neighbor interpolation, each sample image with a dimension of 224 × 224 pixels (px) and 3 color channels (224 × 224 × 3) is reduced to a 1 × 2048 vector (represented in Figure 2 as $f_1, f_2, \ldots, f_{2048}$).

The module-based structure allows the network to grow dynamically while still being able to cope with the scalability-related issues that usually accompany dynamic approaches. This is possible because the modular structure makes it possible to make classifications without needing to use the entire network (see Section 3.1). In addition, the fact that certain parts of the network are able to work independently from one another means that it is possible to add new modules or modify existing ones without affecting others. In short, the modularity of the networks' sub-sections and each one being responsible for a given class or group of classes allow for parts of the networks' knowledge base to be safely added, removed or altered without affecting the rest.

Finally, it is important to mention that other viable options could substitute the chosen feature extraction method (ResNet50), such as VGG16, Inception, or EfficientNet [25–27]. In future work, tests will be done using feature extraction sections from different networks to compare them to each other. In the following sections, the modular dynamic classification component is presented in detail, as well as how the network is trained and how the stored data are used.

3.1. Modular Dynamic Classification

As mentioned, there are two different types of modules present in the presented arquitecture: (a) *endpoints*, which are responsible for storing feature data extracted from a single class during training, and (b) *nodes*, which contain references to two or more sub-modules along with a binary classifier (BC) for each one. Each sub-module can either be an *endpoint* or a *node*. The top part of Figure 3 shows the difference between *nodes* and *endpoints* with a simple example of a potential network structure with three classes.

Each *node* module has its own set of binary classifiers, which, in this case, are simple NN with two output values: a certainty value for "true" and a certainty value for "false" (both between 0 and 1), where the definition of what is true or false depends on the module in question and its location in the network. Each binary classifier represents one of a *nodes'* sub-modules. The binary classifiers that represent *endpoints* define "true" as the only class they are responsible for, and "false" as the classes present in all the other modules and sub-modules in parallel with that *endpoint*. The binary classifiers representing *nodes* consider all of their sub-module classes as "true" and all the classes present in the other parallel modules and sub-modules as "false". During the classification process, these binary classifiers are used to select which sub-module to continue with and end up defining a type of path that eventually leads to a final prediction. During the training process, they are used similarly to determine the best positions in which to "insert" new modules (see Section 3.2).

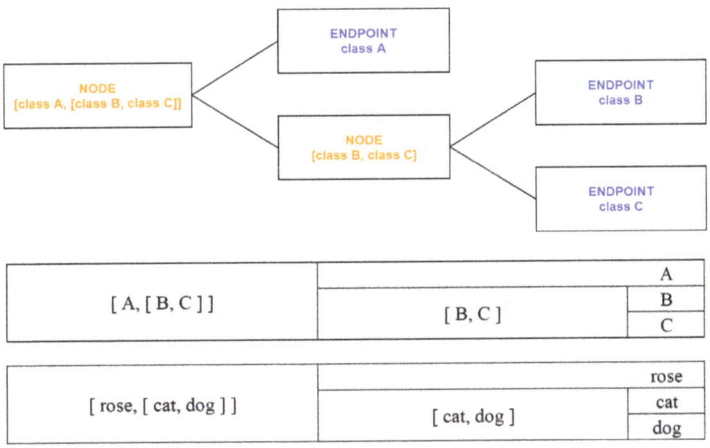

Figure 3. Illustration of the difference between *node* modules and *endpoint* modules, using a demonstration of a potential network structure with three classes. The two bottom rows show a different representation of the same example used in the top row, including an analogy with real-world objects.

Going back to Figure 3, the bottom two rows show the same example as shown at the top but using a different representation, with the last one including an analogy with real-world objects. Classes B and C are joined as children of a *node* (node [B, C]) and class A is on its own. This means that when this network was constructed, the algorithms in the training process calculated that classes B and C were similar to each other, grouped them together and trained new binary classifiers to distinguish between A and B ∪ C. Afterwards, a pair of sub-classifiers were trained to specialize in differentiating between classes B and C. To clarify, in this example, there are 4 binary classifiers present: A (against [B, C]), [B, C] (against A), B (against C), and C (against B).

Going into further detail, each binary classifier has the same structure, but naturally, different trained weights. In other words, the number of neurons per layer, the number of layers, the connections, the activation functions, etc., are all the same for every BC; the only difference between them is their trained weights. The main reason for this is to ensure consistency and provide all the BCs with an equal chance of success and avoid "favoritism". This is important because both the classification and training processes have moments where comparisons are made between the predictions of the different classifiers.

The composition of the binary classifiers consists of six fully connected layers with the following numbers of neurons, 128, 64, 64, 32, 32 and 2, respectively, with rectified linear activation functions (ReLU), except for the last layer. The last layer's activation function is a Sigmoid function. This is because the algorithms that use the output of these classifiers only make use of the true values, so there is no reason to let the false values interfere with the true values, which is what would happen in the case of an activation function, e.g., SoftMax. The architecture of the binary classifier, including the numbers of layers and neurons per layer, are determined empirically. In the future, this classifier will be the object of further studies in a way to improve the scalability of the network and save memory. It is important to justify at this point, the reason behind having a neuron representing the false value when it is not used. The reason is quite trivial: it is much more straightforward to implement and train a binary classifier this way, as this allows for the use of standard backpropagation for the true and false classes as if they were two normal classes. Nevertheless, in the future, the false values will be taken into account and also become part of the decision-making algorithms.

Figure 4 shows a more detailed view of the MDNN architecture (complementing Figure 2): the left side shows the ResNet50 [28] feature extraction, with the 2048 vector of extracted features, and the right side shows the network's flexibility and how the number of sub-modules that a *node* can have is not limited, including sub-modules of sub-modules. The figure also shows the use of the maximum value for deciding which sub-module should be used to continue the process. For the proposed network to meet its purpose of distinguishing between classes, the minimum number of classes for initialization is two. Therefore, the root module will always be a *node*, as *endpoints* only ever represent one class and *nodes* are the only modules that can contain more modules.

It is worth remembering that whenever a binary classifier is called upon, no matter its position in the network, the input values will always be the same feature values (2048 length vector) which are extracted once from the data sample being classified. The positions of modules within other modules are symbolic and serve only to decide if and which other modules are to be used, but the extracted feature data are never altered. This is important to clarify because, when visualizing tree-like structures that are based on NN, one might easily mistake the presented architecture for a single network with a tree format, similar to what is shown in [29], where everything is backwards dependent. Therefore, it must be emphasized that the "tree" serves only as a representation of the order in which things are done and which data are used by each binary classifier. The main takeaway from this is that the feature data are not altered intermediately, i.e., the input data for the last modules are the same as those for the first modules.

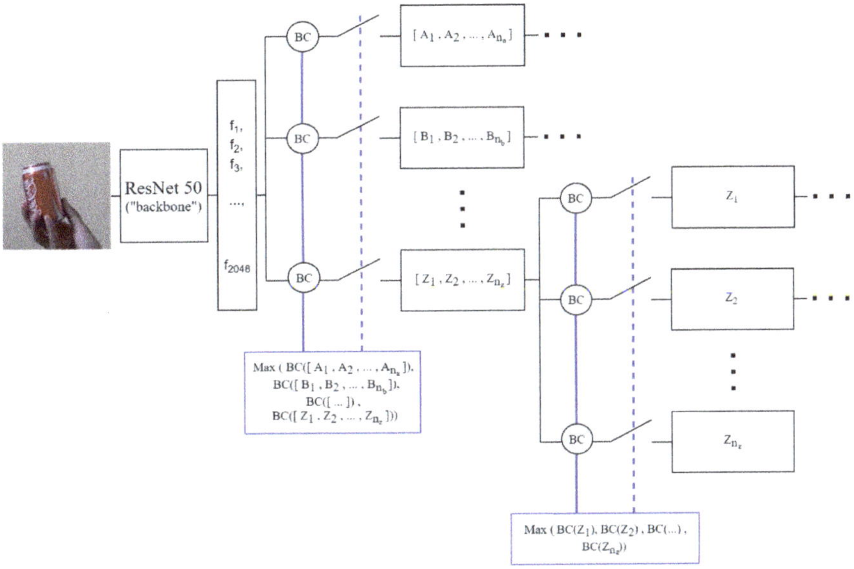

Figure 4. MDNN architecture: on the left, the ResNet50-based feature extraction component [28], with the 2048 vector of extracted features; on the right, the modular dynamic classification component.

The nesting of modules within modules means that the classification process can be recursive. This is because, when applied to a *node*, the results dictate whether it should be re-applied to another sub-*node*, i.e., the classification process is (i) initially applied to the network's root *node* and then, potentially, (ii) recursively applied to more *nodes* depending on the intermediate results.

So, to make a classification, the first step is for all the binary classifiers in the current *node* to process the extracted features and then for all their outputs to be evaluated. Keeping in mind that these binary classifiers indicate the certainty that the input sample data belong to the module they represent, the application of the input sample's extracted feature values to each of the binary classifiers (displayed in Figure 4 as BC, with $BC(X) \in [0, 1]$), results ina group of values between 0 and 1 (because of the final layer's sigmoid activation function) that represent the likelihood of the input sample belonging to each of the modules in the *node* being analyzed. Then, the highest value from these results is sought out, which indicates which module is the most likely to include the correct class. This selected module can then either be an *endpoint* or a *node*, where each one has a different next step. If the chosen module is of the *endpoint* type, then the classification process ends here, and the output prediction is the class represented by that *endpoint*. If, on the other hand, the chosen module is of the *node* type, then it is necessary to enter that *node* and keep repeating this process until eventually an *endpoint* module is selected, resulting in a final classification.

This implementation plays a big role in one of the mentioned objectives, which is to avoid or at least attempt to minimize scalability issues. The main reason this implementation contributes toward this is because only the binary classifiers from the top-scoring *sub-nodes* are used, meaning that, for many cases, classifications are made using only a small percentage of the overall network. This means that, as the network grows, the classification speed is only slowed down for classes that get grouped with lots of other similar classes because they require a bit more time to be distinguished between them; for non-similar classes, only a very small fraction of time is required to make distinctions between them. In future work, the optimization of the classification process will be explored.

The next section is fundamental in understanding the architecture and explains how the network is trained (the second component; we stress that the first component, feature extraction, always remains unchanged—frozen), as it is not similar to the most "traditional" networks.

3.2. Modular Dynamic Training

The process for the addition of a new class starts (as usual) with the network being fed a set of data samples along with a label. With this, the network will (i) process the data samples, (ii) decide where best to place the new *endpoint*, (iii) make any necessary adjustments to the existing modules, and (iv) train the necessary classifiers so that when the network is presented with data from the same class, in the future, it can identify them.

Following the scheme presented in Figure 3, the right side of Figure 5 shows an example of how the network changes as classes are added over time. Here, it is shown how it can classify three classes (shown in Figure 3) (A—rose, B—cat, and C—dog), where B and C are grouped together, and then another four classes are added sequentially(D—tree, E—fish, F—bush, and G—frog), making the network capable of classifying a total, in this example, of seven classes.

The positions in which new modules are placed within the network are not random. There are several procedures behind the calculation of the optimal position for a new module (some of which can be recursive). Learning a new class essentially means that the network can distinguish the new class from the ones which were previously learned.

The most main goal of this class placement process is to avoid conflict or confusion between classes. In avoiding confusion between classes, the goal is to (i) group them by their similarities and then (ii) focus on their differences. This notion of grouping classes by similarities is referring to the parallel placement of modules within *nodes*, as shown in Figure 5, where, among other examples, in the network's initial state, class B is grouped in a node with class C.

Figure 5. The right side shows the network growing progressively as modules are added and grouped. The modules highlighted in green are new and the modules highlighted in blue are pre-existing ones that needed to be retrained either to reduce confusion or to ensure that the parent *nodes* became aware of the new module. The left side describes the steps taken for the addition of each new module (see text for details).

The binary classifiers are used to see if any existing classes share any resemblance with the class being added. The procedure for checking if classes are similar is relatively similar to the classification process, but instead of just searching for the most similar class, here, the goal is to search for any classes that could be partially similar, i.e., it looks for "somewhere" to place the new class.

Upon entering a *node*, there are n possible paths to follow, which can be a mix of *endpoints* and/or *nodes*. The idea here is that any *node* or *endpoint* binary classifiers that, without any alteration, might accidentally mistake the new class for their own should be grouped with that new class and retrained so that, in future, when presented with samples of the new class, they do not repeat the same mistake. To achieve this, it is necessary to first verify which *nodes* and or *endpoints* share any resemblance with the new class. Therefore, the first step here is to classify all the data samples (extracted feature values) of the new class with each of the binary classifiers in the current *node* and then use these results to decide how best to proceed.

After obtaining all the current *node's* classifier results from all the new data samples, the next step is to calculate the average confidence per classifier, which returns a single value for each *node/endpoint*, representing the likelihood of mistaking the new class with the existing class (or classes if it represents a *node*). It should be noted that the use of the average for this task was a natural choice, but in the future, other solutions will be investigated in order to attempt to achieve even better discrimination between classes.

Next, the obtained average values are analyzed to see which classes are the most similar to the new one. Each value is compared with an established threshold value, AV_c, which makes it possible for a final decision to be made on whether a class should be considered similar, where this threshold value dictates the necessary confidence value for two classes to be considered similar.

To establish this threshold value (AV_c), logic says it should not be too high because some similar classes could be missed, but should also not be too low so that only reasonably similar classes are considered. Different values between 0.1 and 0.5 were tested, and the best results were achieved with $AV_c = 0.3$. These tests were done with three different datasets or subsets of those datasets, namely Core50 [9], ImageNet [12] and CIFAR-10 [30]. The ideal value depends on how accurate the binary classifiers are and it essentially dictates how grouped or separated the final network is. Future work for this component includes the use of a dynamic threshold value that is re-calculated as the network expands.

After comparing the averages of the classifications of all the input samples with the threshold value (AV_c), it is known which, if any, modules are similar to the input class. This results in one of five possible outcomes in the function of how many similar *nodes* and/or *endpoints* are found (see also Figure 5 left side): (a) 0 similar modules (0 similar *nodes* and 0 similar *endpoints*); (b) 1 similar *node* and 0 similar *endpoints*; (c) 0 similar *nodes* and 1 similar *endpoint*; (d) more than 1 similar *node* and/or *endpoint* but not all, and (e) all *nodes* and *endpoints* are similar. Each outcome is dealt with differently, as detailed next for each case:

(a) **0 similar modules (0 similar *nodes* and 0 similar *endpoints*)**: Place a new *endpoint* in the current *node* for the new class. Train the classifier for the new *endpoint* with true data as data from the new class and false data as a balanced distribution of data (see Section 3.3 for the explanation) from the other *endpoints* and *nodes* present within in the current *node*.

(b) **1 similar *node* and 0 similar *endpoints***: Enter the similar *node* and repeat the positioning process on its children.

(c) **0 similar *nodes* and 1 similar *endpoint***: Create a new *node* and place a new *endpoint* inside it for the new class as well as the *endpoint* that was matched with the new class. Train the classifier for the new *endpoint* with true data as data from the new class and false data as data from the class it was matched with (see Section 3.3). Retrain the classifier for the pre-existing *endpoint* that was moved into the new *node* with true data as the data it has stored and false data as data from the new class.

(d) **More than 1 similar *node* and/or *endpoint***: Create a new *node* and place inside it a new *endpoint* for the new class as well as all the *endpoints/nodes* that were matched with the new class. Train the classifier for the new *endpoint* with true data as data from the new class and false data as a balanced distribution of the data from all the other *nodes/endpoints* it was matched with (see Section 3.3). Retrain the classifiers for all the pre-existing *nodes* and *endpoints* which were moved into the new *node* using balanced distributions of their own data as true data and balanced distributions of all their sibling's data as false data. The child modules of the *nodes* that were moved into the new *node* do not need to be touched, as they are not dependent on their parents and only relate to each other.

(e) **All *nodes* and *endpoints* are similar**: Place a new *endpoint* in the current *node* for the new class. Train the classifier for the new *endpoint* with true data as data from the new class and false data as a balanced distribution of data from the other endpoints and *nodes* present in the current *node*. Retrain the classifiers for all the pre-existing *nodes* and *endpoints* in the current *node* using balanced distributions of their data as true data and balanced distributions of all their sibling's data as false data.

Out of these processes, case (b) is the only one that does not involve the placement of the new class or the training of any networks, but it repeats the entire process applied to a selected sub-*node* and, eventually, there is a point where there are no more sub-*nodes* and case (b) is no longer be a viable option; therefore, the new class is guaranteed to be placed somewhere in the network eventually.

After the new module is placed within the network, the parent *node* has to be re-trained so that it considers the new class as true, thereby increasing the chance of success for the classification process in production. This is because it increases the probability of future samples of the new class reaching the correct *endpoint*. The retraining of parent *nodes* must be recursive, i.e., the process must then also be applied to the parent of the parent until the root *node* of the network is reached. This process helps to ensure that the initial *nodes* are more likely to send samples of the new class down the correct path during production.

It is important to stress that the use of the same classifiers during the training process as those used during the classification of new data immediately makes huge improvements to the overall classification process. This is because with this strategy, it is possible to predict and correct the modules most likely to cause errors related to the addition of the new class before they are even a problem. The next section is dedicated to explaining how the data are balanced when train or re-training the binary classifiers.

3.3. Balanced Training Data

Research shows that neural networks have higher success rates when trained with balanced data [31], meaning that they should be trained with a similar number of samples per class. As this architecture deals with binary classifiers (true or false), it should aim to use an equal number of true and false samples, leaving us with the problem of deciding what to do when there is a different number of each.

The algorithm implemented for the selection of training samples starts by calculating the maximum number of samples per class that can maintain an ideal distribution based on where they are positioned in the network. In this situation, where various groups of classes have their own sub-groups, an ideal distribution does not mean using an equal number of samples per class. It means the same number of samples should be used from each of a *nodes'* children, i.e., if one of these children is an *endpoint* and one is a *node*, then the same number of samples should be used from each, where the *nodes'* samples are a mix of their children's samples and so on.

For example, recalling Figure 3, if a classifier were trained to recognize an *endpoint* A as true and a *node* [B, C] as false, and A, B and C all had m samples each, the balanced distribution of samples for this classifier would be m samples of A and m samples of [B, C]. Then, to continue aiming for an equal distribution, the m samples of [B, C] would consist of $m/2$ samples of B and $m/2$ samples of C.

This example makes the problem appear straightforward, but with different numbers of samples per class and a more complex network structure (more known classes and more nested *nodes* on the true and false sides), an algorithm is needed for calculating the ideal distribution in any situation and making use of as many samples as possible. Using this equal distribution means classes with fewer samples will reduce the number of samples that can be used from other classes with more samples. So, to minimize this effect, it is recommended to establish a minimum number of samples (MNS) for when classes are added to the network. The currently applied MNS value was determined empirically and was set to 175; nevertheless, when using the MDNN architecture in a real-world situation, there is no knowledge of how many samples will be available per new class, and so a minimum value must be defined. Using MNS = 175 presented good results over the various tests performedand, during intermediate tests with the three datasets mentioned before, increased MNS values provided better results. Data augmentation would normally be a solution for increasing the amount of data available for the classes with fewer samples, but as the stored data consist of feature vectors and not raw images (as mentioned), the produced samples would not be reliable.

There are two main steps involved in computing the maximum number of samples to obtain a balanced training set: (i) calculate the largest number of samples that allows for the use of the distribution explained above and (ii) recursively divide this number by the number of children in a *node* until all *endpoints* are reached.

For the first step, (i), the number of sub-modules in the *node* is multiplied by the number of samples of the sub-module with the largest number of samples. The numbers to be calculated for the sub-modules which are also *nodes* are calculated in the same way. This means that the algorithm is applied recursively until all the sub-modules belonging to the *node* in question have been calculated and the *node* in question itself has also been calculated. When this process is complete, it results in the maximum number of samples that allows for an even distribution of samples for the *node* in question.

So, with M_p representing the maximum possible number of samples that can allow for an equal distribution for the *node* being calculated, $M_{p,i}$ represents the maximum number of samples that can be used by one of the *node's* sub-modules, with p representing the path created by the indexations that lead to the location of the *node* in question and i being the index of a sub-module. For example, in the final state of the network shown in Figure 5, the path required to reach the *node* "[cat, dog]" would be the second sub-module of the root *node* and then the first sub-module of that *node*. Meaning that for the *node* "[cat, dog]", the indexations represented by p would be "2,1", and the number of samples present in the endpoint "dog" would be "$M_{2,1,2}$". The value of n_p represents the number of sub-modules present in the *node* being calculated, and M_p the number of samples which is computed (sometimes recursively) as follows:

$$M_p = n_p \times \max(M_{p,1}, M_{p,2}, \ldots, M_{p,n_p}).$$

The second step (ii) is to keep dividing M_p equally between the children in *nodes* until all the *endpoints* are reached, resulting in the numbers of samples to use per class. For that, the value obtained from the first step is used, M_p (the maximum possible number of samples that allows for an even distribution), and is progressively divided throughout the network. The *node* being calculated distributes its number evenly between its children. The children that are also *nodes* then do the same thing with their values to their children, resulting in sub-divided values. This is repeated until all the sub-modules of the initial *node* are reached and eventually results in a final number of samples to be used for each class (E_p). So, with the value of M_p calculated in part one, it is possible to compute $E_{p,i}$, which is essentially the value of M_p evenly distributed between each of the node's sub-modules.

While in step (i) the parent's values depended on the the children's values, in step (ii) the children's values depend on the parent's values. To initialize this process, the value for the *node* being calculated is set with $E_p = M_p$ and then its sub-modules values are calculated as $E_{p,1} = E_{p,2} = \cdots = E_{p,n_p} = E_p/n_p$. Afterwards, each sub-modules' values are calculated the same way until all the *endpoints* that are descendants of the *node* being calculated are reached, which eventually results in a series of final numbers of samples to be used from each *endpoint*.

4. Tests and Results

One of the most common image classification datasets is ImageNet [12]. In the literature, numerous benchmarks use this dataset, and others like it, where the networks/methods are trained to learn all the classes at once. Although their leaderboards show outstanding accuracy in their results, these are not the datasets and benchmarks used to validate CL methods. As mentioned, to validate continual learning architectures, we need to feed images (data/classes) to the network incrementally. To properly test a networks' capacity for learning continually, a dataset and benchmark with pre-established CL-related rules and guidelines is needed; one of the most used datasets of this type is CORe50 [9].

The CORe50 dataset consists of 50 classes (10 categories with 5 classes each). The 10 categories are as follows: plug adapters, mobile phones, scissors, light bulbs, cans, glasses, balls, markers, cups and remote controls. Figure 6 shows an example of each class and how they are divided into the mentioned categories. The data were collected over 11 distinct sessions (8 indoor and 3 outdoor) characterized by different backgrounds and lighting. For each session and each object, a 15-s video (at 20 fps) was recorded using a Kinect 2.0 sensor producing 300 RGB-D frames. The objects were handheld by the operator, and the camera's point-of-view is that of the operator's eyes. The full dataset consists of 164,866 128 × 128 RGB-D images obtained from 11 sessions × 50 objects × (around 300) frames per session. More details can be seen in [9].

Figure 6. Example images of the 50 classes present in the CORe50 dataset (one image of each class). The 10 columns correspond to the 10 categories in which the classes are divided (5 classes per category).

In summary, each class has around 2398 training images and 900 test images which are split into various test/train batches (the number of batches and their contents depend on the CL scenario being considered). Here, batches represent sets of data (images) that are fed to the network in blocks [9]. The dataset considers three continual learning scenarios: (a) new instances (NI), where all the classes are learned in the first batch and then new data for each class are added over the following batches; (b) new classes (NC), where the first batch includes 10 classes and then the following batches introduce 5 classes at a time until reaching the total 50; and (c) new instances and classes (NIC), where both new instances and classes are presented over time.

We opted to test our framework on the NC scenario, as its implementation is the one most comparable with the paper's objectives. However, of course, we do plan in the future to implement the use of new instances to improve knowledge of already known classes, which will allow us to show results on NI and NIC.

As mentioned, the NC test begins with learning 10 classes and then continuing to learn more classes in groups of 5 until reaching a total of 50; this makes for 9 total steps. The first batch (the 10 initial classes) consists of one class from each of the 10 categories, and the rest of the batches consist of 5 random classes each (different for each run). We executed 30 runs of this test using the algorithms and parameter values described throughout this paper, where the feature extractor (ResNet50) maintained all its default parameters and pre-trained weights, the binary classifiers used the layer structure mentioned in Section 3.1, and the minimum number of samples (MNS), which we typically set to 175, was not needed, as each class had a relatively large number of samples (approximately 2398 as mentioned above). The hyperparameters for the binary classifiers were as follows: RMSprop [32] as the optimizer, the loss was set to sparse categorical cross-entropy, the metrics were set to only sparse categorical accuracy, 4 epochs, and validation split equal to 10%. The 30 runs were all that were executed, i.e., there was no preferential selection (cherry-picking). The tests were made on a 64-bit 3.20 GHz i5-3470 CPU accompanied by 8 GB RAM, and each run took between 12 h and 13 h (approximately 15 min per class). In the future, the whole process will be converted to be processed by GPUs, where it is expected that by using state-of-the-art GPUs, the framework will be able to learn new classes at a much faster rate, potentially approaching real time.

The graphs shown in Figure 7, left and right, are adapted from [9] and show the MDNN results obtained from the tests mentioned in the previous paragraph (black line) compared with three other methods. The copy weight with reinit (CWR) results (in green) are from the CL method presented in [9]; naive (red line) simply continues training as batches become available; and cumulative (blue line) uses the current batch and all previous batches to demonstrate the potential accuracy of a network without CL restrictions and behaves as a kind of upper bound, but not quite, as Lomonaco and Maltoni [9] state that "in principle a smart sequential training approach could outperform a baseline cumulative training". Going back to the figure, the results we are comparing MDNN with on the left are based on a CaffeNet [33] and the ones on the right are based on a VGG-CNN-M model [34], but our results are the same on both graphs (the presented MDNN network was not based on a CaffeNet or a VGG, but simply structured as discussed in the paper).

When it comes to test data, Lomonaco and Maltoni [9] considered what they refer to as a "full test set" where, at any moment during the continual learning process, the tests include all the classes from the dataset, so with that in mind, as the total dataset includes 50 classes and is initialized with 10 classes, the maximum possible accuracy is 20% and not 100%. So, where the x-axis in the graphs in Figure 7 shows the 9 steps mentioned above and the y-axis represents the test accuracy at that point, after training on the first set of classes (step 1–10 classes) all the methods hover around 20%, which means that, based on the data available, they are all performing at close to 100% accuracy. The results shown in the last step (step 9–50 classes) are the accuracies achieved after adding 5 classes at a time and reaching 50 total classes. Here, when using the full test set (all 50 classes), as the networks were presented with all the training data, the accuracy values on this last step are a "normal" accuracy representation, where MDNN finished with 69.8% accuracy, and CWR ended just over 40% accuracy when based on a CaffeNet and just under 40% accuracy when based on a VGG. The cumulative results are similar to MDNN in both instances, showing that, in some cases, the presented network is capable of keeping up with (and outperforming) a method with no CL restrictions, i.e., demonstrating that based on the quote from before, "in principle a smart sequential training approach could outperform a baseline cumulative training", and the presented approach can be considered "smart".

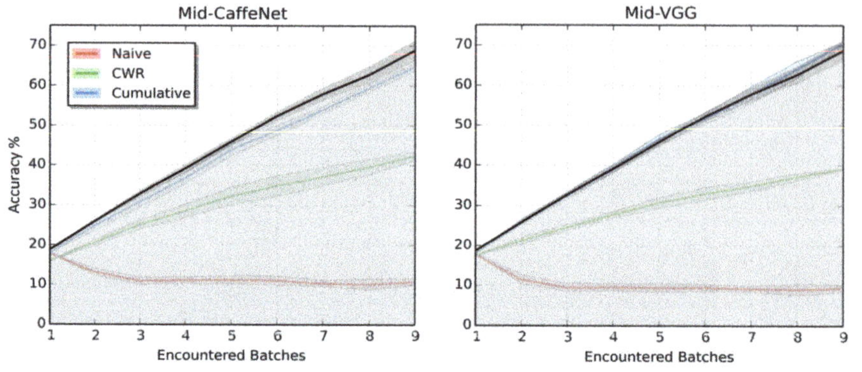

Figure 7. Average accuracy and standard deviation from 30 runs compared with CaffeNet-based approaches (**left**) and VGG-based approaches (**right**). Adapted from [9].

The left graph in Figure 8 shows the average accuracy and standard deviation for each batch over time, where the tests at each batch are only of the classes supposedly learned, meaning that, in the first batch where 10 classes are learned, the network is only tested on its capacity to recognize those same 10 classes. In the same figure, the graph on the right shows the same results but considering the tests always being made with all classes (so when only 10 classes of the total 50 are known, the best possible achievable accuracy is 20%). The results are shown in both formats because the former (left) is how we think the results are the most understandable and gives a more clear idea of the networks' overall performance over time, while the latter (right) is how the results were demonstrated in [9], allowing us to make a more direct comparison.

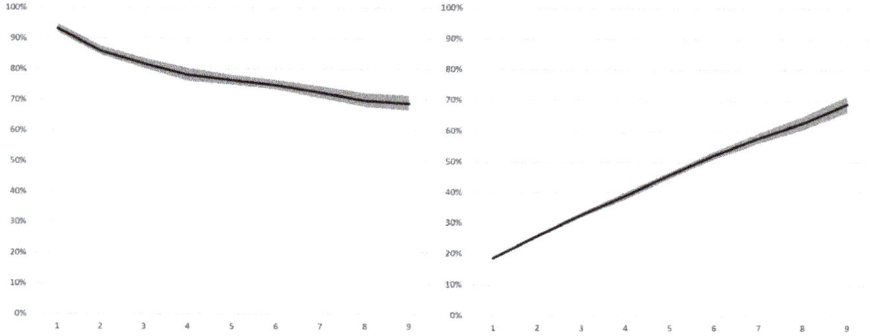

Figure 8. On the left: the average accuracy and standard deviation from 30 runs (only testing with learned classes). On the right: the average accuracy and standard deviation from 30 runs (always testing with all classes).

The table on the left in Table 1 gives a more detailed insight into the performance of the network by showing the individual results from each run and their accuracies over time after each set of 5 classes was added. The initial accuracy results from each run where 10 classes are learned sit between 84.4% and 96.5%, and the final accuracy results after all 50 classes are learned incrementally sit between 57.3% and 76.2%. It is also interesting to note that in some cases, the accuracy can actually be seen to increase when more classes are added (e.g., run 8 going from 35 to 40 known classes, there is a 7.4% increase in accuracy). Additionally, in Table 1, the table on the right showcases all the increases and decreases in accuracy between each batch by using the same row/column representation but instead of showing the accuracy at each step, it shows the increase/decrease (delta) between the current accuracy and that obtained in the previous batch. Increases in accuracy are highlighted in green and decreases in accuracy are highlighted in red.

Table 1. On the left, a table of the average test results obtained by each run after learning each new set of classes, where each row is a run, each column is an increment of 5 new classes and the percentage values within the cells are the accuracy values at each point. On the right, differential results between the accuracy results in the columns are shown on the left where increases in accuracy are green and decreases in accuracy are red.

Run number	Number of Learned Classes								
	10	15	20	25	30	35	40	45	50
1	93.9%	87.5%	82.3%	79.2%	76.9%	76.1%	74.0%	74.5%	72.7%
2	95.4%	89.4%	82.6%	79.5%	76.4%	76.6%	76.2%	72.6%	71.1%
3	94.7%	84.6%	83.5%	78.8%	73.8%	76.8%	76.7%	70.7%	71.8%
4	94.0%	84.3%	78.0%	72.6%	72.5%	67.4%	64.8%	61.6%	62.9%
5	94.6%	86.9%	81.3%	78.4%	75.9%	73.8%	70.0%	71.4%	71.9%
6	93.7%	88.3%	83.0%	79.4%	75.6%	75.4%	74.2%	72.4%	69.3%
7	96.4%	89.8%	83.2%	83.7%	77.1%	73.3%	62.7%	60.8%	63.0%
8	90.9%	83.0%	78.3%	72.0%	68.6%	67.5%	74.9%	71.1%	57.3%
9	91.4%	86.6%	79.5%	65.5%	68.2%	73.8%	73.0%	69.9%	72.1%
10	93.9%	90.2%	88.0%	81.9%	78.9%	80.0%	66.9%	63.5%	63.5%
11	95.6%	86.9%	80.5%	80.6%	77.1%	76.6%	73.5%	73.8%	71.5%
12	85.7%	76.7%	73.8%	74.6%	76.9%	74.0%	72.9%	74.5%	76.2%
13	86.1%	83.0%	78.0%	81.6%	78.5%	76.4%	77.9%	77.0%	76.0%
14	90.7%	85.7%	81.1%	78.2%	77.9%	71.2%	63.6%	60.9%	59.8%
15	84.4%	87.1%	85.6%	83.8%	82.0%	75.4%	74.8%	72.9%	71.4%
16	94.3%	83.4%	81.3%	72.1%	71.7%	68.1%	73.3%	60.7%	60.3%
17	95.4%	82.7%	84.0%	80.5%	78.9%	76.7%	72.2%	70.1%	71.1%
18	90.1%	91.0%	87.6%	85.7%	80.5%	79.3%	72.4%	71.5%	69.1%
19	94.4%	88.7%	88.9%	83.7%	82.7%	79.0%	72.8%	66.6%	68.8%
20	96.0%	81.2%	76.0%	72.9%	78.2%	72.8%	70.8%	67.2%	67.7%
21	93.3%	80.7%	80.0%	79.1%	77.1%	78.4%	76.2%	70.3%	75.1%
22	95.4%	86.5%	83.5%	80.0%	77.8%	79.2%	78.3%	75.2%	74.3%
23	93.7%	84.5%	80.1%	77.0%	72.8%	72.9%	62.9%	59.8%	57.3%
24	95.6%	88.6%	83.4%	71.1%	75.6%	73.0%	70.0%	70.9%	71.2%
25	96.0%	91.2%	83.3%	81.9%	76.4%	75.7%	74.4%	71.8%	70.9%
26	94.6%	87.1%	84.9%	83.7%	81.8%	80.8%	75.3%	74.2%	73.1%
27	94.6%	87.8%	77.6%	76.4%	74.9%	75.1%	75.6%	72.1%	65.6%
28	92.8%	85.6%	84.2%	79.2%	73.4%	69.7%	70.0%	70.7%	69.0%
29	95.9%	83.5%	75.4%	71.3%	71.1%	70.3%	71.2%	68.3%	68.7%
30	93.7%	87.7%	83.9%	77.0%	79.3%	76.1%	76.1%	73.5%	70.2%

Run number	Number of Learned Classes									
	10	15	20	25	30	35	40	45	50	
1	93.9%	−6.5%	−5.2%	−3.1%	−2.2%	−0.9%	−2.1%	+0.5%	−1.8%	72.7%
2	95.4%	−6.1%	−6.8%	−3.1%	−3.1%	+0.3%	−0.4%	−3.6%	−1.5%	71.1%
3	94.7%	−10.0%	−1.1%	−4.7%	−5.1%	+3.0%	−0.1%	−6.0%	+1.1%	71.8%
4	94.0%	−9.8%	−6.3%	−5.4%	−0.1%	−5.1%	−2.6%	−3.3%	+1.3%	62.9%
5	94.6%	−7.6%	−5.6%	−2.9%	−2.5%	−2.1%	−3.8%	+1.4%	+0.5%	71.9%
6	93.7%	−5.4%	−5.3%	−3.6%	−3.8%	−0.2%	−1.3%	−1.7%	−3.1%	69.3%
7	96.4%	−6.6%	−6.6%	+0.5%	−6.6%	−3.8%	−10.6%	−1.9%	+2.2%	63.0%
8	90.9%	−7.9%	−4.7%	−6.3%	−3.4%	−1.1%	+7.4%	−3.8%	−13.8%	57.3%
9	91.4%	−4.8%	−7.1%	−14.0%	+2.7%	+5.6%	−0.8%	−3.1%	+2.2%	72.1%
10	93.9%	−3.6%	−2.2%	−6.1%	−3.0%	+1.1%	−13.0%	−3.5%	−0.0%	63.5%
11	95.6%	−8.7%	−6.4%	+0.0%	−3.5%	−0.5%	−3.1%	+0.2%	−2.2%	71.5%
12	85.7%	−9.0%	−2.9%	+0.8%	+2.4%	−3.0%	−1.1%	+1.6%	+1.7%	76.2%
13	86.1%	−3.1%	−5.0%	+3.6%	−3.1%	−2.1%	+1.5%	−0.9%	−1.0%	76.0%
14	90.7%	−5.1%	−4.5%	−2.9%	−0.4%	−6.6%	−7.6%	−2.7%	−1.1%	59.8%
15	84.4%	+2.7%	−1.5%	−1.8%	−1.8%	−6.6%	−0.5%	−1.9%	−1.5%	71.4%
16	94.3%	−10.9%	−2.1%	−9.2%	−0.4%	−3.6%	+5.2%	−12.6%	−0.4%	60.3%
17	95.4%	−12.7%	+1.4%	−3.6%	−1.5%	−2.2%	−4.5%	−2.1%	+0.9%	71.1%
18	90.1%	+0.9%	−3.4%	−1.9%	−5.2%	−1.2%	−6.9%	−0.9%	−2.4%	69.1%
19	94.4%	−5.7%	+0.2%	−5.2%	−1.0%	−3.8%	−6.2%	−6.2%	+2.2%	68.8%
20	96.0%	−14.8%	−5.2%	−3.1%	+5.3%	−5.4%	−2.0%	−3.6%	+0.4%	67.7%
21	93.3%	−12.6%	−0.7%	−0.9%	−2.0%	+1.3%	−2.2%	−6.0%	+4.8%	75.1%
22	95.4%	−8.9%	−3.0%	−3.5%	−2.1%	+1.3%	−0.8%	−3.1%	−1.0%	74.3%
23	93.7%	−9.2%	−4.4%	−3.1%	−4.2%	+0.1%	−10.0%	−3.1%	−2.5%	57.3%
24	95.6%	−7.0%	−5.2%	−12.3%	+4.6%	−2.7%	−3.0%	+0.9%	+0.3%	71.2%
25	96.0%	−4.8%	−8.0%	−1.4%	−5.5%	−0.7%	−1.3%	−2.7%	−0.8%	70.9%
26	94.6%	−7.6%	−2.1%	−1.2%	−1.9%	−1.0%	−5.5%	−1.0%	−1.1%	73.1%
27	94.6%	−6.9%	−10.2%	−1.2%	−1.5%	+0.2%	+0.5%	−3.6%	−6.5%	65.6%
28	92.8%	−7.2%	−1.4%	−5.0%	−5.8%	−3.7%	+0.3%	+0.7%	−1.7%	69.0%
29	95.9%	−12.4%	−8.1%	−4.1%	−0.2%	−0.8%	+0.9%	−2.9%	+0.4%	68.7%
30	93.7%	−5.9%	−3.9%	−6.8%	+2.3%	−3.3%	+0.0%	−2.6%	−3.3%	70.2%

Figure 9 also shows the MDNN accuracy (dashed black line) on the CORe50 dataset, but this time compared with GoogLeNet [35] using the following methods: cumulative (black line), naive (blue line), AR1 (turquoise line), CWR (purple line), CWR+ (pink line), LWF (green line), EWC (orange line), and SI (red line) (see [14]). The dataset, context, and method of testing are all identical to those discussed in relation to Figure 7. In the GoogLeNet comparison, our method no longer reaches as close to the cumulative result but is more in between AR1 and CWR+.

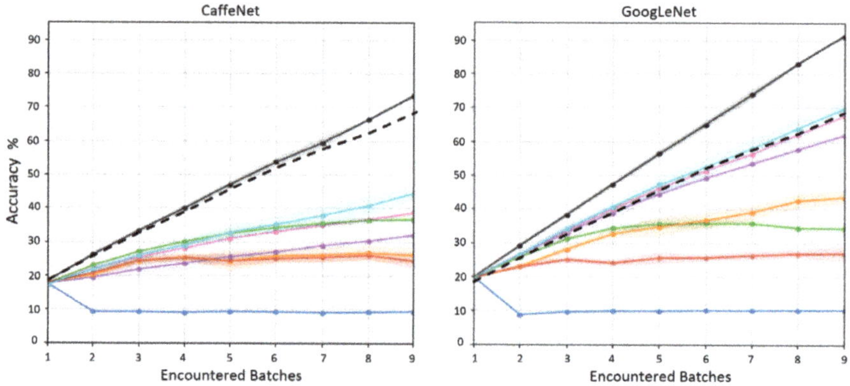

Figure 9. On the left: MDNN accuracy over time on CORe50 compared with cumulative, naive, AR1, CWR, CWR+, LWF, EWC and SI based on CaffeNet (adapted from [14]). On the right: the same as the left but the other methods (not MDNN) are based on GoogLeNet.

Before performing tests on CORe50, MDNN was also tested on ImageNet; although the dataset is not designed for testing CL methods, the test still provided some insights into the method's performance. The test, shown in [36], was made using 11 randomly selected classes. It consisted of the addition of one class at a time and evaluated the overall accuracy after the addition of each class. The test ended with 81% accuracy, where the average was brought down by two similar looking classes that generated confusion between each other (alyssum and astilbe). On top of that test, two additional classes (not from ImageNet) were added to the same network to measure the performance of the network when learning from multiple sources of data. The overall accuracy after the additional two classes were added was 82% and none of the pre-learned classes appeared to be negatively affected.

As the bottom line, a framework is presented that can learn new classes without the need to retrain its entire network of classifiers, with features that help it to cope with the scalability problems that generally come with expanding architectures. The test results on the CORe50 dataset [9] show accuracy results just under 70% and in the case of comparisons against VGG and CaffeNet based models, MDNN was even able to practically match and even surpass results obtained on the same data without CL restrictions. The tests show that the framework has the ability to learn classes incrementally and to maintain its accuracy on old classes as new ones are added, making sure they are not forgotten.

5. Conclusions and Future Work

This paper presents a modular dynamic neural network framework: a continual learning solution that allows new classes to be learned incrementally, growing and structuring itself as it learns while re-training only specific sub-modules as needed. This way, parts of the network are not altered, allowing them to retain their knowledge.

The idea of constantly adding new smaller neural networks to a global main network creates initial concerns for scalability, but the network is structured in such a way that different parts are used for different tasks, which considerably reduces the severity of this problem. In addition to this, the architecture functions in a completely different way compared to other CL methods. The most similar one is LR [22], where the main differences are that (i) their feature extraction component is not static (its alterations are sometimes slowed down during training), where for MDNN, that component is permanently fixed so that the MDNN binary classifiers maintain validity. (ii) When they train new classes, they re-use their stored samples from already known classes. This means that their high-level feature extraction is not affected too badly for old classes by the slow training of their low-level feature extraction. The presented architecture also re-uses samples of already known classes during training but only from particular classes in specific cases, where it depends on the location of the module in question. Finally, (iii) the classification part of their network is made up of one NN of fixed size, whereas the presented architecture is made up of multiple NNs with more being added progressively as needed.

In terms of future work, as mentioned, the framework is made up of many different components, which leaves us with several areas that can be improved in terms of performance and adaptability. The main six (future works) are presented in detail:

(a) *Scalability and memory usage.* With the elimination of catastrophic forgetting being the main focus (and the consideration of memory and network size constraints as secondary goals) priority is given to accuracy over scalability. While the presented approach does make efforts toward scalability and performance-related issues by using only the necessary sub-modules in both the training and classification processes, there is still more that can be done by attempting to reduce the dimension/quantity of the stored features and by attempting to reduce the algorithmic complexity of each module.

(b) *Feature extraction.* The currently implemented feature extractor (ResNet50), while very performant, is not the only option for this task. As mentioned, there are other viable solutions that could substitute it, such as VGG16, Inception, or EfficientNet [25–27]. The type of feature extractor also depends greatly on the data being used. One of the ideas behind the structural separation between the feature extractor and the modular dynamic component is that the feature extractor can be altered to whatever is best suited to the data being dealt with, so, for instance, to apply MDNN to problems other than image-related ones (such as audio, NLP, etc.), theoretically, one would only need to alter the feature extractor (and, of course, the binary classifiers would need to be structured to have the same input format as the feature extractor's output). With this in mind, there is much work to be done to achieve and prove/validate this claim.

(c) *Binary classifiers.* The binary classifiers implemented here are relatively simple neural networks, and, like most other neural network applications, their structure and parameters could be subject to a lot of alterations and still perform equally, making it hard to know the "ideal" structure for any given situation. However, while an ideal structure for each application is not necessarily known, there is still room for work to be done and tests to be made concerning the binary classifiers seen here and to also look into alternative classification methods (instead of neural networks).

(d) *New module placement.* The placement of new modules considers the average accuracy value obtained by the binary classifiers using all the training data provided. This method has proven to be functional, but some factors could be taken into account to improve this algorithm. For instance, when the resulting similarity results from a set of training data show a large standard deviation, this could indicate that only a sub-sample of the class being learned is considered similar, and the current algorithm would not account for this fact.

(e) *Usage of new data of known classes.* Currently, the architecture is not prepared to make use of new data of an already known class. Once a class is learned from a certain set of training data, that is it. The implementation of this feature will consist of preparing the binary classifiers to resume training, the selection of which binary classifiers to continue training (and how much they should be trained) and also the calculation of when to re-arrange/re-structure the network in face of new data, e.g., upon continuing training parts of the network with new data, new intermediate test/validation results could show that some of the classes it considered to be similar are, in fact, not as similar as initially predicted and this situation could benefit from a re-structure. The implementation of this feature will also allow us to make comparisons with other methods' results achieved on the NI and NIC benchmarks from the CORe50 dataset.

(f) *More tests using different datasets.* Finally, after doing the total or partial improvements mentioned in (a–e) we will also test the method using different CL datasets.

In a final summary, this paper presents an extension to an initial proof of concept for MDNN [36] that showed test results from ImageNet, which revealed it to be a viable continual learning approach. Now, the present framework is detailed, improved, validated and tested on a CL-specific benchmark. It is possible to conclude that the framework presents similar or better results when compared with the state-of-the-art CL architectures. Nevertheless, the framework still has space for improvement and the potential to achieve even better performance results with real-time data, CL datasets and standard datasets.

Author Contributions: Conceptualization, D.T., P.J.S.C. and J.M.F.R.; methodology, D.T.; software, D.T.; validation, D.T., P.J.S.C. and J.M.F.R.; formal analysis, D.T., P.J.S.C. and J.M.F.R.; investigation, D.T.; resources, D.T., P.J.S.C. and J.M.F.R.; data curation, D.T., P.J.S.C. and J.M.F.R.; writing—original draft preparation, D.T.; writing—review and editing, D.T., P.J.S.C. and J.M.F.R.; visualization, D.T., P.J.S.C. and J.M.F.R.; supervision, P.J.S.C. and J.M.F.R.; project administration, J.M.F.R.; funding acquisition, P.J.S.C. and J.M.F.R. All authors have read and agreed to the published version of the manuscript.

Funding: This research was funded by the Portuguese Foundation for Science and Technology (FCT), project LARSyS—FCT Project UIDB/50009/2020.

Data Availability Statement: Publicly available datasets were analyzed in this study. This data can be found here: ImageNet-https://image-net.org/download.php (accessed on 15 October 2021) and CORe50-https://vlomonaco.github.io/core50/index.html#download (accessed on 15 October 2021).

Conflicts of Interest: The authors declare no conflict of interest.

References

1. He, K.; Zhang, X.; Ren, S.; Sun, J. Deep residual learning for image recognition. In Proceedings of the IEEE Conference on Computer Vision and Pattern Recognition, Las Vegas, NV, USA, 27–30 June 2016; pp. 770–778.
2. She, Q.; Feng, F.; Hao, X.; Yang, Q.; Lan, C.; Lomonaco, V.; Shi, X.; Wang, Z.; Guo, Y.; Zhang, Y.; et al. OpenLORIS-Object: A dataset and benchmark towards lifelong object recognition. *arXiv* **2019**, arXiv:1911.06487.
3. Parisi, G.I.; Kemker, R.; Part, J.L.; Kanan, C.; Wermter, S. Continual lifelong learning with neural networks: A review. *Neural Netw.* **2019**, *113*, 54–71. [CrossRef] [PubMed]
4. Ebrahimi, S.; Meier, F.; Calandra, R.; Darrell, T.; Rohrbach, M. Adversarial continual learning. In Proceedings of the 16th European Conference Computer Vision, Glasgow, UK, 23–28 August 2020; pp. 386–402.
5. Aljundi, R.; Rohrbach, M.; Tuytelaars, T. Selfless sequential learning. *arXiv* **2018**, arXiv:1806.05421.
6. Chen, C.P.; Liu, Z. Broad learning system: An effective and efficient incremental learning system without the need for deep architecture. *IEEE Trans. Neural Netw. Learn. Syst.* **2017**, *29*, 10–24. [CrossRef] [PubMed]
7. Chen, Z.; Liu, B. *Lifelong Machine Learning*, 2nd ed.; Synthesis Lectures on Artificial Intelligence and Machine Learning; Morgan & Claypool Publishers: San Rafael, CA, USA, 2018, Volume 12, pp. 1–207. [CrossRef]
8. Delange, M.; Aljundi, R.; Masana, M.; Parisot, S.; Jia, X.; Leonardis, A.; Slabaugh, G.; Tuytelaars, T. A continual learning survey: Defying forgetting in classification tasks. *IEEE Trans. Pattern Anal. Mach. Intell.* **2021**. [CrossRef] [PubMed]
9. Lomonaco, V.; Maltoni, D. CORe50: A new dataset and benchmark for continuous object recognition. In Proceedings of the Conference on Robot Learning, Mountain View, CA, USA, 13–15 November 2017; pp. 17–26.
10. Turner, D.; Cardoso, P.J.; Rodrigues, J.M. Continual Learning for Object Classification: Consolidation and reconsolidation. *Perception* 2021, in press.
11. Lesort, T.; Lomonaco, V.; Stoian, A.; Maltoni, D.; Filliat, D.; Díaz-Rodríguez, N. Continual learning for robotics: Definition, framework, learning strategies, opportunities and challenges. *Inf. Fusion* **2020**, *58*, 52–68. [CrossRef]
12. Deng, J.; Dong, W.; Socher, R.; Li, L.J.; Li, K.; Fei-Fei, L. ImageNet: A large-scale hierarchical image database. In Proceedings of the IEEE Conference on Computer Vision and Pattern Recognition, Miami, FL, USA, 20–25 June 2009; pp. 248–255. [CrossRef]
13. Krizhevsky, A.; Nair, V.; Hinton, G. CIFAR-10 and CIFAR-100 Datasets. 2009. Available online: https://www.cs.toronto.edu/~kriz/cifar.html (accessed on 15 October 2021).
14. Maltoni, D.; Lomonaco, V. Continuous learning in single-incremental-task scenarios. *Neural Netw.* **2019**, *116*, 56–73. [CrossRef] [PubMed]

15. Kirkpatrick, J.; Pascanu, R.; Rabinowitz, N.; Veness, J.; Desjardins, G.; Rusu, A.A.; Milan, K.; Quan, J.; Ramalho, T.; Grabska-Barwinska, A.; et al. Overcoming catastrophic forgetting in neural networks. *Proc. Natl. Acad. Sci. USA* **2017**, *114*, 3521–3526. [CrossRef] [PubMed]
16. Li, Z.; Hoiem, D. Learning without Forgetting. *IEEE Trans. Pattern Anal. Mach. Intell.* **2018**, *40*, 2935–2947. [CrossRef] [PubMed]
17. Zenke, F.; Poole, B.; Ganguli, S. Continual learning through synaptic intelligence. In Proceedings of the International Conference on Machine Learning, Sydney, Australia, 6–11 August 2017; pp. 3987–3995.
18. Rebuffi, S.A.; Kolesnikov, A.; Sperl, G.; Lampert, C.H. iCaRL: Incremental Classifier and Representation Learning. In Proceedings of the 2017 IEEE Conference on Computer Vision and Pattern Recognition (CVPR), Honolulu, HI, USA, 21–26 July 2017; pp. 2001–2010. [CrossRef]
19. Mallya, A.; Lazebnik, S. PackNet: Adding Multiple Tasks to a Single Network by Iterative Pruning. In Proceedings of the 2018 IEEE/CVF Conference on Computer Vision and Pattern Recognition, Salt Lake City, UT, USA, 18–22 June 2018; pp. 7765–7773. [CrossRef]
20. Requeima, J.; Gordon, J.; Bronskill, J.; Nowozin, S.; Turner, R.E. Fast and flexible multi-task classification using conditional neural adaptive processes. *Adv. Neural Inf. Process. Syst.* **2019**, *32*, 7959–7970.
21. van de Ven, G.M.; Tolias, A.S. Three continual learning scenarios and a case for generative replay. In Proceedings of the ICLR 2019, New Orleans, LA, USA, 6–9 May 2019.
22. Pellegrini, L.; Graffieti, G.; Lomonaco, V.; Maltoni, D. Latent Replay for Real-Time Continual Learning. In Proceedings of the 2020 IEEE/RSJ International Conference on Intelligent Robots and Systems (IROS), Las Vegas, NV, USA, 24–29 October 2020; pp. 10203–10209. [CrossRef]
23. Mai, Z.; Li, R.; Jeong, J.; Quispe, D.; Kim, H.; Sanner, S. Online continual learning in image classification: An empirical survey. *arXiv* **2021**, arXiv:2101.10423.
24. Sharma, N.; Jain, V.; Mishra, A. An analysis of convolutional neural networks for image classification. *Procedia Comput. Sci.* **2018**, *132*, 377–384. [CrossRef]
25. Simonyan, K.; Zisserman, A. Very deep convolutional networks for large-scale image recognition. *arXiv* **2014**, arXiv:1409.1556.
26. Szegedy, C.; Vanhoucke, V.; Ioffe, S.; Shlens, J.; Wojna, Z. Rethinking the Inception Architecture for Computer Vision. In Proceedings of the 2016 IEEE Conference on Computer Vision and Pattern Recognition (CVPR), Las Vegas, NV, USA, 27–30 June 2016; pp. 2818–2826. [CrossRef]
27. Tan, M.; Le, Q. Efficientnet: Rethinking model scaling for convolutional neural networks. In Proceedings of the International Conference on Machine Learning, Long Beach, CA, USA, 9–15 June 2019; pp. 6105–6114.
28. Fang, W.; Wang, C.; Chen, X.; Wan, W.; Li, H.; Zhu, S.; Fang, Y.; Liu, B.; Hong, Y. Recognizing global reservoirs from Landsat 8 imageries: A Deep Learning Approach. In Proceedings of the AGU Fall Meeting, Washington, DC, USA, 10–14 December 2018.
29. Wan, A.; Dunlap, L.; Ho, D.; Yin, J.; Lee, S.; Jin, H.; Petryk, S.; Bargal, S.A.; Gonzalez, J.E. NBDT: Neural-backed decision trees. *arXiv* **2020**, arXiv:2004.00221.
30. Krizhevsky, A.; Hinton, G. *Learning Multiple Layers of Features from Tiny Images*; Technical Report; University of Toronto: Toronto, ON, Canada, 2009.
31. Hensman, P.; Masko, D. *The Impact of Imbalanced Training Data for Convolutional Neural Networks*; Degree Project in Computer Science; KTH Royal Institute of Technology: Stockholm, Sweden, 2015.
32. Tieleman, T.; Hinton, G. Lecture 6.5-rmsprop: Divide the gradient by a running average of its recent magnitude. *COURSERA Neural Netw. Mach. Learn.* **2012**, *4*, 26–31.
33. Jia, Y.; Shelhamer, E.; Donahue, J.; Karayev, S.; Long, J.; Girshick, R.; Guadarrama, S.; Darrell, T. Caffe: Convolutional Architecture for Fast Feature Embedding. In Proceedings of the 22nd ACM International Conference on Multimedia, Orlando, FL, USA, 3–7 November 2014. [CrossRef]
34. Chatfield, K.; Simonyan, K.; Vedaldi, A.; Zisserman, A. Return of the devil in the details: Delving deep into convolutional nets. *arXiv* **2014**, arXiv:1405.3531.
35. Russakovsky, O.; Deng, J.; Su, H.; Krause, J.; Satheesh, S.; Ma, S.; Huang, Z.; Karpathy, A.; Khosla, A.; Bernstein, M.; et al. ImageNet Large Scale Visual Recognition Challenge. *Int. J. Comput. Vis.* **2015**, *115*, 211–252. [CrossRef]
36. Turner, D.; Cardoso, P.J.; Rodrigues, J.M. Continual Learning for Object Classification: A Modular Approach. In *Universal Access in Human-Computer Interaction. Access to Media, Learning and Assistive Environments*; Antona, M., Stephanidis, C., Eds.; LNCS 12769; Springer: Cham, Switzerland, 2021. [CrossRef]

applied sciences

MDPI

Article

Community Detection Based on Graph Representation Learning in Evolutionary Networks

Dongming Chen, Mingshuo Nie, Jie Wang, Yun Kong, Dongqi Wang * and Xinyu Huang

Software College, Northeastern University, Shenyang 110169, China; chendm@mail.neu.edu.cn (D.C.); niemingshuo@stumail.neu.edu.cn (M.N.); 1971161@stu.neu.edu.cn (J.W.); 1971157@stu.neu.edu.cn (Y.K.); huangxinyu@mail.neu.edu.cn (X.H.)

* Correspondence: wangdq@swc.neu.edu.cn; Tel.: +86-136-2403-9571

Abstract: Aiming at analyzing the temporal structures in evolutionary networks, we propose a community detection algorithm based on graph representation learning. The proposed algorithm employs a Laplacian matrix to obtain the node relationship information of the directly connected edges of the network structure at the previous time slice, the deep sparse autoencoder learns to represent the network structure under the current time slice, and the *K*-means clustering algorithm is used to partition the low-dimensional feature matrix of the network structure under the current time slice into communities. Experiments on three real datasets show that the proposed algorithm outperformed the baselines regarding effectiveness and feasibility.

Keywords: evolutionary networks; graph representation; community detection; deep sparse autoencoder

Citation: Chen, D.; Nie, M.; Wang, J.; Kong, Y.; Wang, D.; Huang, X. Community Detection Based on Graph Representation Learning in Evolutionary Networks. *Appl. Sci.* **2021**, *11*, 4497. https://doi.org/10.3390/app11104497

Academic Editors: João M. F. Rodrigues, Pedro J. S. Cardoso and Marta Chinnici

Received: 12 April 2021
Accepted: 12 May 2021
Published: 14 May 2021

1. Introduction

The evolutionary network refers to a network in which the nodes and edges change over time. As a typical evolutionary network, social networks are very common in modern life, such as Flickr, Facebook, Twitter, and so on. In the above-mentioned social networks, an individual or a group of users may join and exit the network, and the intimacy with friends is increased through obtaining to know friends of friends. Community detection in an evolutionary network is a complex and challenging problem. The development and change of the network structure under the adjacent time slices are slow, however, the statistical characteristics of the network will change significantly over a long period.

Therefore, it is necessary to re-identify the community structure. As the organization of a dynamic community changes constantly, a dynamic community tends to be more complex. Network representation has gradually become the main supporting technology of dynamic community detection, which provides great convenience for dynamic community detection. The application of network representation technology can greatly improve the detection performance [1,2].

Traditional community detection algorithms are mostly based on static networks, and the influence of time on the network structure is ignored. For example, network communication operators hope to reduce the communication costs in the same social circle in different periods, proposing a flexible billing activity for their customers. The static community is uncertain about the actual network connectivity, and this marketing policy is obviously not effective. If considering the change of time for community detection, the real-time community structure can be dynamically represented, which is beneficial to providing individualized services for users.

Thus, it can be seen that the evolution of the network structure has an important impact on community detection, which is likely to bring commercial value [3]. Early research on the division of dynamic network communities [4–8] viewed the dynamic network as a series of snapshot networks in chronological order, and obtained the community structure of each snapshot by using the static network community detection algorithm; however, this kind of thinking does not make full use of the evolution information of the network structure, and a fundamental drawback rooted in such schemes is that most of traditional algorithms are sensitive to tiny changes in the network structure [9].

The DeepWalk [10] algorithm uses a random walk method to obtain the neighbors of nodes in the network to generate a fixed-length node sequence, and the node sequence is fed into the Skip-Gram model for learning representation to apply to other tasks. The SDNE [11] algorithm uses a deep sparse autoencoder [12] to simultaneously optimize the first-order and second-order similarity objective functions, conducting semi-supervised learning representation, and it is applied to visualization, multi-label classification, link prediction, etc. In machine learning research, scholars have proposed learning methods for graph embedding [13–15]; however, these methods usually only work well on certain networks.

To overcome these limitations, we propose a novel community detection algorithm based on evolutionary network representation learning (Learning Community Detection Based on Evolutionary Network, LCDEN). The LCDEN algorithm uses a Laplacian matrix to obtain the node relationship information of the directly connected edges in the network structure of the previous time slice, learns the network structure under the current time slice through the deep sparse autoencoder, and uses a K-means clustering algorithm to divide the community of the obtained low-dimensional feature matrix of the network structure under the current time slice. Our contributions can be summarized as follows:

(1) We propose a novel algorithm for community detection in evolutionary networks, which solves the limitation that traditional community detection algorithms were unable to handle the temporal information of a network structure.
(2) The proposed algorithm can effectively use the historical temporal information of a network structure and apply a deep learning model to the research of evolutionary network representation learning.
(3) The proposed algorithm has advantages in different datasets and has higher detection performance and computational efficiency.

2. Related Work

Community detection is one of the most active topics in the field of graph mining and network science. With the continuous expansion of the real-world network scale and the introduction of temporal information, the research on community detection in evolutionary networks can explore community detection algorithms based on the network structure and network information conforming to the real world.

Network representation learning. In the research of network representation learning, one of the foremost requirements of network representation learning is preserving network topology properties in learned low-dimension representations [16]. Researchers often use methods based on matrix decomposition or random walking to learn the graph representation, such as LLE [17], Laplacian Eigenmaps [14], Deepwalk [10], and LINE [18]. However, these methods typically have high computational complexity and poor generalization ability. In addition, they lack the use of temporal information in evolutionary networks.

Evolutionary network. The research on the dynamic evolution of networks has been an ongoing hotspot. The dynamic information in networks has proven to be crucial for understanding networks. Many works have attempted to combine the research of evolutionary network representation learning with the basic research of complex networks. To reveal the dynamic mechanism of information dissemination networks, Rodriguez et al. [19] modeled the information propagation process as several discrete networks with continuous-time slices occurring at different rates.

Fathy et al. [20] realized the joint learning of the graph structure information and time dimension information by combining the graph attention mechanism with the convolution of the time dimension, and processing the dynamic graph structure data by using the self-attention layer as time passes. It is of great significance to learn and fuse a network structure, multi-modal information, and temporal information based on effective methods, which can obtain more accurate representations of evolutionary networks.

Community detection. In recent years, community structure detection has received extensive attention, and communities play an important role in complex systems [21]. However, most of the traditional community detection algorithms for static networks are not suitable for research on evolutionary network structures. Chen et al. [22] proposed a community detection algorithm based on the minimum change granularity for evolutionary networks (DWGD), which can complete dynamic community detection with time slices as the minimum granularity. Wang et al. [23] proposed that Markov chain-based community detection algorithm for evolutionary networks had better computational efficiency and detection performance. In addition, many researchers also studied community detection in evolutionary networks by the Coherent Neighborhood Propinquity of dynamic networks [24], building compressed graphs [25], etc.

3. Preliminaries

3.1. Definitions

Given a network $G = (V, E)$, where $V = \{v_1, v_2, \ldots, v_n\}$ is the set of nodes, $E = \{e_1, e_2, \ldots, e_n\}$ is the set of edges. The adjacency matrix $A = [a_{ij}]_{n \times n}$ represents the connection relationship between nodes, the value of the corresponding element in the matrix indicates whether the edge exists, if there is an edge between node v_i and node v_j, then $a_{ij} = 1$; otherwise $a_{ij} = 0$. In this paper, we take the adjacency matrix A as the closest relation matrix in the network to describe the similarity between nodes in the graph and their proximity relation.

(1) Node proximity

Given a network $G = (V, E)$, $u,\ v \in V$, the proximity $Nei(u, v)$ between node u and node v is defined as

$$Nei(u, v) = e^{\alpha(2-l)}, \tag{1}$$

where l is the shortest path length from node u to node v, $l \geq 2$, α is the degree of attenuation, and the value range is $(0, 1)$. If the path length l increases, the proximity decreases, and α controls the degree of proximity attenuation. The larger the value of α, the faster the proximity relationship between the nodes decays.

(2) Proximity matrix

Given a network $G = (V, E)$, $M = [m_{ij}]_{n \times n}$ is a matrix corresponding to the network G, where we use the proximity formula to calculate the proximity between the two corresponding nodes v_i and v_j in M, $m_{ij} = Nei(v_i, v_j)$, with $v_i, v_j \in V$; then, we call M the proximity matrix of G.

3.2. Data Preprocessing

The data preprocessing mainly converts the adjacency matrix into a proximity matrix. The breadth-first traversal algorithm is employed to obtain the path length, then we calculate the neighbor relationship between nodes that are not directly connected by the attenuation, and the topology of the community can be well exhibited. However, when the path length is greater than a certain threshold, node pairs that are not in the same community may also have a certain proximity value, which will lead to an ambiguous partition of the community. Therefore, we set the threshold *Le* of the path length. By adjusting the parameters empirically, we only calculate the proximity value between nodes that are mutually reachable within the length of *Le*.

At the same time slice, we construct a loss function for directly connected nodes in the network, this is formulated as

$$L_1 = \sum_{i,j=1}^{n} A_{i,j} ||(y_i - y_j)||_2^2 = 2tr\left(Y^T L Y\right) \tag{2}$$

where i, j represent two adjacent nodes, n is the number of nodes, A_{ij} is the adjacency matrix, y_i and y_j are the representations of two adjacent nodes corresponding to the presentation layer, and L is the Laplacian matrix of the network, obtained from the corresponding diagonal matrix and adjacency matrix. The Laplacian matrix is calculated as

$$L = D - A, \tag{3}$$

where D is a diagonal matrix, A is the adjacency matrix of the network. To ensure that the evolutionary network can smoothly evolve with time, the relational information of the directly connected nodes of the network structure of the last time slice are obtained by using the Laplacian matrix. By employing the Laplacian matrix, the node information loss function of directly connected nodes is constructed as

$$L'_1 = 2tr\left(Y_t^T L_{t-1} Y_t\right), \tag{4}$$

where Y_t is the representation of the network structure of time slice t, L_{t-1} is the Laplacian matrix of the network structure of time slice $t-1$.

Under different time slices, the scale of the evolutionary network varies. A zero-filling operation is usually performed for training the data calculations in neural networks. However, there may be such cases where non-zero elements are far less than the zero elements in the data matrix. In this way, the increase of zero elements will affect the sparsity of the matrix, and it is easy to increase the error of zero elements due to the reconstruction of the deep autoencoder. Therefore, a sparsity constraint based on the deep automatic encoder is beneficial to control the reconstruction difference of the deep autoencoder. In the proposed algorithm, the sparsity parameter is added to the depth autoencoder during the encoding process, and we obtain a deep sparse autoencoder, which is helpful to express the information of the original data without loss.

The first layer to the second layer in the deep sparse autoencoder is an encoding process as well as a dimensionality reduction process. According to the size of the data, we set the number of data nodes in each layer, which is also the data dimension. By setting the data dimension size in each layer, the deep sparse autoencoder will obtain the low-dimensional vector corresponding to the input data after encoding. From the second layer to the third layer, this is a decoding process. The deep sparse autoencoder decodes the low-dimensional vector obtained from the input data and then outputs the vector of the same dimension as the input data.

This process employs the backpropagation algorithm to train the data by adjusting the parameters related to the encoder and decoder—for example, weights, deviations, etc., to minimize the reconstruction difference. Consequently, the output expression vector approximates the input data information. Finally, the resulting representation of the coding layer is exactly the low-dimensional Eigenmatrix that needs to be output.

In the proposed algorithm, a deep sparse autoencoder is used to represent and learn the network structure under the current time slice of the evolutionary network. We adopt a similar idea to calculate the reconstruction error of building deep sparse autoencoders [11], and a backpropagation algorithm is employed to obtain the node information expression of the network at the next time slice. The reconstruction error of the deep sparse autoencoder is established, which is formulated as

$$L_2 = \sum_{i=1}^{n} \left|\left| (x'_i - x_i) * b_i \right|\right|_2^2, \tag{5}$$

where x'_i and x_i are the input data and reconstructed data, respectively. $b_i = x_i * (\beta - 1) + 1$, b_i is a sparsely constrained parameter, and β is a tunable parameter.

To satisfy the temporal smooth transition of the evolutionary network structure, the node information of the directly connected edges in the network of the last time slice is employed to construct the loss function, and the reconstruction error of the deep sparse autoencoder is incorporated to construct the overall loss function.

In the proposed algorithm, to preserve part of the historical node information of the network structure of the last time slice and, simultaneously, effectively present the feature of the network structure of the next time slice, the overall loss functions are jointly optimized to effectively train the data. The loss function on constructing the learning model of the overall graph presentation is defined as

$$L_{min} = \theta L'_1 + \gamma L_2, \tag{6}$$

where θ and γ are two tunable parameters.

4. Methods

The proposed algorithm employs a Laplacian matrix to map the nodes with directly connected edges in the network structure of the last time slice to the deep sparse autoencoder representation layer. It is beneficial to retain the node information of the network of the last time slice, and it is not easy to lose the historical evolution information of the network. Meanwhile, the temporal smoothness affects the structure of the evolutionary network and, thus, is taken into account in the graph representation learning algorithm.

Furthermore, the deep sparse autoencoder is used to represent the network structure under the current time slice of the evolution network, to obtain a better expression of the characteristics of the network structure, and then the representation vector of the node is achieved. Finally, the *K*-means clustering algorithm is used to cluster the low-dimensional feature matrix of the network structure under the current time slice to obtain the community structure.

For the network structure of continuous-time slices, we successively utilize the information between the node pairs of the network structure under the last time slice. The network structure of the current time slice is learned and presented through the deep sparse autoencoder, and then the low-dimensional representation matrix of the network structure of the current time slice is obtained.

The following shows the process of the proposed algorithm:

(1) Initialize the relevant parameters and load the dataset.
(2) Based on the breadth-first traversal algorithm to traverse the network structure adjacency matrix of the current time slice.
(3) Calculate the degree by using the path length to obtain the proximity matrix.
(4) Use a Laplacian Matrix to obtain the information of directly connected nodes in the network of a time slice, and the features of the proximity matrix of the network under the current time slice are extracted by employing a deep sparse autoencoder.
(5) Use the *K*-means algorithm to obtain the network community structure under the current time slice.

The proposed algorithm (Algorithm 1) is based on the TensorFlow framework, and its pseudocode is as follows.

In this pseudocode, the Breadth-first traversal algorithm is used to convert the adjacency matrix A_t into the proximity matrix M_t, and then to input it into the deep sparse autoencoder for encoding. Simultaneously, we use the adjacency matrix A_{t-1} to calculate the Laplacian matrix, to map the node information with directly connected edges to the presentation layer of the deep sparse autoencoder, and to obtain the result of the presentation layer—that is, the required low-dimensional matrix representation vector Y_t. Finally, the *K*-means clustering algorithm is used to cluster the low-dimensional matrix representation vector Y_t, and the community results are obtained.

Algorithm 1. LCDEN algorithm

Input: Evolution graph G, adjacency matrix A_{t-1}, adjacency matrix A_t, number of communities k, path length threshold L_e, degree of attenuation α, deep sparse autoencoder with 3 layers, number of nodes in each layer of the deep autoencoder $d^{(i)}$, $i = 1, 2, 3$.
Output: Community results $Com_t = \{c_1, c_2, \ldots, c_{k'}\}$.

```
for i in G.nodes do
    Initialize each node in graph G to an unvisited state
    Initialize the queue, que
    Initialize traversal path length set, length_dict
    Mark i as the accessed state, initialize the path_length(i) = 0
    Record path_length to length_dict
    Add i to the que
    while que != None do
        Remove node j from que
        for u from 0 to Length(v) do
            if v ≤ L and u is unvisited state then
                Mark u as the accessed state
                path_length(i) = Σ_{l=i}^{v+1} path_length(l)
                Record path_length(u) and Σ_{l=i}^{v} path_length(l) to length_dict
                Add u to que
            end if
        end for
      Set v to be accessed end state
      end while
      for u in G.nodes do
            Calculate Nei(i,u) = e^{α(2−l)}
      end for
  end for
  Obtain neighbor degree matrix M_t
  Build deep sparse autoencoder
  Input neighbor degree matrix M_t to deep autoencoder
Calculate loss function L_2
Obtain hidden layer representation y
Calculate Laplacian matrix L_{t−1} = D_{t−1} − A_{t−1}
Calculate local node information loss function L'_1
Calculate L_min = θL'_1 + γL_2
  repeat
    Joint optimization L_min
  until converge
Obtain a low dimensional feature matrix Y_t
  Run K-means clustering
  Return community results Com_t = {c_1, c_2, ..., c_k'}
```

5. Experiments

In this section, we evaluate the computational efficiency and detection performance of our proposed algorithm on three different datasets. We use the Silhouette Coefficient [26] and time-consumption as evaluation metrics to evaluate the performance of different models.

5.1. Datasets

The following three network datasets were employed for the experiments.

(1) Superuser temporal network [27]

The stack exchange website super user network dataset is an interactive temporal network on the stack exchange website. The relevant metadata stored are users and the interactions between users. Here, nodes represent users, and edges indicate interactions between users.

(2) Wiki-Talk-Temporal dataset [27]

The Wikipedia network is a temporal network, and Wikipedia users edit each other's "conversation" pages. The directed edge (u, v, t) indicates that a user u edited the conversation page of user v at time t.

(3) Twitter network dataset [28]

The data were purchased by The Numerical Analysis and Scientific Computing research group in the Department of Mathematics and Statistics at the University of Strathclyde, using a grant made available by the University of Strathclyde through their EPSRC-funded "Developing Leaders" program. In the dataset, node i and node j represent the node numbers.

5.2. Evaluation Metrics

We employed the Silhouette Coefficient and time-consumption to evaluate the experimental results. The Silhouette Coefficient effectively combines the cohesion and separation of the clustering for evaluation. The advantage of employing the Silhouette Coefficient as a cluster result evaluation is that no real data information is needed for comparison. The range of the Silhouette Coefficient is [−1, 1], and the larger the value is, the better the clustering effect.

The calculation formula of the Silhouette Coefficient of each sample point is as follows:

$$s(i) = \frac{b(i) - a(i)}{\max\{a(i), b(i)\}} \tag{7}$$

where i represents a sample point, $a(i)$ represents the average distance between each sample i and all other sample points in a cluster. It is used to quantify the cohesion of the cluster. We select a cluster b, which does not contain sample point i, and then calculate the average distance between sample point i and all sample points in cluster b. Similarly, we traverse all other clusters until the nearest average distance is found. This is represented as $b(i)$, which is used to quantify the degree of separation between clusters.

We calculate the Silhouette Coefficient of the overall sample—that is, we find the average value as the overall Silhouette Coefficient of the data cluster, to denote the closeness of the data cluster.

5.3. Baselines

To verify the effectiveness of the proposed algorithm, we chose the DWGD algorithm [22] for our comparative experiments. The DWGD algorithm can partition communities for dynamic weighted networks. The weight of the edges may increase, decrease, or remain unchanged with the change of time. The DWGD algorithm can well deal with the increase and decrease of the edge weight. Experiments on certain datasets verified the effectiveness of the algorithm. The DWGD algorithm can perform community detection for evolutionary networks and weighted networks. Therefore, the algorithm proposed in this paper was compared with the DWGD algorithm on the problem of community detection in dynamic weighted networks.

5.4. Experimental Analysis

The proposed algorithm (abbreviated as LCDEN) and the DWGD algorithm were used to divide the network structure of the data set containing a time stamp under different time slices, and the obtained Silhouette Coefficient and running time were compared. The experimental results and analysis are as follows.

(1) The comparative experimental results on the Superuser temporal network are shown in Figure 1. As shown in Figure 1a, we can conclude that, in the same time slice, the Silhouette Coefficient of the proposed algorithm is higher than that of the DWGD

algorithm, that is, the clustering effect is better. Figure 1b demonstrates that, in the same time slice, the running time of the proposed algorithm is less than the DWGD algorithm.

(2) The comparative experimental results based on the Wiki-Talk-Temporal network (Wikipedia network dataset) are shown in Figure 2.

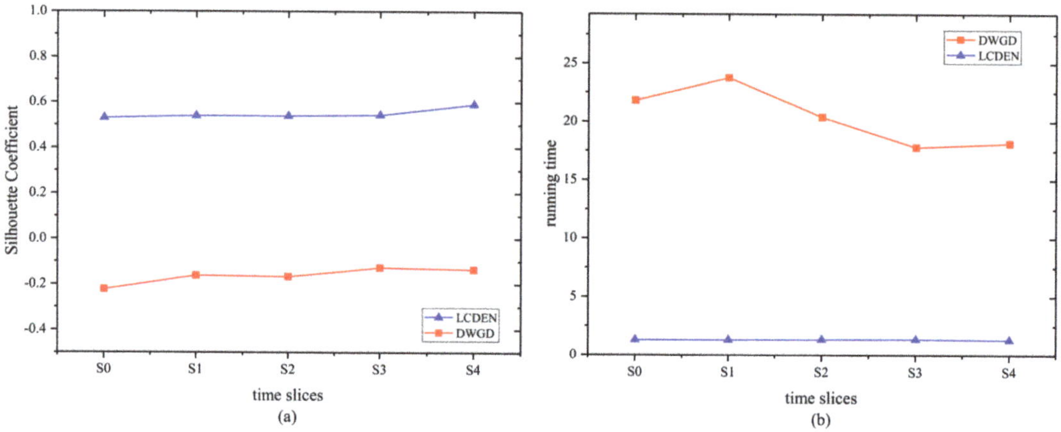

Figure 1. Comparison of the Silhouette Coefficient (**a**) and running time (**b**) on the Superuser temporal network.

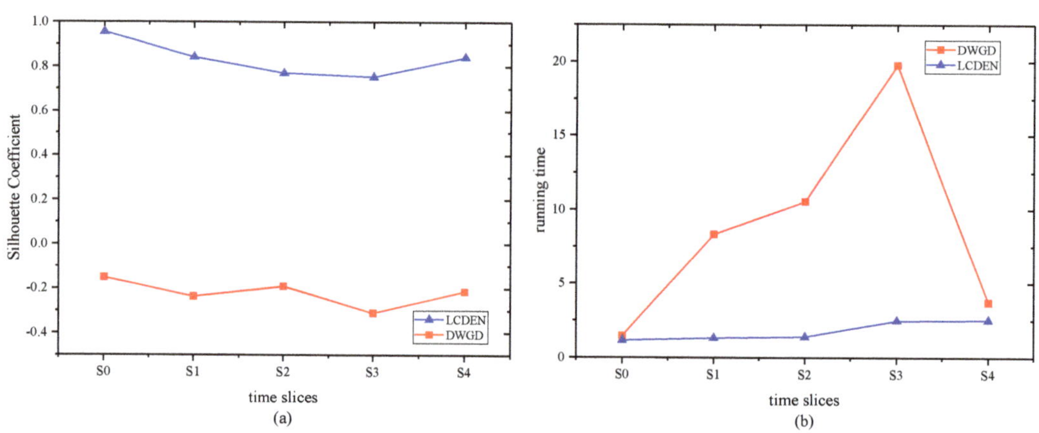

Figure 2. Comparison of the Silhouette Coefficient (**a**) and running time (**b**) on the Wiki-Talk-Temporal network.

Figure 2a also shows that the Silhouette Coefficient of the proposed algorithm is higher than the DWGD algorithm at the same time slice, indicating a better clustering effect of the proposed algorithm. From Figure 2b, we can see that in the time slices of S0 and S4, the running time of the proposed algorithm and the DWGD algorithm are close to each other, and, in other time slices, the running time of the proposed algorithm is less than that of the DWGD algorithm.

(3) The comparative experimental results based on the Twitter dataset are shown in Figure 3.

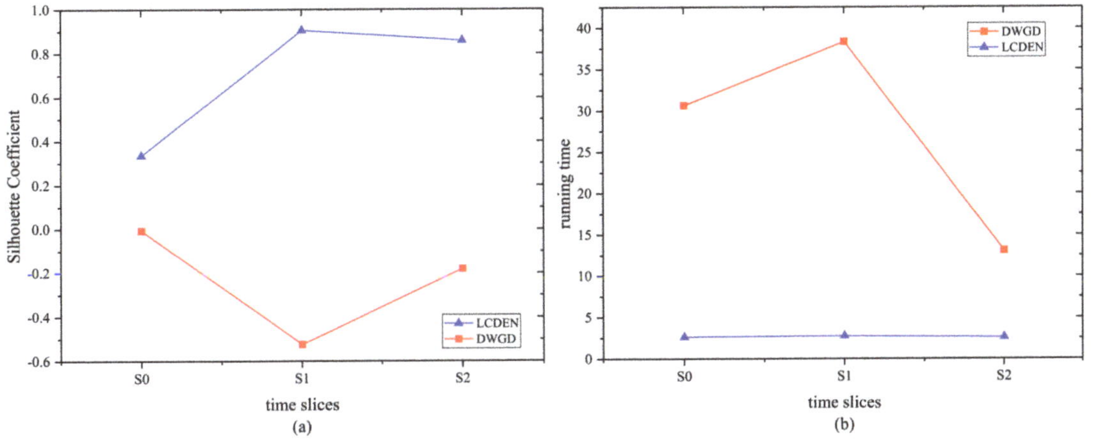

Figure 3. Comparison of the Silhouette Coefficient (**a**) and running time (**b**) on the Twitter network.

Figure 3a illustrates that the Silhouette Coefficient of the proposed algorithm is higher than that of the DWGD algorithm, indicating a better clustering effect of the proposed algorithm. Figure 3b discloses that the running time required for the proposed algorithm is less than that of the DWGD algorithm.

6. Results and Discussion

The community structures are groups of nodes that are more strongly or frequently connected among themselves than with others. Community detection is proposed to find the most reasonable partitions of a network via the observed topological structures [29]. However, traditional community detection methods are usually oriented to static networks and cannot deal with large-scale evolutionary networks, while the most common ones in the real world are complex systems that change with time.

The proposed algorithm makes full use of the historical structure of the network and uses the deep learning model to process the temporal information of the evolutionary network, thus, obtaining the most reasonable communities in different time slices of the given evolutionary network. By comparing with the baselines, the proposed algorithm demonstrated advantages in computational efficiency and detection performance.

This is because the use of the deep learning model uses the historical structure of the network as the prior knowledge of the community detection at the current time, and the deep learning model obtains more input data when compared with the baselines. The research in community detection will further promote the development of the optimization of transportation network structures [30,31], the analysis of social networks [32,33], and research of biological systems [3,34].

7. Conclusions

In this paper, we propose an evolutionary network community detection algorithm that embeds nodes into vectors and uses the K-means algorithm for community member clustering. To process the temporal data, the proposed algorithm employs a Laplacian matrix to represent the historical network structural information and uses a deep sparse autoencoder to encode the prior and the current information. The experimental results on representative datasets showed that the proposed algorithm outperformed the baselines in computational efficiency and detection performance.

Author Contributions: Conceptualization, D.C., Y.K. and D.W.; Formal analysis, M.N., J.W. and X.H.; Funding acquisition, D.C. and D.W.; Methodology, M.N. and Y.K.; Project administration, D.C. and D.W.; Resources, M.N. and X.H.; Software, M.N., J.W. and Y.K.; Supervision, D.C. and D.W.;

Validation, D.C. and D.W.; Visualization, M.N. and J.W.; Writing–original draft, M.N., J.W. and Y.K.; Writing–review & editing, D.C., D.W. and X.H. All authors have read and agreed to the published version of the manuscript.

Funding: This work is partially supported by the Natural Science Foundation of Liaoning Province under Grant No. 20170540320, the Doctoral Scientific Research Foundation of Liaoning Province under Grant No. 20170520358, and the Fundamental Research Funds for the Central Universities under Grant No. N161702001, No. N172410005-2.

Conflicts of Interest: The authors declare no conflict of interest.

References

1. He, T.; Hu, L.; Chan, K.C.C.; Hu, P. Learning Latent Factors for Community Identification and Summarization. *IEEE Access* **2018**, *6*, 30137–30148. [CrossRef]
2. Zhang, X.; Zhang, J.; Yang, J. Dynamic Community Recognition Algorithm Based on Node Embedding and Linear Clustering. In *Innovative Computing*; Springer: Singapore, 2020; pp. 829–837.
3. Martinet, L.-E.; Kramer, M.A.; Viles, W.; Perkins, L.N.; Spencer, E.; Chu, C.J.; Cash, S.S.; Kolaczyk, E.D. Robust dynamic community detection with applications to human brain functional networks. *Nat. Commun.* **2020**, *11*, 1–13. [CrossRef] [PubMed]
4. Watts, D.J.; Strogatz, S.H. Collective dynamics of 'small-world' networks. *Nat. Cell Biol.* **1998**, *393*, 440–442. [CrossRef]
5. Asur, S.; Parthasarathy, S.; Ucar, D. An Event-Based Framework for Characterizing the Evolutionary Behavior of Interaction Graphs. In Proceedings of the 13th ACM SIGKDD International Conference on Knowledge Discovery and Data Mining—KDD '07, San Jose, CA, USA, 12–15 August 2007; pp. 913–921. [CrossRef]
6. Kumar, R.; Novak, J.; Raghavan, P.; Tomkins, A. On the Bursty Evolution of Blogspace. *World Wide Web* **2005**, *8*, 159–178. [CrossRef]
7. Kumar, R.; Novak, J.; Tomkins, A. Structure and Evolution of Online Social Networks. In *Link Mining: Models, Algorithms, and Applications*; Springer: New York, NY, USA, 2010; pp. 337–357.
8. Lin, Y.-R.; Sundaram, H.; Chi, Y.; Tatemura, J.; Tseng, B.L. Blog Community Discovery and Evolution Based on Mutual Awareness Expansion. In Proceedings of the IEEE/WIC/ACM International Conference on Web Intelligence (WI'07), Fremont, CA, USA, 2–5 November 2007; pp. 2–5.
9. Yu, L.; Li, P.; Zhang, J.; Kurths, J. Dynamic community discovery via common subspace projection. *New J. Phys.* **2021**, *23*, 033029. [CrossRef]
10. Yang, C.; Liu, Z.; Zhao, D.; Sun, M.; Chang, E.Y. Network Representation Learning with Rich Text Information. In Proceedings of the 24th International Conference on Artificial Intelligence, Buenos Aires, Argentina, 25–31 July 2015; pp. 2111–2117.
11. Wang, D.; Cui, P.; Zhu, W. Structural Deep Network Embedding. In the Proceedings of 22nd ACM SIGKDD International Conference on Knowledge Discovery and Data Mining, San Francisco, CA, USA, 13–16 August 2016; pp. 1225–1234.
12. Bengio, Y.; Lamblin, P.; Popovici, D.; Larochelle, H. Greedy Layer-Wise Training of Deep Networks. In *Advances in Neural Information Processing Systems 19, Proceedings of the 20th Annual Conference on Neural Information Processing Systems (NIPS 2006), Vancouver, BC, Canada, 4–5 December 2007*; MIT Press: Cambridge, MA, USA, 2007; p. 153.
13. Cox, M.A.; Cox, T.F. Multidimensional Scaling. In *Handbook of Data Visualization*; Springer: Berlin/Heidelberg, Germany, 2008; pp. 315–347.
14. Belkin, M.; Niyogi, P. Laplacian Eigenmaps and Spectral Techniques for Embedding and Clustering. In Proceedings of the 14th International Conference on Neural Information Processing Systems: Natural and Synthetic, Vancouver, BC, Canada, 3–8 December 2001; pp. 585–591.
15. Tenenbaum, J.B.; De Silva, V.; Langford, J.C. A global geometric framework for nonlinear dimensionality reduction. *Science* **2000**, *290*, 2319–2323. [CrossRef]
16. Sun, H.; Jie, W.; Loo, J.; Chen, L.; Wang, Z.; Ma, S.; Li, G.; Zhang, S. Network Representation Learning Enhanced by Partial Community Information That Is Found Using Game Theory. *Information* **2021**, *12*, 186. [CrossRef]
17. Hou, Y.; Zhang, P.; Xu, X.; Zhang, X.; Li, W. Nonlinear Dimensionality Reduction by Locally Linear Inlaying. *IEEE Trans. Neural Netw.* **2009**, *20*, 300–315. [CrossRef] [PubMed]
18. Tang, J.; Qu, M.; Wang, M.; Zhang, M.; Yan, J.; Mei, Q. LINE: Large-scale Information Network Embedding. In Proceedings of the 24th International Conference on World Wide Web, Florence, Italy, 18–22 May 2015; pp. 1067–1077.
19. Rodriguez, M.G.; Balduzzi, D.; Schölkopf, B. Uncovering the temporal dynamics of diffusion networks. *arXiv* **2011**, arXiv:1105.0697 2011.
20. Fathy, A.; Li, K. TemporalGAT: Attention-Based Dynamic Graph Representation Learning. In Proceedings of the Pacific-Asia Conference on Knowledge Discovery and Data Mining, Singapore, 11–14 May 2020; pp. 413–423.
21. Li, M.; Liu, J.; Wu, P.; Teng, X. Evolutionary Network Embedding Preserving both Local Proximity and Community Structure. *IEEE Trans. Evol. Comput.* **2019**, *24*, 523–535. [CrossRef]
22. Chen, D.M.; Wang, Y.K.; Huang, X.Y.; Wang, D.Q. Community Detection Algorithm for Complex Networks Based on Group Density. *J. Northeast. Univ. Nat. Sci. Ed.* **2019**, *40*, 186–191.

23. Wang, Z.; Wang, C.; Li, X.; Gao, C.; Li, X.; Zhu, J. Evolutionary Markov Dynamics for Network Community Detection. *IEEE Trans. Knowl. Data Eng.* **2020**, 1. [CrossRef]
24. Chen, N.; Hu, B.; Rui, Y. Dynamic Network Community Detection with Coherent Neighborhood Propinquity. *IEEE Access* **2020**, *8*, 27915–27926. [CrossRef]
25. Seifikar, M.; Farzi, S.; Barati, M. C-Blondel: An Efficient Louvain-Based Dynamic Community Detection Algorithm. *IEEE Trans. Comput. Soc. Syst.* **2020**, *7*, 308–318. [CrossRef]
26. Rousseeuw, P.J. Silhouettes: A graphical aid to the interpretation and validation of cluster analysis. *J. Comput. Appl. Math.* **1987**, *20*, 53–65. [CrossRef]
27. Paranjape, A.; Benson, A.R.; Leskovec, J. Motifs in Temporal Networks. In Proceedings of the 10th ACM International Conference on Web Search and Data Mining, Cambridge, UK, 6–10 February 2017; pp. 601–610.
28. Yang, J.; Leskovec, J. Patterns of Temporal Variation in Online Media. In Proceedings of the 4th ACM International Conference on Distributed Event-Based Systems—DEBS '10, Cambridge, UK, July 12–15 2010; pp. 177–186.
29. Huang, X.; Chen, D.; Ren, T.; Wang, D. A survey of community detection methods in multilayer networks. *Data Min. Knowl. Discov.* **2021**, *35*, 1–45. [CrossRef]
30. Ding, R.; Ujang, N.; Bin Hamid, H.; Manan, M.S.A.; He, Y.; Li, R.; Wu, J. Detecting the urban traffic network structure dynamics through the growth and analysis of multi-layer networks. *Phys. A Stat. Mech. Appl.* **2018**, *503*, 800–817. [CrossRef]
31. Liu, R.-R.; Jia, C.-X.; Lai, Y.-C. Remote control of cascading dynamics on complex multilayer networks. *New J. Phys.* **2019**, *21*, 045002. [CrossRef]
32. Rozario, V.S.; Chowdhury, A.; Morshed, M.S.J. Community detection in social network using temporal data. *arXiv* **2009**, arXiv:1904.05291.
33. Al-Garadi, M.A.; Varathan, K.D.; Ravana, S.D.; Ahmed, E.; Mujtaba, G.; Khan, M.U.S.; Khan, S.U. Analysis of online social network connections for identification of influential users: Survey and open research issues. *ACM Comput. Surv.* **2018**, *51*, 1–37. [CrossRef]
34. Sanchez-Rodriguez, L.M.; Iturria-Medina, Y.; Mouches, P.; Sotero, R.C. A method for multiscale community detection in brain networks. *bioRxiv* **2019**, 743732. [CrossRef]

MDPI
St. Alban-Anlage 66
4052 Basel
Switzerland
www.mdpi.com

Applied Sciences Editorial Office
E-mail: applsci@mdpi.com
www.mdpi.com/journal/applsci